AF632605

SUSTAINABLE FARMING

NEW INDIA PUBLISHING AGENCY
New Delhi-110 034

SUSTAINABLE FARMING

Partha Sarathi Patra
Assistant Professor
Uttar Banga Krishi Viswavidyalaya
Pundibari, Cooch Behar, West Bengal, India

Kausik Pradhan
Associate Professor and Head
Department of Agricultural Extension
Uttar Banga Krishi Viswavidyalaya
Pundibari, Coochbehar, West Bengal, India

Shyamal Kheroar
Department of Agronomy
Bidhan Chandra Krishi Viswavidyalaya
West Bengal, India

Amrit Tamang
Assistant Professor
Soil Science and Agricultural Chemistry
Uttar Banga Krishi Vishwavidayalaya
(UBKV), Cooch Behar, India

Pabitra Adhikary
Subject Matter Specialist (Agronomy)
Farm Science Center, WBUAFS
Kolkata, West Bengal, India

nipa®

NEW INDIA PUBLISHING AGENCY
New Delhi-110 034

NEW INDIA PUBLISHING AGENCY
101, Vikas Surya Plaza, CU Block, LSC Market
Pitam Pura, New Delhi – 110 034, India
Email: info@nipabooks.com
Web: www.nipabooks.com

For customer assistance, please contact
Phone: + 91-11-27 34 17 17 Fax: + 91-11- 27 34 16 16
E-Mail: feedbacks@nipabooks.com

© 2021, Publisher

ISBN: 978-93-90175-05-5

All rights reserved. No part of this publication may be reproduced, stored in a retrieval system or transmitted in any form or by any means, including electronic, mechanical, photocopying recording or otherwise without the prior written permission of the publisher or the copyright holder.

This book contains information obtained from authentic and highly reliable sources. Reasonable efforts have been made to publish reliable data and information, but the author/s, editor/s and publisher cannot assume responsibility for the validity, accuracy or completeness of all materials or information published herein or the consequences of their use. The work is published with the understanding that the publisher and author/s are not attempting to render any professional services. The author/s, editor/s and publisher have attempted to trace and acknowledge the copyright holders of all material reproduced in this publication and apologize to copyright holders if permission and/or acknowledgements to publish in this form have not been taken. If any copyrighted material has not been acknowledged, please write to us and let us know so that we may rectify the error, in subsequent reprints.

Trademark Notice: NIPA, the NIPA logos and their presentations (the way they are written/ presented) in this book are the trademarks of the publisher and hence may not be used without written permission, if copied or used without authorization, the infringer will be prosecuted as per law.

NIPA also publishes books in a variety of electronic formats. Some content that appears in print may not be available in electronic books, and vice versa.

Composed and Designed by NIPA.

Dedicated to
Farmer's of North Bengal

Preface

In the third world developing countries like India, the agriculture and allied sector plays a pivotal role in case of ensuring food security, nutritional security and economic development of the pro-poor people. This sector is also responsible for generating income and employment of a large number of people involved in agricultural food value chain. Agriculture sector is not only associated with mere food production but it is also associated with the food processing industry, wholesale and retail marketing of the agricultural products. Consequently, to meet these responsibilities the agriculture sector is very much dependent on the efficiency of the primary agricultural producers for quantitative and qualitative production of food to meet the growing demand. But still the primary producers are facing several challenges at the farm level which are adversely affecting the ability to produce the agricultural products to meet the requirement of the value chain actors and consumers. Among the challenges the major challenges embedded with the agricultural sector are gigantic population pressure, whims and fences of weather parameters like rainfall, temperature, natural disaster, rapid environmental degradation, risk in case of using synthetic external inputs, depletion of natural resources, poor accessibility of updated scientific information, market intelligence etc. Under such a situation, there is need to make the agricultural producers more concern about the delineation and adoption of new ways for mitigating and averting the discourses of climate change and associated uncertainties with their farming enterprises. There is need for a movement on evergreen revolution now days wherein the food security would be maintained by considering the ecological imperatives of the environment to make it sustainable with a prolonged impact. So, there is an immense role to be played by the agricultural scientists, the agricultural producers and value chain actors by conceptualising the concept of sustainable agriculture in a holistic manner.

Through this book a variety of erudite scholars explore the concept, dimension, opportunities, strategies, innovative technologies, challenges and potential solution in the light of sustainability of agriculture and allied sectors to send a message to the planners, policy makers regarding the practices towards sustainable agriculture.

We, the editors of this book owe our deepest sense of gratitude to the chapter contributors of this book without whose strong diligence and determination it is hardly possible to complete.

Editors

Contents

List of Contributors

Abha Manohar K, *Research Scholar, Department of Forestry, UBKV, Cooch Behar, West Bengal, India*

Amrit Tamang, *Assistant Professor, Department of Soil Science and Agricultural Chemistry Faculty of Agriculture, UBKV, Cooch Behar, West Bengal, India*

Arindam Kundu, *Research Fellow, Department of Agronomy, Faculty of Agriculture, BCKV Mohanpur, Nadia-741252, West Bengal, India*

Arju Sayed Ahmed, *Research Scholar, Department of Agronomy, UBKV, Cooch Behar, West Bengal, India*

Ashok Choudhury, *Professor, Department of Soil Science & Agril. Chemistry, Uttar Banga Krishi Viswavidyalya, Pundibari, Cooch Behar, West Bengal, India*

Asit Mandal, *Scientist, Division of Soil Biology, Indian Institute of Soil Science, Nabibagh Berasia Road, Bhopal, India*

Bappa Paramanik, *Subject Matter Specialist (Soil Science), DakshinDinajpurKrishiVigyan Kendra,UBKV, Majhian, Patiram, Dakshin Dinajpur, West Bengal, India*

Biplab Tudu, *Assistant Professor, Agricultural Entomology, RRRS, Hill Zone, Pedong Kalimpong, UBKV, West Bengal, India*

Biplov C. Sarkar, *Research Scholar, Department of Forestry, UBKV, Cooch Behar, West Bengal, India*

Biswajit Patra, *Assistant Professor, Agricultural Entomology, RRS, Hill Zone, Kalimpong UBKV, West Bengal, India*

Bisweswar Mahato, *Subject Matter Specialist (Soil Science), KrishiVigyan Kendra, Kalyan Vivekenandanagar, Purulia-723126, West Bengal, India*

Dilip Kumar Hajra, *Assistant Professor, Animal Science, Department of Agronomy, UBKV Cooch Behar, West Bengal, India*

F.H. Rahman, *Principal scientist, ICAR-ATARI Kolkata, Salt Lake, Sector-III Kolkata-700097,West Bengal, India*

Hriday Komal Tarafdar, *Assistant Professor, Soil Science and Agril. Chemistry, Regional Research Station, Hill Zone, Kalimpong, UBKV, West Bengal, India*

Indranil Das, *Assistant Agricultural Chemist, Fertilizer Control Laboratory, Govt. of West Bengal, Abas, Midnapore-721102, West Bengal, India*

Kausik Pradhan, *Associate Professor, Department of Agricultural Extension, Uttar Banga Krishi Viswavidyalya, Pundibari, Cooch Behar, West Bengal, India*

Lakshmi Hijam, *Assistant Professor, Department of Genetics and Plant Breeding, UBKV, Cooch Behar, West Bengal, India*

Moumita Chakraborty, *Assistant Professor, Department of Genetics and Plant Breeding, UBKV, Cooch Behar, West Bengal, India*

Megna Rashid, *Faculty of Forestry, SKUAST, Kashmir, Jammu and Kashmir, India*

Nabakishore Mahato, *Field Assistant, Farm Section, Regional Research Station (OAZ) UBKV, Majhian, Dakshin Dinajpur, West Bengal, India*

Nazir A. Pala, *Assistant Professor, Faculty of Forestry, SKUAST, Kashmir, Jammu and Kashmir, India*

Nonigopal Shit, *Assistant Professor, Animal Science, Regional Research Station, RRS UBKV, Cooch Behar, West Bengal, India*

Pabitra Adhikary, *Subject Matter Specialist, Agronomy, North 24 Parganas Krishi Vigyan Kendra, Ashokenagar, North 24 Parganas, India*

Parimal Panda, *Technical Assistant, Soil Microbiology Laboratory, Regional Research Station,UBKV, Pundibari, Cooch Behar-736165, West Bengal, India*

Partha Sarathi Patra, *Assistant Professor (Agronomy), RSS Terai Zone, Directorate of Research, UBKV Cooch Behar, West Bengal, India*

Rajesh Saha, *Research Scholar, Department of Agronomy, UBKV Cooch Behar, West Bengal, India*

Ranajit Panda, *Research Associate, IMD, GKMS-Project, Regional Research Station (OAZ) UBKV, Majhian, Dakshin Dinajpur-733133, West Bengal, India*

Sagar Tamang, *(Scientist, Plant Protection), Krishi Vigyan Kendra, Purba Mednipur, BCKV West Bengal, India*

Sandip Hembram, *Young Professional-II, ICAR-NICRA, Nimpith,Ramkrishna Ashram KrishiVigyan Kendra, South 24 Pargana-743338, West Bengal, India*

Satyajit Hembram, *Assistant Professor, Plant Pathology, RRS, Terai Zone, UBKV, Cooch Behar, West Bengal, India*

Shyamal Kheroar, *Assistant Professor (Agronomy), AINP on Jute and Allied Fibres Directorate of Research, UBKV, Cooch Behar, India*

Sibdas Baskey, *Assistant Professor, Plant Pathology, RRS, Hill Zone, Kalimpong, UBKV West Bengal, India*

Subhra Shil, *Subject Matter Specialist, Horticulture, Krishi Vigyan Kendra, Khowai, Tripura India*

Sumit Chakraborty, *Professor, Department of Forestry, UBKV, Cooch Behar, West Bengal India*

Suvendu Kumar Roy, *Professor, Department of Genetics and Plant Breeding, UBKV, Cooch Behar, West Bengal, India*

Tarik Mitra, *Soil Scientist, National Remote Sensing Centre, Dept. of Space, Govt. of India Hyderabad-500037, Andhra Pradesh, India*

1

Basic Elements of Sustainable Agriculture

Pabitra Adhikary, Partha Sarathi Patra, Shyamal Kheroar
Amrit Tamang and *Rajesh Saha*

Sustainable agriculture's benefit to farm and community economies is grounded in four well-established economic development principles and a fifth, concern for the community:

Lower input farming system (LIFS)

Lower input farming system means reduced production levels with larger area of land used in order to maintain the same revenue; or reduced dependency on external inputs with similar production structure but improved valorization of the product through transformation and or direct marketing, cutting the length of the production chain is a sustainability concept (produce and consume locally). LIFS are seeking optimization of on-farm resources, minimization of off-farm resources use. It also leads to mixed farming system, characterized by animal valorization of vegetal production on farm and organic manure restitution to farm soil from farm husbandry. Mixed character of farming system optimally combines annual and perennial crops and animal production on the same farm. Lower input farming also maintains ecological infrastructures (at least 5% in integrated farming)-which prove to be very beneficial for biodiversity – can provide shelter for organisms providing biological control and also be a source of raw material (wood) for producing on-farm energy.

Diversification

A wider deployment of agricultural biodiversity is an essential component for sustainable and secure food production. Diversity of kingdoms, species and gene pools can increase the productivity of farming systems in a range of growing conditions. Diversity can maintain and increase soil fertility and mitigate the impact of pests and diseases. Diversity of diet, founded on diverse farming systems, provides better nutrition and quality health, with additional benefits for human productivity and livelihoods. Agricultural biodiversity will also be absolutely essential to mitigate the impacts of climate change, not simply as a source of traits but as the underpinnings of more resilient farm ecosystems.

Conservation of natural resources

Conservation of natural resources plays an important role in decreasing food safety risks on the farm. Stream-side vegetation, grassed filter strips, and wetlands help keep our water supply clean by reducing the movement of pathogens, nutrients and pesticides into streams, rivers and lakes. Windbreaks and hedge rows reduce the amount of dust and other airborne contaminates blowing onto produce fields. Conservation practices also serve as wildlife habitat which can support beneficial insects that prey upon pests, raptors that serve as on-farm rodent control, or other species that help to reduce the need for toxic chemicals to control agricultural pests. In sustainable agriculture, economic value is created by maintaining the productivity of land and water resources while enhancing human health and the environment.

Capturing value-added

As farmers struggle to find ways to increase farm income, interest in "adding value" to raw agricultural products has grown tremendously. The value of farm products can be increased in endless ways: by cleaning and cooling, packaging, processing, distributing, cooking, combining, churning, culturing, grinding, hulling, extracting, drying, smoking, handcrafting, spinning, weaving, labeling or packaging. For producers, capturing value usually means capturing some of the value added by processing and marketing. Marketing directly to the consumer can be done on a small or large scale and in a variety of ways. Options for the producer who enjoys direct contact with consumers include selling at farmers' markets and through community supported agriculture systems. Other options include sales directly to restaurants and local institutions, as well as mail order and Internet sales (Born and Bachmann, 2006). Therefore, these value-added strategies will require the formation of a coop of local farmers and a collaborative relationship with the local community.

Community

Social sustainability relates to the quality of life for those who work and live on the farm, as well as those in the local community. Fair treatment of workers, positive farm family relationships, personal interactions with consumers, and choosing to purchase supplies locally are just some of the aspects considered in social sustainability. Community supported agriculture (CSA), farmers markets, U-pick, cooperatives, and on-farm events are just some of the ways a sustainable farm can have a positive impact on the local community. In essence, the farm supports the community and the community supports the farm. A sustainable farming system include shared commitment to profitability, food security, food safety, open space for water recharge, natural habitats for flora, fauna and recreation and a cooperative and supportive social and economic community infrastructure. Recognition of the role farming has played in stabilizing our community is critical or we shall continue to disintegrate our rural fabric and preferred standards of living. In other words, we must rekindle a sense of caring about the welfare of our neighbors in order for viable rural and urban communities to survive.

Reference

Holly Born and Janet Bachmann, 2006. Adding Value to Farm Products: An Overview. ATTRA-National Sustainable Agriculture Information Service. Pp: 1–3.

2

Farming System and Sustainable Agriculture

Shyamal Kheroar, Partha Sarathi Patra and *Pabitra Adhikary*

Introduction

Nature is dynamic; it causes inheritable changes in all living organisms. For thousands of years, farmer around the world have been selecting and conserving varieties of different crop plants that they cultivated. This process has generated a rich wealth of varieties in each crop plant, seen to be most abundant in countries near the equator and India is no exception. In the world there are Mega centers of biodiversity of which aware present in India, this is because in India, the different soils and agro-climatic situations are present, so Indian farmers grow a large number of crops. Generation of Indian farmers, with their continued selection and conservation, has created a rich wealth of varieties in many crops. Therefore India is the original home of many crops such as rice, little and kodo millets, red gram, moth bean, jute, pepper, cardamom, many vegetables and fruit species. These plants were identified from the wild, selected and cultivated by Indian farmers over hundreds of years. The present wealth of varieties in India includes both crops that originated in the country and those that were introduced from the other countries during the distant and recent paste are soybean, sunflower, oil palm and kiwi fruit.

The Indian economy is predominantly rural and agriculture oriented. In agriculture, 85% of the holdings are less than two hectares and the declining trend in the average size of the farmer holdings, poses a serious problem. Majority of them are dry lands, which depend on erratic monsoon rains. The rest of the area is cultivated with supplemental irrigation. The farmers concentrate mainly on crop production, which is invariably subjected to a high degree of uncertainly income and employment.

In India the cultivable land is 143.8 million hectares and there is very little possibility of extending it further. Therefore, to meet the requirement of food grains for increasing population, the only option open is through time and effective space utilization in agriculture. The time concept relates to increasing the intensity of cropping under assured irrigated conditions, whereas space utilization pertains to building up of vertical dimension through multi-tier cropping and farming system approach. Thus by making use of these time and space concept either in irrigated or in rained areas, the productivity per unit area per unit time can be substantially enhanced. Therefore the only way to increase an agricultural production in the small/marginal units of farming is to increase the productivity per unit time and area. This may be achieved of quicker maturing varieties with equal yields or by improving techniques of culture, fertilizer use, weed and pest control.

The green revolution in mid sixties stirred by research based new technological developments involving new materials, models and ways of organizing farm inputs (water, fertilizer, chemical etc) and government's policies transformed the agriculture dramatically. The outcome chewed a many fold increase in production and productivity viz. food grain production of 211 million tones (which was only 74.23 million tones in 1966-67) and food grain productivity of 1697 kg/ha (which was only 644 kg/ha in 1966-67). The country thus became self sufficient in food production despite tremendous pressure to sustain 16% of world human population and 10% of cattle population with just 2.4% of total land.

However modern crop production technology has considerably raised output but has residues in farm produce, gene erosion of fossil fuel based inputs such as fertilizer, pesticide, herbicides and labor spacing but energy intensive farm machinery.

Land water and plant (bio diversity) are the natural resources and conservation of these natural resources is vital important. As far as resources degradation is concentrated, India is facing top soil depletion, falling ground water table and contamination of water, ecological imbalance due to deforestation and healthy hazards linked to increase use of toxic agro-chemical because of increase in food consumption. The adverse soil loss is estimated to be over 16 tones/ha/year, which is layer of top soil. Annual top soil loss is 6000 million tones, which carries about 8.4 million tones of nutrients and it is more than what we are using today. Similarly Tata Energy Research Institute (TERI) reported that more than 10% of GDP is being lost annually on account of environment and around 11-26% of country's agricultural output is lost on account of soil degradation.

The rate of human induced land and degradation is very high. Out of total geographical area of 329 million ha, 187 million ha, representing almost 57% of total geographical area of the country are reported to have degraded of which 162 million ha is way of wind and water related degradation.

What is farming system?

Farming systems represent an appropriate combination of farm enterprises, viz. cropping systems, livestock, fisheries, forestry, poultry and the means available to the farmer to raise them for profitability. It interacts adequately with the environment without dislocating the ecological and socio-economic balance on one hand and attempts to meet the national goals on the other.

Integrated farming system (IFS)

Integration of various agricultural enterprises, viz. cropping, animal husbandry, fishery, forestry etc., a judicious mix of any one or more with cropping complements the cropping enterprise.

Farming system research

It is highly location-specific research which is multi and interdisciplinary in nature and uses whole farm approach for improved technologies to enhance and stabilize agricultural production. The research strategy includes base data analysis, on-centre research and on-farm research. This is the final evaluation of system in the real world situation of the farmer.

Farming system concept

Sustainability is the objective utilization of inputs without impairing the quality of environment with which it interacts. Therefore, it is clear that farming system is a process in which sustainability of production is the objective.

The overall objective is to evolve technically feasible and economically viable farming system models by integrating cropping with allied enterprises for irrigated, rained, hilly and coastal areas with a view to generate income and employment from the farm.

Specific objectives

1. To identify existing farming systems in specific areas and access their relative viability.
2. To formulate farming system. Model is involving main and allied enterprises for different farming situations.
3. To ensure optional utilization and conservation of available resources and effective recycling of farm residues within system and
4. To maintain sustainable production system without damaging resources/ environment.
5. To rise over all profitability of farmhouse hold by complementing main/ allied enterprises with other.

Scope of farming system

Farming enterprises include crop, livestock, poultry, fish, free, sericulture etc. A combination of one or more enterprises with cropping, when carefully chosen, planned and executed gives greater dividends than a single enterprise, especially for small and marginal farmers. Farm as a unit is to be considered and planned for effective integration of the enterprises to be combined with crop production activity.

Integration of farm enterprises depends on factors such as

1. Soil and climatic features of the selected area.
2. Availability of the resources, land, labor & Capital.
3. Present level of utilization of resources.
4. Economics of proposed integrated farming system.
5. Managerial skill of farmer.

Characteristics of farming system research

1. It is holistic or system oriented,
2. It is problems solving: involvement of farmers in problem identification and solving process,
3. It is farmer participatory,
4. It envisages location specific technology solutions,
5. It is for specific client group – small/marginal farmer,
6. It adopts bottom up approach,
7. It compasses extensive on farm activities, collaboration between farmer and scientist,
8. It is gender sensitive,
9. It ultimate objective is sustainability,
10. It focuses on actual adoption,
11. It recognizes interdependence among multiple clients.

Components of farming systems

In the integrated farming system, it is always emphasized to combine cropping with other enterprises/activities, many enterprises are available and these includes cattle maintenance sheep or goat rearing, poultry, piggery, rabbit rearing, bee keeping etc.Any one or more can be combined with the cropping.

1. **Buffaloes:** Important dairy breeds of buffalo are murrah, mehsana, zefarabadi, Godavari. Feeding: Cattle feed generally contains fibrous coarse low nutrient straw material. Roughage is basic for cattle ration

and includes legumes non-legume hays, straw and silage of legume and grasses. Per day requirement @ 1 kg concentrate per 2 lit of milk, green fodder (20-30 kg), straw 5-7 kg & water – 32 lit.

2. **Sheep Rearing:** Sheep are well adapted to many areas. They are excellent gleaners and make use of much of waste feed. Consume roughage, converting a relative cheap food into a good cash product. Housing is not expensive. Feeding: 1-2 kg of leguminous hay per day. Protein supplied through concentrate as groundnut cake.
3. **Goat Rearing:** In India, activity of goat rearing under different environments. The activity is also associated with different systems such as crop or animal based, single animal or mixed herd small or large scale. Goat is mainly reared foe meat, milk hide and skin meat preferred in India, A goat on hoof fetches a better price than a sheep on hoof. Feeding: per head nutrients requirement to goat is low. Hence they are suitable for resources poor small farmers with marginal grazing lands they eat plants and leaves of tree, which any other animals not touch. Goat eats 4-5 times that of body weight concentrate of maize, groundnut cake etc. and clean and fresh water.

Figure 1: Different components of farming systems

4. **Poultry:** poultry is one of the fattest growing food industries in the world. Poultry meat accounts for about 27% of total meat consumed worldwide poultry industry in India is relatively a new agricultural industry. Egg production may reach up to 5000 crores and broiler meat production 330 thousand tones (by 2000) the average global consumption is 120 eggs per person/year and in India it is only 32-33 eggs per capita/year. To meet the nutritional requirement the per capita consumption estimated at 180 eggs 9 kg meat/year. Feed: The feed conservation efficiency of the bird is superior to other animals. About 60-70% of the total expenditure on poultry farming is spent on the poultry feed. Hence, use of cheap and

efficient ration will give maximum profit cereals-maize, barley, oats, wheat, rice – broken mineral/salt – limestone, salt manganese.

5. **Duck rearing:** Ducks account for about 7% of the poultry population in India. They are popular in cereal and logged states like west Bengal. Orissa, A.P, T. Nadu, they have production potential of about 130-140 eggs/bird/year. These can rear in marshy riverside Westland. Duck farming can be a better alternative. Feeding: Eating fallen grains in harvested paddy fields, small fishes and other aquatic materials.
6. **Turkey rearing:** Turkey is a robust bird and can be reared in humid tropics. It actively feed on a variety of crop residues and insects in the farm.
7. **Piggery:** pigs are maintained for production of pork.
8. **Rabbit Rearing:** In India is of recent origin though hunting of wild rabbits for meat is not uncommon. Rabbit can be easily reared with relatively less concentrate feed with high production rate.
9. **Bee Keeping:** Bee keeping is one of the most important agro-based industries which do not required any raw material like other industries. Nectar and pollen from flower are the raw material, which available in plenty in nature. Bee keeping can be started with a single colony. Honey collection: Honey should have good quality. Qualities such as aroma, color, consistency and floral sources are important. Honey is an excellent energy food with an average of about 3500 calories per kg. it is directly absorbed into the human blood stream requiring no digestion.
10. **Aquaculture:** Ponds serve as domestic requirement of water, supplementary irrigation to crop and pisciculture with the traditional management, farmer obtain hardly 300 – 400 of wild and culture fish/ha/year. However, polyfish culture with the stocking density of 7500 fingerlings and supplementary feeding will boost the total biomass production. Species: cattle, Rohu, common carp, silver carp, and grass carp (feed on aquatic plants). Management: Pond depth – 1.5-2.0 m, water should be slightly alkaline, pH-7.5-8.5. If the pH less than 6.5, it can be adjusted with addition of lime. Higher pH (> 8.5) can be reduced with addition of Gypsum. Application of fresh dung may also reduce high pH in the water. The fish are to be nourished with supplementary feeding with rice bran and oilseed cakes. This will enable faster growth and better yield. Each variety of crops stocked to 500 fingerlings with the total of 5000-8000/ha. This gives 2000 to 5000 kg/ha of fish annually.
11. **Sericulture:** Definition: the keeping of silk moths and their larvae for the production of silk or Seri culture is defined as a practice of combining

mulberry cultivation; silkworm rearing and silk reeling. Sericulture is a recognized practice in India. The total area under mulberry is 240 thousand ha in the country. It plays an important role in in socio-economic development of rural poor in some areas. Climate condition favorable for mulberry and rearing of silk worms throughout the year, Karnataka is the major silk producing state in India (temp 21 to 30° C), in Kashmir climate suit from May to October.

12. **Moriculture:** Cultivation of mulberry plants is called as Moriculture. The crop yields well for 12 years. Yield of mulberry leaves is 30-40 t/ha/year. Rearing: eggs are allowed to be laid over a cardboard. In Bamboo tray rice husk is spread. Tender chopped mulberry leaves are added to the tray. The hatched out larvae are transferred to the leaves it is important to change the leaves every 2 – 3 hours during the first 2 – 3 days. The cocoon constructed with silk. The cocoons required for further rearing are kept separately and moths are allowed to emerge from them.
13. **Mushroom Cultivation:** Mushroom is an edible fungi great diversity in shape, size and colour. Essentially mushroom is a vegetable that is a cultivated that is cultivated in protected farms in a highly sanitized atmosphere; mushroom contains 90% moisture with in quality protein, fairly good source of vitamin C and B complex. It is rich source of mineral like ca, P, K & Cu. They contain less of fat and CHO and are considered good for diabetic and blood pressure patients. Varieties: (*i*) Oyster mushroom (Pleurotus ostreatus) (*ii*) Paddy straw mushroom-Volvarilla volvacea (*iii*) White button mushroom-Agaricus bisporus

Oyster mushroom

Paddy straw mushroom

White button mushroom

14. **Biogas Plant:** Biogas is a clean, unpolluted and cheap source of energy, which can be obtained by a simple mechanism and little investment. The gas is generated from the cow dung during anaerobic decomposition. Biogas generation is a complex bio-chemical process, celluloytic material are broken down in methane and CO_2 by different group of micro-organism. IT can be used for cooking purpose, buring lamps, etc. Biogas near to kitchen & cattle shed to reduce cost of gas transfer and cow dung transport, sunlight is important for temperature. Biogas slurry: slurry is obtained after the production of biogas. It is enriched manure; another

positive aspect of this manure is that even after weeks of exposure to the atmosphere the slurry does not attract fleas and worms. Dry slurry contains about 1.8% N, 1.10% P and 1.50% K.

Classification of Farming Systems

A. According to the size of the farm

1. Collective farming

It includes the direct collection of plant products from non-arable lands. It may include either regular or irregular harvesting of uncultivated plants; honeying and fishing usually go hand in hand with collection. Actual cultivation is not needed. The natural products like honey, gum, flower etc are collected. Such plant product may be collected from forestry area.

2. Cultivation farming

In this system, farming community cultivates the land for growing crops for obtaining maximum production per unit area.

(i) Small scale farming: In this type, the farming is done on small size of holding and other factors of production are small in quantity and scale of production is also small.

Advantages

1. Intensive cultivation is possible
2. Labour problem do not affect the production.
3. It is easy to manage the farm
4. There is less loss due to natural calamities like frost, heavy rainfall, and diseases.
5. Per unit output increases

Disadvantages

1. Small-scale farming cannot take advantages of various economic measures:
2. Cost of production per unit is more
3. Mechanization of agriculture is not possible.
4. Farmer does not get employment round the year.

(ii) Large scale farming: when farming is done on large size holding with large amount of capital, large labour force, large organization and large risk are called large-scale farming.

In other words when the factors of production are large in quantity, the small of farming is said to be large. In India 40 to 50 hectares land holding may be said large scale farming but in countries like America, Canada and Australia even 100 ha. farms are also called as small farms.

Advantages

1. Production of large scale farming is more economical. The cost of production per unit is less.
2. Per unit production is increased.
3. Better marketing of agricultural products is possible. Processing, transportation, storage, packaging of produce is economical.
4. Costly machine like tractor, combined harvester can be maintained on the farm.
5. Subsidiary occupation such as dairy, poultry, bee keeping based on maintained on the farm.
6. Proper utilization of factors of production is possible.
7. Research work is possible.
8. It increases bargaining power of people.

Disadvantages

1. If demand of produce decreases and production exceeds the market demand there will be more loss to the large farm.
2. In case of labour strike there will be more loss on the farm.
3. Due to natural calamities like frost, drought, flood, insects and diseases the large farm will suffer a lot.
4. It will be difficult to manage large scale farm.

B. According to production of land, labour and capital investment

(1) Intensive cultivation

In intensive cultivation, more labour and capital are used in the same piece of land. In other words land remains fixed in quantity while other factors are increased. If the same land is rare due to population pressure, while labour and capital are comparatively cheap, intensive cultivation is preferred than extensive cultivation. The application of intensive cultivation method depends mainly upon-

(i) Increasing population and
(ii) Technical improvement.

In the earlier stages of development population was small and technical knowledge of agriculture was also limited hence extensive method was adopted but, as population increases intensive cultivation become necessary and improvement in technique make its adoption is possible.

(2) Extensive cultivation

When more area is brought under cultivation to increases the output it is termed as extensive cultivation. In extensive cultivation land is chiefly available but availability of other factors increases less proportionately. A cultivator wishing to increases his output may follow either intensive method or extensive method but the selection of these two methods is based on cost. If following extensive cultivation than by following intensive cultivation can raise the additional output more cheaply, extensive method of cultivation will be useful. If on the other hand intensive cultivation seems to be the cheaper method he will naturally adopt it. If land is cheaper and it can be had at a normal cost while labour and capital are comparatively costlier, extensive cultivation will be cheaper method of obtaining increased output. In early times when land was plentiful extensive cultivation was followed. The extensive and intensive cultivation go side by side in a country for a certain period of time and afterwards intensive cultivation may become more important method. In most of the countries extensive and intensive methods of cultivation generally go hand in hand.

C. According to the value of products or income or on the basis of comparative advantages

(1) Specialized farming

The farm in which 50% or more income of total crop production is derived from a single crop is called specialized farming or The farm in which only single crop is cultivated for selling in the market and the income of the farm depends mainly on that crop is called specialized farming by Hopkins.

According to the definition if 50% income is derived from paddy from any farm this is called paddy farm similarly sugarcane farm wheat farm, vegetable farm, orchard farm etc.

Advantages

1. Better use of land: More profitable to grow crops on land best suited to it. *E.g.,* jute growing or cultivation on swampy land in west Bengal.
2. Better marketing: it allows grading, processing, storing, transporting and financing the produce.
3. Less equipment and labour.
4. Costly and efficient machinery can be kept: A wheat harvester thresher can be maintained in a highly specialized wheat farm.

5. The efficiency and skill of the labor increased: Specialization allows a man to be more efficient and expert at doing a few things.
6. Farm records can be maintained easily.
7. Intensity of production leads to relatively large amount of output.
8. Better management: fewer enterprises on the farm are liable to be less neglected and sources of wastage can easily be detected.

Disadvantages

1. There is greater risk: When failure of crop and decreasing market price of the product, demand in market of product.
2. It is not possible to maintain soil fertility-lack of crop rotation.
3. The productive resources *i.e.,* land; labor and capital are not fully utilized.
4. Irregular income of the farm as they get income only once or twice in a year.
5. Proper utilization of resources is not possible.
6. By product of crop are not property utilized, as numbers of livestock's are less in number.
7. Due to specialization of a single enterprise, the knowledge about other enterprises vainness.
8. Does not help in supplying all the food needs of the family members of the farmer.

(2) Diversified Farming

A diversified farm is one that has several production enterprises or sources of income but no source of income equal as much as 50% of the total income from that source on such farm farmers depends on several sources of incomes. It is also called as general farming.

Advantages

1. Better use of land, labour and capital: Better area land through adoption of crop rotations, steady employment of farm and family labour and more profitable use of equipment are obtained in diversified farming.
2. The farmer and labour engaged all the year round in different activities.
3. Less risk to crop failure and market price of the product.
4. The by-products of this farm can utilize properly as cattle, poultry, birds, etc. are reared with crop production.
5. Regular and quicker return is obtained from various enterprises.
6. Soil erosion can be checked as land kept under cultivated throughout the year

7. Soil fertility can be checked as land kept under cultivated throughout the year.
8. Diversified farming is less risky than specialized farming.
9. Best use of all equipments.

Disadvantages

1. Do not fetch desirable profit so long as co-operative marketing facility is not there.
2. Proper inspection of different enterprises is difficult.
3. It is not possible to farmer to maintain all types of machinery required for different crops.
4. The wastage of farm in any farm is difficult to detect.

D. According to the water supply

(1) Rainfed Farming

Agriculture mainly depends on the rainfall in most part of the country. 80% of the total cultivated arable land is rainfed. Rain fed farming is very risky system of farming where the success of the crop depends on the cycle of the monsoon. Timely rainfall is the pre-requisite of this farming. The uneven rainfall is quite detrimental to crop production.

Characteristics of rainfed farming

1. Crop and varieties, which can withstand moisture stress should be cultivated.
2. Kharif crops sown after receiving monsoon.
3. Not possible to adopt improved methods of cultivation only one or two crops are grown.
4. Crops rotation is not followed.
5. Soils of these areas are deficits in nutrient.
6. Mixed cropping should be practiced and adopt deep-rooted crops.
7. Short duration varieties fit well in rain fed areas.
8. The crops that are tolerant to drought should be cultivated.
9. Soil moisture should be preserve by mulching, FYM application.
10. Soil erosion which may be called "Creeping death" of the soil is a worldwide problem, so necessary measures should adopt to keep the soil productive.

(2) Irrigated Farming

The crop can be grown throughout the year; moisture is not a limited factor.

Characteristics

1. The round the year cropping pattern becomes possible.
2. Intensive cropping is possible.
3. Production can be increased by proper utilization of productive resources.
4. Crop rotation can be executed properly due to adequate irrigation facility.
5. Manuring is safely done in irrigated crop.
6. The field experiment is possible, because of timely irrigation facility.

E. According to type of rotations

(1) Lay System

In this system, several years of arable farming are followed by several years of grassed and legumes utilized for livestock production.

(i) Unregulated lay farming: In this system natural vegetation grasses, bushy growth on pasture is allowed to grow during the period of fallow. This is an improved managed pasture.

(ii) Regulated lay system: During the period of fallow, certain types of grasses are grown or planted. These are the well managed pasture with fencing and adopting rotational grazing system.

(2) Perennial crop system

The crop, which covers the land for many years *e.g.*, Tea, Coffee, sugarcane, In some cases tree crops (oil palm, rubber) are alternated with fallow in other with arable farming, grazing etc.

(i) Type of rotation: The world rotation has two meanings according to the time period involved. There is long term alteration between various retypes of land use such as arable farming, tree farming, grassland use etc. in this rotation means the sequence of this basic type of land use on a given field. Within arable farming there is also the term crop rotation which means the short-term sequences of different arable crops on one field.

(ii) Intensity of the rotation: It is denoted by "R" simple and appropriate criteria for classification, which gives the true relationship between crop cultivation and following within the total length of one cycle of utilization.

$$R = \frac{\text{Length of the cycle of land utilization}}{\text{No. of years of cultivation}} \times 100$$

The length of the cycle is the sum of the number of years of arable in farming the number of fallow years "R" indicates the production of the area under cultivation in relation to a total area available for arable farming.

(a) Shifting cultivation: R < 33% in the case there is more number of years of fallow than actual cultivation.

(b) Lay or fallow farming: In this case R < 66% and 33%

(c) Permanent cultivation; in this large area is cultivated and small area is left fallow. R> 66%.

(d) Multiple cropping: where R = > 100%. If R = 150% means 50% area is under two crops in a year. If R = 300% means three crops in a year are being grown.

F. According to degree of commercialization

(1) Commercialized farming

More than 50% of the produce is for sale.

(2) Partly commercialized farming

More than 50% of the value of produce is for home consumption.

(3) Subsistence farming

Virtually there is no sale of crop and animal products, but used for home consumption. Subsistence farming is a type of farming where the farmers of our country cultivate the crop in their land for the livings. Hence, the holding is small in size; so improved method of cultivation is not possible. They fail to meet the total requirement. They reared cattle, poultry, along with crop cultivation in limited land meet their requirement.

Advantages

1. Utilizing productive resources profitably.
2. Farmers with their family members engaged though the year as they rearing cattle, poultry etc.
3. Farmer meet their demand from the income from cattle, poultry etc.
4. By-product used properly.

Disadvantages

1. Fails to adopt improved crop cultivation technique do to small holding.
2. Cultivation mainly depends on monsoon rain.
3. Procurement of seed, fertilizer as and when required is difficult.
4. Income of this farm is very low.

G. According to degree of nomadic

(1) Total nomadic

In this system, the animal owners do not have permanent place of residence. They do not practice regular cultivation. Their families move with the herds.

(2) Semi nomadic

Animal owners have a permanent place of residence, while Supplementary cultivation is practiced. However, for long periods of time, they travel their herd to distant grazing areas.

(3) Transhumant

Means seasonal migration of livestock to suitable grazing ground or it is the situation in which farmer with a permanent residence sends their herd with herdsman for long period of time to distant grazing areas.

(4) Partial nomadic

Farmer has permanent residence and who have herds at their disposal, which remains in the vicinity.

(5) Stationary animal husbandry

It occurs where the animals are remained on the holding or in the village throughout the entire year.

Sustainable agriculture

In India progress in agriculture was through exploitation of recourses to serve the interest of rules in the power. In the post-independence era, we were faced with the difficult task of feeding an increasing population and were burned by recurring import of food grains. Attending self-sufficient in food grain production was the challenges and a goal before Agriculture scientist. The element of strategy to achieve self-sufficiency in food production within a short time were to expand the area under cropping, make large investments in development of water resources, develop input supplies and market infrastructure. Developments and promotion of high yielding verities responsive to high input was the cornerstone for green revolution of 1970.

1. We are self-sufficient in food grain production. Now the question arises whether to follow the same strategies, which resulted in green revolutions or to redefine them. Consideration of our past achievement new challenges and the related development worldwide can answer this question.
2. Although we are now self-sufficient in food grain production, the per capital availability has not increased significantly and a large fraction of population are still below poverty line due to lack of purchasing capacity.

3. At the present, growth rate of 2% per annum, we will be about 1.3 billion by 2025 and 1.6 billion by 2060. Hence there is need to produce more from shrinking resources base.
4. Major gains in productivity and production in the past three decades have been from areas, which had no serious limitation to production (adequate irrigation productive soils, ideal climate etc.) productivity and production increases was moderate in eastern and central India and Deccan plateau.

Even in areas with no serious limitation for production, technology adopted for enhancing productivity has also simultaneous weekend the resource base resulting in a series of second – generation problems.

Modern crop production technology has considerably raised output but has created problems of land degradation pesticide residue in farm produce, gene erosion, and atmosphere and water pollution. The natural resources base is degraded and diminished and the quality of the environment sustaining human life is adversely affected. Agricultural production has sustained man and great civilization. The history of the world reveals that great civilization flourished along irrigation sources and mismanagement of these resources saw the extinction of these civilizations. With expanding population and rapid depletion and degradation of the natural resource base, sustainable agriculture has assumed very great significance. The task of meeting the needs of the present generation without eroding the ecological assets of the future generation is receiving up priority by environmental planners.

The world sustainability is now widely used in developments circles. But what does it really mean? According to a dictionary meaning sustainability refers to keeping an effort going continuously, the ability to last out and keep from falling. In the context of agriculture, sustainability basically refers to the capacity to remain productive while marinating the resources base. For example, the Technical Advisory Committee og the Consultative group on International Agricultural Research (TAC/CGIAR)- States: Sustainable agriculture is the successful management of resources for agriculture to satisfy human changing needs, while maintaining of enhancing the quality of the environments and conserving natural resources.

Judging agricultural to be sustainable if it is

1. **Ecological sound:** which means that the quality of natural resources is maintained and the vitality of the entire agro-ecosystem form humans, crop and animal to soil organism is enhanced.
2. **Economically viable:** which means that farmers can produce enough for self-sufficient and or income is measured not only in terms of direct farm produce (yield) but also in terms of function such as conservation

resources and minimizes risks.

3. Socially just, which means that resources and power are distributed in such a way that the basic needs of all the members of society are met and their rights to land use, adequate capital technical assistant and market opportunities are assured.
4. **Humane:** Means that all forms of life (plant, animal, human).
5. **Adaptable:** Means that rural communities are capable of adjusting to the constantly changing condition for farming populating growth, policies, market demand. Etc.

Definition: There are many definition of sustainable agriculture. Simply stated, "Sustainable agriculture is a form of agriculture aimed at meeting the needs of the present generation without enhancing the resource base of the future generation".

In order to feed the burgeoning population more food has to be produced and this to be done without enhancing the resources base. Expanding agriculture to ecologically fragile area means Greater threat to environment. According to some workers sustainable agriculture is minimal dependence on synthetic Fertilizer, pesticide and antibiotics. It is also considered as a system of cultivation with the use of manure, crop rotation and minimal tillage.

Principal of sustainable agriculture

1. No fight with nature, but co-operation with them.
2. Living soil-maintain the soil healthy.
3. Lessen the use of outside and distant Resources.
4. Diversity and adjustment: The diversity of all plants and animal should be preserved and increased.
5. Durable livelihood: durable lemmatization about how much resource can be utilized by man in an area.
6. Self-dependence: The new knowledge and technology should be increase self-confidence, which help for effecting planning of resources.

Role of IFS in sustainable agriculture

A wide chain of critical environment threats is posing challenges to the mankind all over the world. Some of these threats are degeneration of soil, water, forest and marine resources essential for increasing production of food, fodder, fiber, fuel, timber etc. Degradation of the natural resources is the main issue threading sustainable developments of agriculture. Causes of resources base degradation: following are the main reasons of derogation of natural resources.

(i) Soil and water conservation to prevent degradation of soil productivity and lengthening crop growing season for optimum crop production.

(ii) Efficient use of limited irrigation water: without leading to the problems of oil, salinity, alkalinity and high ground water tables.

(iii) Crop rotation to mitigate weeds, diseases and insect problems to increase soil productivity and minimizes soil erosion.

(iv) Integrated nutrient management that reduces the need for chemical fertilizer improves the soil health and minimizes environmental pollution by conjunctive use of organics, in organics and bio fertilizer.

(v) Integrated pest management that reduce the need for agro-chemical by crop rotation, weather monitoring, use of resistant cultivators, planning time and biological pest control.

(vi) Management systems to control weeds by preventive measure, tillage, timely Intercultivation and crop rotation to improve plant health.

All such ecologically sound practices require more information, trained labour and management skills per unit production than current input intensive agriculture.

Difference between Sustainable and Modern Agriculture

Particulars	Sustainable agriculture	Modern agriculture
Plant Nutrient	Farm yard manures, compost, vermicompost, green manure, bio-fertilizer and crop rotation are used.	Chemical fertilizers are used.
Pest control	Cultural methods, crop rotation and biological methods are used.	Toxic chemical are used.
Inputs	High diversity, renewable and biodegradable inputs are used	High productivity and low diverse chemicals are used fragile ecology.
Ecology	Stable ecology	easily broken
Use of resources	The rate of extraction from forests, fisheries, underground water source other renewable do not exceed the rate of regeneration.	The rate of extraction exceeds threat of regeneration. Falling of trees, deforestation, overgrazing and pollution of water-bodied takes.
Quality of food material	Food material are safe	Food material contain toxic residue.

Types of enterprises

1. Independent enterprises

Independent enterprises are those which have no direct bearing on each other, an increase in the level of one neither help not-hinders the level of other. In such cases each product should be treated separately *e.g.,* production of wheat and maize independently.

2. Joint enterprises

Joint products are those which are produced together *e.g.*, cotton and cottonseed, wheat and straw etc. the quantity of one product decides the quantity of the other products. In case of joint products there is no economic decision the make with respect to the combination of products and two products can be treated as one.

3. Competitive enterprises

Competitive enterprises are those which complete for use of the farmer limited resources, use of resources to produce more of necessitates a sacrifice in the quantity of other product.

When enterprises are competitive three things determine the exact combination of the product, which would be most profitable.

1. The rate at which one enterprises substitute for the other.
2. Prices of the products and
3. The cost of producing the product.

The rate of which, one product substitute for another is known as the marginal rate of substitution. The products can:

1. Constant rates of substitution.
2. Decreasing rate of substitution
3. Increasing rates of substitution. *e.g.*, paddy-sorghum, paddy-groundnut

4. Supplementary enterprises

Two products are said to be supplementary when an increase in the level off one does not adversely affect the production of the other but adds to the income of the farm i.e enterprise which do not complete with each other but adds to the total income. For example, on many small farms daily enterprise or a poultry enterprise may be supplementary to the to the main crop enterprises because they utilize surplus family labour and shelter available and perhaps even some feeds, which would otherwise go to waste. When two products should be product up to the end of supplementary stage, sometime enterprises are supplementary for one resource but competitive foe another. In such cases the relationship should be treated as one of competitive, even though, they are supplementary to one another in respect of other sources, *e.g.*, mixed crops.

5. Complementary enterprises

Complementary enterprises are those, which add to the production of each other *e.g.*, Berseem and maize crops. Two products are complementary when the transfer of available input for the production of the one product to the production of the other results in increases in the production of both products. Then two crops are complementary enterprises, the use of resources for the two crops result in the increases production on both the crops.

Two enterprises do not remain complementary over all possible combinations. They become competitive at some stages. When both complementary and competitive relationship occurs, the complementary relationship occurs; the complementary relationship occurs first and then is followed by competitive relationship.

Merits or advantages of IFS

1. **Productivity:** IFS provides an opportunity to increase economic yield per unit area per unit time by virtue of intensification of crop and allied enterprises.
2. **Profitability:** Use waste material of one component at the least cost. Thus reduction of cost of production and form the linkage of utilization of waste material, elimination of middleman interference in most input used. Working out net profit B/C ratio is increased.
3. **Potentiality or Sustainability:** Organic supplementation through effective utilization of by-products of linked component is done thus providing an opportunity to sustain the potentiality of production base for much longer periods.
4. **Balanced Food:** We link components of varied nature enabling to produce different sources of nutrition.
5. **Environmental Safety:** In IFFS waste materials are effectively recycled by linking appropriate components, thus minimize environment pollution.
6. **Recycling:** Effective recycling of waste material in IFFS.
7. **Income Rounds the year:** Due to interaction of enterprises with crops, eggs, milk, mushroom, honey, cocoons silkworm provides flow of money to the farmer round the year.
8. **Adoption of New Technology:** Resources farmers (big farmer) fully utilize technology, IFS farmers, linkage of dairy/mushroom/sericulture/ vegetable. Money flow round the year gives an inducement to the small/ original farmers to go for the adoption technologies.
9. **Saving Energy:** To identify an alternative source to reduce our dependence on fossil energy source within short time. Effective recycling technique the organic wastes available in the system can be utilized to generate biogas. Energy crisis can be postponed to the later period.
10. **Meeting Fodder crisis:** Every piece of land area is effectively utilized. Plantation of perennial legume fodder trees on field borders and also

fixing the atmospheric nitrogen. These practices will greatly relieve the problem of non-availability of quality fodder to the animal component linked.

11. **Solving Fuel and Timber Crisis:** Linking agro-forestry appropriately the production level of fuel and industrial wood can be enhanced without determining effect on crop. This will also greatly reduce deforestation, preserving our natural ecosystem.
12. **Employment Generation:** Combing crop with livestock enterprises would increase the labour requirement significantly and would help in reducing the problems of under employment to a great extent IFS provide enough scope to employ family labour round the year.
13. **Agro – industries:** When one of produce linked in IFS are increased to commercial level there is surplus value adoption leading to development of allied agro – industries.
14. **Increasing Input Efficiency:** IFS provide good scope to use inputs in different component greater efficiency and benefit cost ratio.

LISA or BISA

The term LISA (low-input sustainable agriculture) was coined in developed countries in the light of the environmental hazards caused by high or, more often, excessive inputs of fertilizers and pesticides. Not only do these overdoses constitute a threat to the environment, but they also represent a costly production factor. It is against this background that low inputs' were equated with sustainable agriculture'. In the case of industrialized countries, low inputs actually means a reduction of the high levels current in modem' agriculture. It is misleading, however, to apply this same concept to agriculture in developing countries, where corresponding input levels are generally low. Here sustainability problems are generally those of depletion rather than of pollution. Loss of soil fertility results from a continuous mining of plant nutrients without their adequate replenishment. The age-old method of shifting cultivation-which made it possible to grow crops for a few years on nutrients derived from burning the forest-can no longer meet the demands of the present populations, let alone those of future generations.

Clearly, maximum use should be made of local resources such as crop residues, green manures and biological nitrogen fixation, but it must also be realized that residual organic matter cannot replace the harvest that is removed. Furthermore, large areas of acid soils in the tropics are inherently poor, so that even the organic matter they produce is very low in plant nutrients. Therefore, a combination of organic materials and mineral fertilizers, sufficient to replenish the plant nutrients removed by the crops and to enhance soil fertility, should be promoted.

FAO is actively supporting the development of Integrated Plant Nutrition Systems (IPNS) as an approach that is economically viable, environmentally sound and conducive to the urgently needed increases in food production. Naturally, IPNS has to be 'tailored' specifically to the resources available in different agro-ecological conditions, and also adapted to various farming systems. It would be preferable, therefore, to advocate BISA-a balanced inputs sustainable agriculture-rather than LISA. The term balanced' refers to an adequate and feasible combination of organic and mineral inputs, taking into account the availability of organic material and labor. Balanced' also points to the need to ensure a supply of the three basic plant nutrients, N. P and K, since organic material supplies only a substantial amount of N.

Integrated Pest Management (IPM) is another key component in sustainable agricultural systems. By keeping the use of chemical pesticides to a minimum and applying principles based on pest ecology (*e.g.,* pest life cycles, predator/prey relationships), methods of cultivation and production can be adjusted so that farmers reduce their costs while avoiding the risks of excessive use of pesticides.

In the developing countries, LISA is often broadened to LEISA (low-external-input sustainable agriculture), to take account of the cost of purchased inputs which small farmers can afford rarely. The solution is not, however, to deny farmers the use of inputs. Such an attitude would be comparable to solving the problem of the hungry by recommending an even lower food intake. It is the task of public policy to create the economic motivation, and the infrastructure, necessary to give people access to the inputs they need.

Since the early 1980s, FAO has focused on the development of agricultural production and farming systems in a holistic manner and to this end a farming systems analysis and development approach was adopted in the Regular Programme and field work. Initial activities were targeted on the development of methodologies and procedures for the implementation of this approach in field projects and programmes Such methodologies include: production and farm data analysis, rapid rural appraisal techniques, analysis of indigenous production practices, the economics of cropping systems and the preparation of long term multidisciplinary programmes for on-farm technology testing. Emphasis is also given to training, through guidelines for seminars, informal workshops with universities, and the implementation of regional training programmes. Existing methodologies will be further developed to enable household and production systems to be integrated with multiple family objectives that include probability, equitability and sustainability. A new focus is given to the integration of traditional and modern farm and crop management practices and to the reduction of risks through on-farm and off-farm diversification of rural incomes, particularly in fragile ecosystems such as mountains, semi-arid areas and forests.

Indicator of sustainability

An indicator is something that helps you understand where you are, which way you are going and how far you are from where you want to be. A good indicator alerts you to a problem before it gets too bad and helps you recognize what needs to be done to fix the problem. Indicators of a sustainable community point to areas where the links between the economy, environment and society are weak. They allow you to see where the problem areas are and help show the way to fix those problems.

Indicators of sustainability are different from traditional indicators of economic, social, and environmental progress. Traditional indicators-such as stockholder profits, asthma rates, and water quality-measure changes in one part of a community as if they were entirely independent of the other parts. Sustainability indicators reflect the reality that the three different segments are very tightly interconnected, as shown in the figure below:

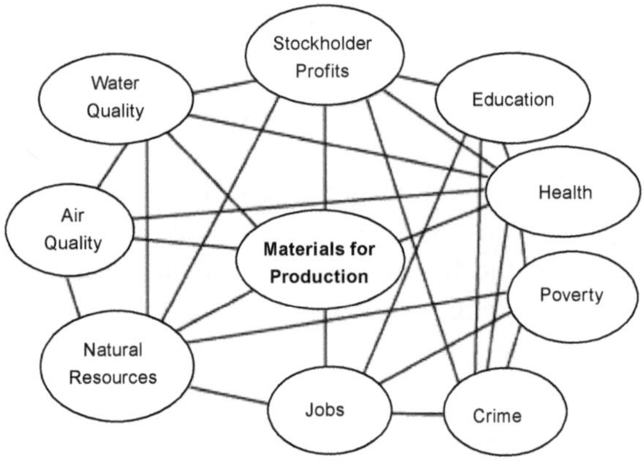

Figure 2: Communities are a web of interactions among the environment, the economy and society.

As this figure illustrates, the natural resource base provides the materials for production on which jobs and stockholder profits depend. Jobs affect the poverty rate and the poverty rate is related to crime. Air quality, water quality and materials used for production have an effect on health. They may also have an effect on stockholder profits: if a process requires clean water as an input, cleaning up poor quality water prior to processing is an extra expense, which reduces profits. Likewise, health problems, whether due to general air quality problems or exposure to toxic materials, have an effect on worker productivity and contribute to the rising costs of health insurance.

Sustainability requires this type of integrated view of the world-it requires multidimensional indicators that show the links among a community's economy, environment, and society. For example, the Gross Domestic Product (GDP), a well-publicized traditional indicator, measures the amount of money being spent in a country. It is generally reported as a measure of the country's economic well-being: the more money being spent, the higher the GDP and the better the overall economic well-being is assumed to be. However, because GDP reflects only the amount of economic activity, regardless of the effect of that activity on the community's social and environmental health, GDP can go up when overall community health goes down. For example, when there is a ten-car pileup on the highway, the GDP goes up because of the money spent on medical fees and repair costs. On the other hand, if ten people decide not to buy cars and instead walk to work, their health and wealth may increase but the GDP goes down.

In contrast, a comparable sustainability indicator is the Index of Sustainable Economic Welfare. In order to get a more complete picture of what is economic progress, the ISEW subtracts from the GDP corrections for harmful bases or consequences of economic activity and adds to the GDP corrections for significant activities such as unpaid domestic labor. For instance, the ISEW accounts for air pollution by estimating the cost of damage per ton of five key air pollutants. It accounts for depletion of resources by estimating the cost to replace a barrel of oil equivalent with the same amount of energy from a renewable source. It estimates the cost of climate change due to greenhouse gas emissions per ton of emissions. The cost of ozone depletion is also calculated per ton of ozone depleting substance produced. Additionally, adjustments are made to reflect concern about unequal income distribution. The correction for unpaid domestic labor is based on the average domestic pay rate. Some health expenses are considered as not contributing to welfare, as well as some education expenses.

Like the GDP, the ISEW bundles together in one index tremendous amounts of information, but the key difference is that the information takes into account the links between environment, economy and society.

Indicators of sustainable community are useful to different communities for different reasons. For a healthy, vibrant community, indicators help monitor that health so that negative trends are caught and dealt with before they become a problem. For communities with economic, social, or environmental problems, indicators can point the way to a better future. For all communities, indicators can generate discussion among people with different backgrounds and viewpoints, and, in the process, help create a shared vision of what the community should be.

Characteristics of effective indicators

An indicator is something that points to an issue or condition. Its purpose is to show you how well a system is working. If there is a problem, an indicator can help you determine what direction to take to address the issue. Indicators are as varied as the types of systems they monitor. However, there are certain characteristics that effective indicators have in common:

- Effective indicators are **relevant**; they show you something about the system that you need to know.
- Effective indicators are **easy to understand**, even by people who are not experts.
- Effective indicators are **reliable**; you can trust the information that the indicator is providing.
- Lastly, effective indicators are based on **accessible data**; the information is available or can be gathered while there is still time to act.

Note that these are all numeric measurements. Indicators are quantifiable. An indicator is not the same thing as an indication, which is generally not quantifiable, but just a vague clue. In addition to being quantifiable, effective indicators have the four basic characteristics noted above. These characteristics are:

- **Relevant:** An indicator must be relevant, that is, it must fit the purpose for measuring. As indicators, the gas gauge and the report card both measure facts that are relevant. If, instead of measuring the amount of gas in the tank, the gas gauge showed the octane rating of the gasoline, it would not help you decide when to refill the tank. Likewise, a report card that measured the number of pencils used by the student would be a poor indicator of academic performance.
- **Understandable:** An indicator must be understandable. You need to know what it is telling you. There are many different types of gas gauges. Some gauges have a lever that moves between 'full' and 'empty' marks. Other gauges use lights to achieve the same effect. Some gauges show the number of gallons of gasoline left in the tank. Although different, each gauge is understandable to the driver. Similarly, with the report card, different schools have different ways of reporting academic progress. Some schools have letter grades A through F. Other schools use numbers from 100 to 0. Still other schools use written comments. Like the gas gauge, these different measures all express the student's progress or lack of progress in a way that is understandable to the person reading the report card.

  On the other hand, a gas gauge that showed the number of BTU's left in the tank would probably not be very useful to you in deciding when to fill up the tank. Likewise, a report card that gave grades in ancient Greek script would be a mystery to most people. In order for you to know when action is needed, you must be able to understand what an indicator is telling you.

- **Reliable:** An indicator must be reliable. You must trust what the indicator shows. A good gas gauge and an accurate report card give information that can be relied on. A gas gauge that shows the tank is empty when in fact it is half full would make you stop for gasoline before it is needed. A gas gauge that shows the tank is half full when in fact it is empty would cause you to run out of gas in an inconvenient place. Similarly, if a student's grade were reported wrong, an honors student could be sent for remedial work and a student who needs help would not get it. An indicator is only useful if you know you can believe what it is showing you.

  Reliability is not the same as precision. When your gas gauge registers empty, you know there is still a gallon or so of gasoline left as a reserve. The gas gauge reliably under-reports the amount of gasoline. An indicator does not necessarily need to be precise; it just needs to give a reliable picture of the system it is measuring.

- **Accessible Data:** Indicators must provide timely information. They must give you information while there is time to act. For example, imagine a gas gauge that only gave you the amount of gasoline in the tank when the engine was started. After you have been driving for several hours, that reading is no longer useful. You need to know how much gasoline is in the tank at each moment. Similarly, a report card distributed a week before graduation arrives too late to give a student remedial help. In order for an indicator to be useful in preventing or solving a problem, it must give you the information while there is still time to correct the problem.

One of the biggest problems with developing indicators of sustainability is that frequently the best indicators are those for which there is no data, while the indicators for which there is data are the least able to measure sustainability. This has led many communities to choose traditional data sources and measures for indicators. There are several advantages to traditional indicators. First, the data is readily available and can be used to compare communities. Second, traditional indicators can help to define problem areas. Third, traditional indicators can be combined to create sustainability indicators.

However, there is a real danger that traditional data sources and traditional indicators will focus attention on the traditional solutions that created an unsustainable community in the first place. It may be tempting to keep measuring 'number of jobs,' but measuring 'number of jobs that pay a livable wage and include benefits' will lead to better solutions. Discussions that include the phrase 'but you can't get that data' are not going to lead to indicators of sustainability. In fact, if you define a list of indicators and find that the data is readily available for every one of them, you probably have not thought hard enough about sustainability. Try to define the best indicators and only settle for less as an interim step while developing data sources for better indicators.

Agronomic requirement in cropping system

Agronomic considerations for different cropping system are different due to inclusion of more than one crop as in intercropping or sequence cropping system. Thus, principal involved in management of l/c and sequence system are different.

Intercrop is practiced with two objectives

- To get additional yields through and intercrop as bonus and
- To avoid risk.

Management of intercropping systems

In intercropping system crops are grown simultaneously. Management practices aim to provide favorable environment to all the component, exploit favorable interaction among the component crops and minimize competition among the components.

- **Seedbed preparation:** The objective of land preparation is to establish an ideal zone for the seedling that minimizes the stress. Potential stress condition include inadequate or excess moisture, unfavorable temperature for a given species, soil crushing weeds, residue of preceding crop and insect or pathogen attack. Important of seedbed is the same in both conventional (monoculture) and in multiple cropping. Seedbed preparation depends on the crop. Deep rooted crops responds to deep ploughing while for most of cereal shallow tillage is sufficient. The crops with small seed required fine seedbed, cotton, and maize, planted on ridges, certain crops on flat seedbed. Since more than one crop in intercropping, the seedbed preparation is generally prepared as per the needs of base crop. Sugarcane planted in furrow and intercrop sown on ridges, certain crops on flat seedbed. Since more than one crop in intercropping, the seedbed preparation is generally prepared as per the needs of base crop. Sugarcane planted in furrow and intercrop sown on ridges. In Groundnut + red gram intercropping system, flat seedbed is prepared for sowing crops. However, ICRISAT is recommending broad bed and furrow for black soils. In rice + maize intercropping system, ridges and trenches are formed. Maize is planted on ridges and rice in trenches.
- **Varieties:** The varieties of component crop in intercropping system should be less competing with the base crop and peak nutrient demand period should different from the base crop. Maturity period between two components is

minimum of 30 days. Hybrids varieties of sorghum like CSH-6, CSH-9 are suitable for intercropping with long duration variety of red gram like C11 and LRG 30 because of wider gap between maturity periods. The verities selected for intercrop should be have thin leaves, tolerant to shading and less branching. If the base crop is shorter than intercrop, the intercrop should be compact with erect branching and its early growth should be slow. The characteristics of the base crop should be as in sole crop.

- **Sowing:** Practices of sowing are slightly altered to accommodate inter-crop in such a way that it cause less competition to the base crop. Widening inter row spacing of cereal component to accommodate more rows of component legume crop improves legume yield and efficiency of the intercrop system. Sowing of base crop is done either as paired row, paired – winder row or skip row of base crop are brought close by reducing inter row spacing. The spacing between two pairs of rows is increased to accommodate the inter crop. Such row arrangement of component crops within the rows improves the amount of light transmitted tio the lower component crop, which can enhance legume yield in cereal legume intercropping system.

For example–the normal row spacing of Rainfed is 30 cm the row spacing is reduce to 20cm between paired rows and 50cm spacing in two in two pairs. The spacing in paired row planting designed as 20/50 cm indicated that the spacing between two rows in pairs is 20 cm and among the pairs 50cm.

Similarly, pearl millet is planted with row spacing 30/60 cm in paired row planting. These changes in crop geometry do not alter the yield of base crop, but intercrops are benefited to some extent. Or when alternating pairs of sorghum rows 90 cm with two rows of an associated legume, Singh (1972) found that LER was greater compared at 60cm between rows with two rows of the legume in between Planting in fixed ratio of intercrop is most common. The intercropping system of groundnut + red gram is either in 5:1 or 7:1 ratio and sorghum + red gram in 2:1 ratio. In these cases the normal three tined or four seed drill can be used without any modification. The hole pertaining to intercrop row in the hopper is closed with a piece of clh in that row, intercrop is sown with alkali or kera.

For higher yields, base crop population is maintained at its sole crop population and intercrop population is kept at 80 percent of its sole crop population.

Relative sowing time of component crop is important management variable manipulated in cereal-legume I/C system but has not been extensively studied. Sowing may be staggered to increase the temporal difference, which might result in higher yield advantage. (Singh et el.1981).

- **Fertilizer Application:** The nutrient uptake is generally more in intercropping system compared to pure crops. When the legume are associated with cereal crop in intercropping system, legume supplement a portion of nitrogen required of cereal crop; the amount may be of 20kg/ha by legumes. Application of higher dose of nitrogen to the cereal + legume intercropping system not only reduce the nitrogen fixation capacity of legumes, but also growth of legume is suppressed by aggressive cereals owing to fast growth of cereals with increased availability of nitrogen, cereal + legume intercropping, therefore is mainly advantageous under low fertilizer application.

  Considering all the factors, it is suggested that the nitrogen dose recommended for base crop as pure crop is sufficient for intercropping system with cereal + legume or legume + legume. With regards to phosphorus and potassium, one eighth to one fourth of the recommended dose of intercrop is also added in addition to recommended dose of base crops to meet the extra demand. Basal dose of nitrogen is applied to rows of both components in cereal + legume inter crop. Top dressing of nitrogen is done only cereal rows. P & K are applied as basal dose to both crops.

- **Water Requirement:** The technique of water management is the same for sole cropping and intercropping or sequential cropping. However, the presence of an addition crop may have an important effect on evapo-transpiration. With proper water management, it is possible to grow two crops where normally only one crop is raised under rain fed condition. Intercropping system is generally recommended for rain fed crop get the stable yields. The total water requirement of I/c does not increase much compared to sole cropping. At ICRISAT, the water requirement of sole sorghum and intercropping with red gram was almost similar (584 and 585 mm, respectively). However in a more competitive crop like onion as intercropped in groundnut increase the total W.R by about 50mm. the total water used in intercropping system is almost same as in sole crops, but yields are increased. Thus water use efficiency of intercropping is higher than sole crops.

  Scheduling of water: if one of the crop is irrigated based on its requirement, the other crop may suffer due to excess water stress, sometimes leading to total failure of crop. In cotton black gram I/C system, cotton is irrigated once in 15-20 days. The intercrop black gram is often affected by excess water and gives poor yield. In such situations, skip furrow method of irrigation is advocated. Scheduling irrigation at IW/CPE ratio of 0.60 to 0.80 or irrigation at one bar soil moisture tension is suitable for most of the systems.

- **Weed Management:** Generally it is believed that intensive cropping reduces weed problems. Weed infestation depends on the crop, plant density and cultural operation done. Weed problems is less in intercropping system compared to the sole crops. This is due to more complete crop cover and high density available in I/C cause severe competition with weeds and reduce weed growth. The weed suppressing ability of intercrop is dependent upon the component crops selected, genotype used, plant density adopted, proportion of component crops, their spatial arrangement and fertility moisture status of the soil.

Experiment carried out at ICRISAT, Hyderabad, indicated that there was 50-75% reduction in weed infestation by intercropping. Pigeon pea + sorghum I/C system, which is extensively practiced in Karnataka, M.S and A.P is known to reduce weed intensity. The higher plant population used and also complete covering of the soil earlier in intercropping system reduces weed infestation. In late maturing crops that are planted wide rows, presence of early maturing crops helps to cover the maturing crops that are planted wide rows, presence of early maturing crops helps to cover the vacant inters row space and keeps weed under check.

Quick growing non-competitive, compact legumes like green gram and black gram act as another crop due to their good canopy coverage. In certain situations, intercrops are used as biological agents to control weeds, black gram, geen gram, cow pea in sorghum and cowpea in banana reduce weed population and one hand weeding can be avoided by this method, however, in some intercropping system like maize + groundnut, rice+ tapioca, maize + tapioca, weed problem is similar to their sole crops. The growth habit of genotype used in intercropping has a great influence on weed growth. Weed infestation in I/C is influenced by early growth and competitive additives of the component crops. If one or both the component crops are vigorous and cover the land area rapid ally, weed infestation is greatly reduced. Early crop canopy to cover the soil is more important than rapid increase in plant height.

It is well known that, different species of weeds are associated with different crops, but weeds present in sole crops are different than those present in intercropping system. At Hyderabad, in pearl millet as sole crop mixed weed flora was observed as Celosia, Digit aria and Cupreous in sole crop of groundnut. In pearl millet + groundnut intercropping system, type of weeds changes with proportion of component crops. As more rows of groundnut are introduced in place of pearl millet of rows, there is a striking increase in both numbers and biomass of the tall and competitive Celosia, especially in groundnut rows.

Weed problem is less; weed control is necessary in intercropping system. But labour required for weeding is less; second weeding is not necessary because of crop coverage and limited weed growth. Normally two hand weeding are required, but it may restrict to one hand weeding under I/C. in sorghum + red gram or sorghum + cowpea, just one weeding is sufficient to get high yield as in weed-free check.

The critical period of weed free condition may be extended a little longer in I/C than in sole cropping. this is because the critical growth stages of the component crops very temporally in I/C. for example, critical weed free period has to be extended to first 7 weeks in sorghum + red gram I/C while sole sorghum crop requires only requires only 2-4 weeks weed free period.

Chemical weed control is difficult in intercropping system because the herbicide may be selective to one crop but non-selective to another. Atrazine control weeds in sole sorghum, but it is not suitable for sorghum + red gram intercropping system, as it is toxic to red gram. Herbicides suitable for intercropping systems as-

- Maize + green gram & Maize + cowpea-Butachlor (pre-emergence) (Machete)
- Sorghum + pulse – fluchloriam (PPI) (Basalin) or Alacholor (pre-emergence) (loss)
- Sorghum + red gram – prometryne (pre-emergence)
- Sugarcane + groundnut – nitro fen (pre-emergence) (to KE-25).

- **Pest and Disease in Intercropping System:** Pest and diseases are believed to be less in intercropping system due to crop diversity than sole crops. Some plant combination may enhance soil fungicide and antibiotics through indirect effects on soil organic matter content. The spread of the diseases is altered by the presence of different crops. Little leaf of Brinjal is less when Brinjal is sheltered by maize or sorghum, as the insect-carrying virus first attacks maize or sorghum; virus infestation is less on Brinjal. Non – host plant in mixtures may emit chemicals or odor that affects the pests, thereby protecting host plants.

The concept of crop diversification for the management of nematode population has been applied mainly in the form of decoy and trap crops. Decoy crops are non-host crops, which are planted to make nematode waste their infection potential. This is affected by activating larva of nematode in the absence of hosts by the decoy crops.

Crop	**Nematode**	**Decay crop**
Brinjal	*Meloidogyne incognita*	Sesamum orientale
Tomato	*Meloidogyne pratylenchus alleni*	Caster, ground nut
Soybean	Pratylenchus sp.	Crotalarias spectabills

Trap crops are host crops sown to attract nematode but destined but destined to be harvest or destroyed before the nematode manage to hatch. This is advocated for cryst nematode. The technique involves sowing in pineapple plantations; tomatoes are planted and ploughed in to reduce root knot nematodes. There is also evidence that, some plants adversely affect nematode population through toxic action. Marigold reduces the population of Pratylenchus species worms.

Management of sequential cropping system

Unlike intercropping crops are grown one after another in sequential cropping and hence, management practices are different.

- **Seedbed preparation:** Seedbed prepare as per crops, viz. Puddling for rice, ridges and furrow for vegetable, maize and cotton and flat seedbed for several other crops. However, two problems are encountered in seedbed preparation in sequential cropping system.
  - The time available for seedbed preparation is less in high intensity cropping system. Frequent rain interference the land preparation.
  - Due to prevent crop field may be in condition. For example field preparation after rice is difficult, it is mainly because soil structure is destroyed during puddling.

The time between harvesting of crop and sowing of next crop is more, if rice is the preceding crop. To avoid this problem minimum tillage or zero tillage is adopted. It is the common practices to sow pulse crop just before or immediately after harvesting. In rice-wheat system the stubbles are killed by spraying Paraquat and wheat is sown in plough furrows between stubble of rice.

In irrigated agriculture the time available for tillage between two successive crops is minimal leading to minimum tillage. Minimum tillage is applicable for soils with, (1) A course texture surface soil, (2) Good drainage (3) High biological activity of soil fauna. (4) An adequate quantity of soil residue mulch and (5) Favourable initial soil moisture and friable soil consistency over a wide range of soil moisture.

It is not possible to practice zero or minimum tillage in all sequence cropping systems. If sunflower is the preceding crop, ploughing is essential to oxidize the allelo chemical of sunflower. The stubbles of pearl millet and sorghum, which contain high C : N ratio immobilizes nitrogen. It is therefore, necessary to remove them. Stubble also interfere the operations.

In rice-rice-green gram system summer ploughing is done and later when water is available in the rainy season puddling is done and first crop is sown. For the second rice crop minimum tillage as incorporation of crop residues and rice is planted. Green is sown as a relay crop in the second rice crop.

In Cotton-Sorghum-finger millet cropping system in garden lands, thorough field preparation is done and field is laid out into check basins to transplant finger millet. In next season, cotton is planted among the stubbles of finger millet without any kind of preparation. Weeds are controlled by Inter-cultivation operation in Nigeria, no till planting of sorghum into residues of the previous crop maintained the seed at 10°C lower temperature in the seedling zone reaches 41°C

- **Varieties:** Short duration of crops is selected to fit well in the multiple cropping systems. Photo-insensitive varieties are essential for successful sequence cropping system.
- **Sowing:** Sowing is not a problem as provided sufficient time for seedbed preparation. If seedbed not prepared well, difficult the establishment of crop for example-cotton establishment is difficult in black soil after rice. Due tom hard pans in the shallow layer, root penetration is difficult. If field allowed to dry for land preparation, loss of one month, hence seedling are raised either on twisted paddy straw of leaf cups before harvesting of rice crop. After harvesting a cow bar hole is made up to 30 cm. it is partly filled with sand and soil mixture and cotton seedling are planted. The establishment of pulse crop is difficult after rice broadcasting of seeds rice crop or rice subjected results uneven germination and high seed rate is necessary. Crops planted in stubbles are subjected to competition from regenerated stubbles. It can be overcome by spraying preparation or digger on stubbles.

  Delay in sowing is common problems in intensive cropping systems. To reduce yields loss due to transpiration of overage seedlings higher level of nitrogen is applied to induce tillering. In rice – wheat system, wheat yields are reduced considerably when the sowing of wheat are delayed beyond November. In such situation, transpiration of 40 to 50 days old seedlings of wheat is done. FYM is broadcasted over the field to maintain higher soil temperature during December.
- **Soil fertility management:** Soil fertility management become more complex in intensive cropping because of the additional factors such as residual effect of nutrients applied to the previous crops, possible effect of legume in the system, complementary and competitive interaction from the component crops and influence crop residues left in the soil. The modern or chemical agriculture, which includes higher cropping intensity involving improved varieties, heavier inputs of fertilizer and water, increasing yields and accelerated removal of plant nutrients have inputs of fertilizer and water, increasing yields and accelerated removal of plant nutrients have added newer dimensions to the fertility management.

  Fertilizer practices for sequence cropping: based on long term fertility experiments conducted in various parts of India, the following broad conclusions are: system productivity increased with the application of P along with N, and further increased with use of N, P and K, Application of N at recommended dose is advocated to each of the in cropping system.

Phosphorus management in cropping system needs careful adjustment of P fertilizer dose taking into the account type of fertilizer, soil characteristics and their yield level, extent of P removal and growing environment. In cropping system – involving wheat, fertilizer P dose to Kharif crop crop can be reduced if preceding wheat has received P in adequate amounts.

Removal of K in proportion to N is very high in cropping systems, particularly those involving cereal and fodder crops. It is important to apply K fertilizer are recommended dose to maintain soil fertility. In K rich soils of Coimbatore, K application at 50% recommended dose to each crops in the sequence rice-rice-soybean was optimum. Among the secondary nutrients, sulphur application is benefited particularly to sesamum-mustard, soybean-safflower and groundnut-mustard cropping sequence (Tondon, 1991)

Among micronutrients, Zn deficiency is the most common as nearly 50% soils of intensive cultivated areas suffers from Zn deficiency. Rainy season crops like rice, maize and sorghum respond more to applied Zn than winter crops like wheat and chickpea. Under long term experiments at Ludhiana (Punjab), significant decline in productivity of maize was records after 10 to 12 years' annuals cycles, due to fall in Zn status. Use of organics prevents Zn deficiency in intensive cropping system under normal soils.

- **Cropping system and integrated nutrient management (INM):** The concept of integrated nutrients management (INM) involves use of various inorganic, organic, biological sources of nutrient for improvement and maintenance of soil fertility leading to sustained crop production. Crop responses to organic and biological sources of nutrients are not spectacular as to fertilizers, but the supplementary and complementary use of these resources is known to enhance the use efficiency of applied fertilizer besides improving soil physico-chemical properties and preventing emergence of micro – nutrient deficiencies. The major components of the INM are fertilizer, organic manures, green manures, crop residues and Bio-fertilizers. In cereal-based cropping systems, about 25-50% fertilizer NPK dose of rainy season crops could be curtailed with the use of organics such as FYM, green manure and crop residues. In sugarcane based system, integrated use of sulphitation press-mud, cane trash and Biofertilizers each with inorganic fertilizers and green leaf manuring showed 20-25% economy of fertilizers N applied to sugarcane by improving the use efficiency of N, P and other nutrient. Effect of intensive cropping on soil properties: physical properties-continuous ground cover due to intensive cropping minimizes soil erosion, run-off losses and crust formation. Relatively higher amount of crop residues due to intensive cropping improves soil structure. Chemical Properties-Organic residues in intensive cropping systems should be recycled to maintain optimum organic carbon in the soil for sustained production.

Factors for determine the fertilizer schedules are

- **Soil supplying power:** Growing different crops during different seasons alters the soil nutrient status, estimated by soil analysis at the beginning of the season. The soil supplying power increases with legume in rotation. The available nitrogen and potassium in soil after groundnut are higher to initial status of the soil. But after pearl millet, only potassium status in the soil is improved and no changes in P.
- **Nutrient uptake by crops:** The total amount of nutrients taken by the crops in one sequence gives an indication of the fertilizer requirement of the system. The balance is obtained by subtracting the fertilizer applied to crops that nutrient taken by the crops.
- **Residual effect of fertilizers:** The extent of residues left over in the soil depends on the type of fertilizer used. Phosphate fertilizer and FYM have considerable residue in the soil, which is useful for subsequent crops. The residues left by potassium fertilizers are marginal. Legume effect: Legumes add nitrogen to the soil in the range of 15 to 20 kg/ha. The amount of nitrogen added depends on the purpose. Green gram grown for grain, contributes 24 and 30 kg N respectively to the succeeding crop. Inclusion of leguminous green manures in the system add 40 kg to 120 kg N/ha. The availability of phosphorous is also increased by incorporation of green manure crops. Potassium availability to subsequent crop is also increased by groundnut crop residues: crop residues add considerable quantity of nutrients to the soil, cotton planted in finger millet stubbles benefits by 20 to 30 kg/ha due to decomposition of stubbles. Deep rooted crops-cotton, red gram absorbs nutrients from deeper layers. Leaf fall and decomposing add phosphorus to op layers crop residue contain high C: N ratio like stubbles of sorghum, pearl millet temporarily immobile nitrogen. Residue of legume's crop contains low C: N ratio and they decompose quickly and release nutrients. Efficiency of crops: jute is more efficient crop for utilizing of nitrogen followed by summer rice, maize, potato and groundnut in that order. Phosphorus efficient crops, jute > summer rice> Kharif rice> potato > groundnut > maize. Groundnut is more efficient in potassium utilization followed by maize, jute, summer rice, Kharif rice, and potato. Fertilizer recommendation should be based on cropping system *e.g.,* in wheat based cropping system an extra dose of 25% nitrogen is recommended for wheat when it is grown after sorghum, pearl

millet. When wheat, after pulse crop needs 20 to 30 kg less nitrogen per hectare. Phosphate fertilizers are added through green manure crops, not to apply phosphates to succeeding wheat crop. In rice based cropping system consisting of rice-rice in Kharif and rabbi and sorghum, maize, finger millet, soybean in summer it is sufficient to apply phosphorus and potassium to summer crops only while nitrogen is applied to all the crops. Thus, following system approach in fertilizer recommendation can save lot of fertilizer.

- **Water management:** There is no carry over effect of irrigation as in case of fertilizer, rice – rice is efficient cropping system for total yield, but it consume large amount of water especially in summer. If water is scare in summer instead of rice, groundnut is used in cropping system. Method of irrigation: the layout should be so planned that most of the crops can be suitable, in rice-rice-groundnut system; rice is irrigated by flood method, while groundnut by boarder strips. In cotton – sorghum-finger millet system, cotton, sorghum by furrow method while finger millet checks – basin method is adopted. More remunerative and less water consuming crop rotation have standardized have been standardized at different location of India. Rice-mustard-green gram, rice-potato-green gram rotation were found more water efficient systems at memari in W.B under high level of irrigation in tarai region of U.P, rice-lentil and rice-wheat cropping system were found better. Pre monsoon groundnut-rabi sorghum sequence was highly remunerative with high water use efficiency compared to sugarcane alone in M.S when irrigation water is not limiting. Under limited water supply, however, rice – chickpea-green gram and rice-mustard – green gram are more remunerative with high water use efficiency.
- **Weed management:** Weed problems observed individual crop, and weed shift occur and their carry over effect of weed control method on succeeding crop. Weeds are dynamic in nature, generally broad-leaved weeds occur in wheat occur in wheat at later stage and 2, 4 D is applied as post emergence herbicide to control them. In rice-wheat system, canary grass (Phallaris minor) is a menace for wheat crop. Other weed seed species are decomposed and loss viability, but Phataris minor seed do not loss viability. When sown in rice stubble is heavily infested with Phalaris minor. In cotton-sorghum-finger millet sequences cropping with zero tillage weeds are controlled by herbicide in two rotations.

Herbicide applied to the previous crop may be toxic to the succeeding crop. Higher dose of Atrazine applied to sorghum crop affect germination of succeeding pulse crops. Herbicide recommendation should be depends on succeeding crops, ploughing before the planting season helps to kill most of the weeds.

- **Pest and diseases:** Pest and diseases infestation more in sequence cropping due to continuous cropping, carry over effect of insecticides is not observed.
- **Harvesting:** In sequences cropping crop can be harvested at physiological maturity stage instead of harvesting maturity. The field can vacate one week earlier. Because of continuous cropping the harvesting time may coincide with heavy rains and special post-harvest operations, like artificial drying, treating the crop with common salt etc. are practices to save the produce. Unlike industries, agriculture is practiced by 105 million farm families who live in 0.6 million village. More than 40% of them are below the poverty line. Nearly 85 million farm families belong to small and marginal categories. Only 25 to 30% of the modern agricultural technology has reached the farmers. This is often because the technology has not been consistent with the condition with condition of the farm situations. Since there is no further scope for horizontal expansion of land for cultivation, the only alternative left is for vertical expansion on space and time particularly for small and marginal farmers (constituting 76% farming community) who do not have much of reasons, especially in rain fed areas. The new farming system research strategy should, therefore to develop technology with participatory approach of farmers.

Types of enterprises factors governing choice and size of enterprises

(1) Competition effect
Competition between different intercropped species, for higher light, nutrients, water, carbon dioxide and other growth factors.

(2) Complementary effect
It is define as effect of one component on other, which enhance growth and productivity as compared to competition above.

The principle of combing enterprises

A farm manager is often confronted with the problems as to what enterprises to select and the level at which each enterprises should be taken up. How far he can go far or should go in combine enterprises with another depends partly on the inter-relationships, between different enterprises and the prizes of products and inputs.

What is organic farming?

Organic farming system in India is not new and is being followed from ancient time. It is a method of farming system which primarily aimed at cultivating the land and raising crops in such a way, as to keep the soil alive and in good health by use of organic wastes (crop, animal and farm wastes, aquatic wastes) and other biological materials along with beneficial microbes (bio-fertilizers) to release nutrients to crops for increased sustainable production in an eco-friendly pollution free environment.

As per the definition of the United States Department of Agriculture (USDA) study team on organic farming "organic farming is a system which avoids or largely excludes the use of synthetic inputs (such as fertilizers, pesticides, hormones, feed additives etc) and to the maximum extent feasible rely upon crop rotations, crop residues, animal manures, off-farm organic waste, mineral grade rock additives and biological system of nutrient mobilization and plant protection".

FAO suggested that "Organic agriculture is a unique production management system which promotes and enhances agro-ecosystem health, including biodiversity, biological cycles and soil biological activity, and this is accomplished by using on-farm agronomic, biological and mechanical methods in exclusion of all synthetic off-farm inputs".

Need of organic farming

With the increase in population our compulsion would be not only to stabilize agricultural production but to increase it further in sustainable manner. The scientists have realized that the 'Green Revolution' with high input use has reached a plateau and is now sustained with diminishing return of falling dividends. Thus, a natural balance needs to be maintained at all cost for existence of life and property. The obvious choice for that would be more relevant in the present era, when these agrochemicals which are produced from fossil fuel and are not renewable and are diminishing in availability. It may also cost heavily on our foreign exchange in future.

Figure 3: Different components of organic farming

The key characteristics of organic farming

- Protecting the long term fertility of soils by maintaining organic matter levels, encouraging soil biological activity, and careful mechanical intervention
- Providing crop nutrients indirectly using relatively insoluble nutrient sources which are made available to the plant by the action of soil micro-organisms
- Nitrogen self-sufficiency through the use of legumes and biological nitrogen fixation, as well as effective recycling of organic materials including crop residues and livestock manures
- Weed, disease and pest control relying primarily on crop rotations, natural predators, diversity, organic manuring, resistant varieties and limited (preferably minimal) thermal, biological and chemical intervention
- The extensive management of livestock, paying full regard to their evolutionary adaptations, behavioral needs and animal welfare issues with respect to nutrition, housing, health, breeding and rearing
- Careful attention to the impact of the farming system on the wider environment and the conservation of wildlife and natural habitats.

Advantages of organic farming

- Organic manures produce optimal condition in the soil for high yields and good quality crops.
- They supply the entire nutrient required by the plant (NPK, secondary and micronutrients).
- They improve plant growth and physiological activities of plants.
- They improve the soil physical properties such as granulation and good tilt, good giving good aeration easy rot penetration and improved water holding capacity.
- They improve the soil chemical properties such as supply and retention of soil nutrient and promote favorable chemical reaction.
- They reduce the need for purchased inputs.
- Most of the organic manures are wastes of byproduct which accumulated load to pollution.
- Organic fertilizers are considered as complete palnt food.
- Organically grown crop are believed to provide more healthy and nationally superior food for man and animals that those grown with commercial fertilizers.
- Organically grown plants are more resistant to disease and insect and hence only a few chemical sprays or other protective treatment are required.

- There is an increasing consumer are willing to pay more for organic foods.
- Organic farming helps to avoid chain reaction in the environment for chemical spray and dusts.
- Organic farming helps to prevent environment degradation and can be used to regenerate degraded areas.
- Since the basic aim is diversification of crops, much more secure income can be obtained that when they reply on only one crop or enterprise.

Why organic farming not adopted on large scale?

Cause of low adoption:

- Chemical are easy to use and less costly.
- The benefit of organic practices is not seen immediately.
- Large quintiles of organic inputs are required.
- Difficult to get organic fertilizer.
- Unorganized market for organically grown produce.
- Preference to consume organic food is yet not established.
- Economic loss to transition (from traditional agriculture to organic agriculture).
- No experimental evidence on the cost benefit ratio of organic farming.
- Government effort to propagate.
- Scientific research is also scare.

Reference

Balasubramanian P and Palaniappan SP 2006. *Principles and Practices of Agronomy.* Agrobios.

Joshi M and Parbhakarasetty TK 2005. *Sustainability Through Organic Farming.* Kalyani.

Palaniappan SP and Anandurai K 1999. *Organic Farming-Theory and Practice.* Scientific Publ.

Panda SC 2004. *Cropping systems and Farming Systems.* Agribios.

Reddy TY and Reddy GHS 2015. *Principles of Agronomy.* Kalyani Publ.

Sharma AK 2001. *A Handbook of Organic Farming.* Agrobios

3

Basics of Vermicomposting A Viable Technology for Rural Livelihood

Bappa Paramanik, Parimal Panda, Indranil Das, Arindam Kundu Asit Mandal* and *Ashok Choudhury

Introduction

Solid waste management is a worldwide problem and it is becoming more and more complicated day by day due to rise in population, industrialization and changes in our life style. Large quantities of organic wastes are produced from agricultural production and farming systems which include animal manures, sewage biosolids, food wastes, and industrial organic wastes. Globally, these wastes have the potential of increasing soil and water pollution as because they are currently disposed of by land-spreading, incineration, or into landfills. As much as 50%-60% of the total wastes that are disposed into landfills are organic wastes. If these were turned into materials useful in farming system, there would be great savings in primary plant nutrients and metabolic energy.

Vermicomposting has been arising as an innovative eco-technology for the method by which compost or mixed manure of organic origin or conversion of various types of waste is prepared by the use of earthworms. It is a controlled degradation (non-thermophilic, boioxidative process) of the organic wastes for the consumption of earthworms, helps in the recycling of food wastes, reduces the waste bulk density and the final product may contain hormone-like substance which accelerates the plant growth. Vermicompost is humus like, finely granulated and stabilized material which can be used as a soil conditioner to reintegrate the organic matter to the agricultural soils. Industrial wastes remain largely unutilized and often cause environmental problems like ground and surface water pollution, foul odours, occupying vast land areas etc. Non-toxic and organic industrial wastes could be potential raw material for vermitechnology. In the last two decades, vermin-technology has been applied for the management of industrial wastes and sludges and to convert them into vermicompost for land restoration practices. The success of the process depends upon several process parameters like quality of raw material, pH, temperature, moisture, aeration etc., type of vermicomposting system and earthworm species used. Vermicomposting and vermiculture are two interlinked and interdependent processes where vermiculture can be done in the presence of decomposable waste organic matter

Participating organisms of the vermicomposting

Bacteria, micro-organisms and earthworms participate in the vermicomposting process, of which bacteria and micro-organisms are used as food for worms. The main organism for vermicomposting is earthworms. Charles Darwin described earthworms as the 'unheralded soldiers of mankind', and Aristotle called them as the 'intestine of earth', as they could digest a wide variety of organic materials (Darwin and Seward 1903; Martin 1976). Earthworms can also be called as biological indicators of soil health, for soils with earthworms definitely support healthy populations of bacteria, fungi, actinomycetes, protozoans, insects, spiders, millipedes, and a host of other organisms that are essential for sustaining a healthy soil (Ismail, 1997c). Earthworm casts get converted into stable soil aggregates by the action of gums that result from microbial digestion of their organic compounds (Waksman and Martin, 1939), or by the binding effect of fungal hyphae (Parle, 1963 a,b). There are also reports that certain metabolites produced by earthworms may be responsible to stimulate plant growth (Gavrilov, 1962; Neilson, 1965).

Biology of earthworms

The earthworm is a tube shaped segmented invertebrate. Its body holds its shape because it's full of a liquid called coelomic fluid found between the body wall and the alimentary canal. If one were to view a cross section of the worm body it would resemble a target, with the centre representing the internal organs and the outer circle representing the body wall. Earthworm has a long, cylindrical body with a pointed head. In some species the posterior end is slightly flattened, while in others the body is cylindrical throughout. Rings that surround the moist, soft body allow the earthworm to twist and turn, especially since it has no backbone. With no true legs, bristles (setae) on the body move back and forth, allowing the earthworm to crawl. Earthworm breathes through its body surface. Detailed morphological and anatomical description of different species of earthworms is available in Aiyyar and Ananthakrishnan (1993).

Figure 1 shows the alimentary canal of an earthworm. Food is ingested through the mouth into a bag like structure referred to as crop. In some species a distinct crop is absent. Later the food passes through the gizzard, where ingested stones grind it up. After passing through the intestine for digestion, what's left is eliminated as castings.

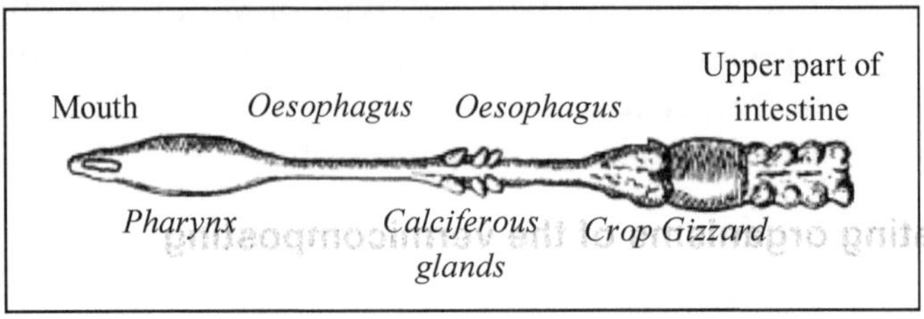

Figure 1: Alimentary canal of an earthworm (Lumbricus).
Source: Martin, 1976

Reproduction of worms

Earthworms that are sexually mature have a prominent band around their body, which is called as the Clitellum. This is usually visible around 8-12 weeks of age. During copulation, the worms will join together at the clitellum (sometimes for quite a long period of time). Reproductive material is exchanged. When the worms separate, a ring of mucus material forms at the clitellum of each worm. This process is known as copulation (Figure 2).

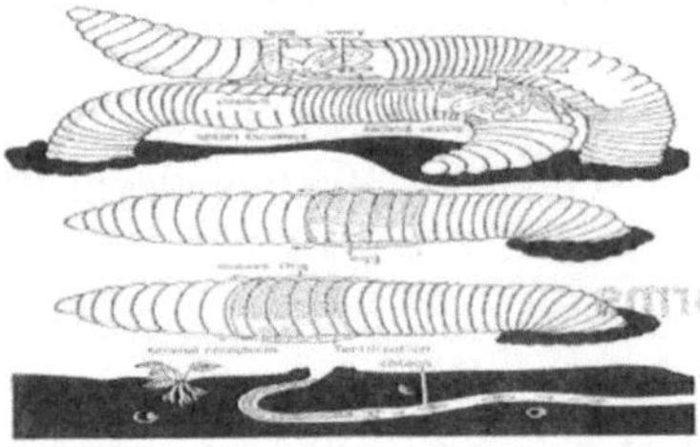

Figure 2: Copulation of earthworms. Adapted from "Vermicomposting as an Eco-tool in sustainable solid waste management " by Anna University

Sperm from the other worm is stored in sacs. As the mucus slides over the worm, it encases the sperm and eggs inside. After slipping free from the worm, both ends seal, forming a lemon-shape cocoon approximately 3.2 mm long. Two or more baby worms will hatch from one end of the cocoon in approximately 3 weeks. Baby worms are whitish to almost transparent and are 12 to 25 mm long. Red worms take 4 to 6 weeks to become sexually mature.

The worm will then wriggle backwards, and the mucus ring slips off over the head. The ring seals, forming a 'capsule' (also called an 'egg'). All the necessary reproductive material is sealed inside. The worm capsule, when first deposited is soft and milky white. This quickly hardens and turns a light lemon colour. The capsule goes through various colour changes through different toning of yellow, then to a rusty brown colour. Capsules are almost ready to hatch when they are the rusty colour.

The capsule is about the size of a grape seed, but size is related to the size of the worm, with larger worms producing larger capsules. Worm capsules have been known to survive drought conditions for 12 months or more, hatching when conditions become suitable again.

Classification of earthworm

Kingdom	:	Animalia
Phylum	:	Annelida
Class	:	Oligochaeta
Order	:	Opisthopora
Family	:	Lumbricidae, Eudriladea & Megascolecidae
Genus	:	A large number of genera have been described in literature
Species	:	Commonly adopted worms in vermiculture are *Bimastos parvus, Dendrobaena rubida, Dendrobaena veneta, Eisenia fetida, Eisenia hortensis* and *Eisenia andrei* belonging to family Lumbricidae; *Eudrilus eugeniae* family Eudrilidae; *Amynthas diffringens, Amynthus morrisi, Lampito mauritii, Metaphire anomala, Metaphire birmanica, Perionyx excavates, Perionyx sansibaricus, Megascolex megascolex, Notoscolex, Pontoscolex corethrurus, Octochaetona serrata, Octochaetona surensis, Pheritima elongate, Pheritima posthuma* belonging to order Megascolecidae.

Types of earthworms

Earthworms are invertebrates. There are nearly 3600 types of earthworms in the world and they are mainly divided into two types: (1) burrowing; and (2) non-burrowing. The burrowing types *Pertima elongata* and *Pertima asiatica* live deep in the soil. On the other hand, the non-burrowing types *Eisenia fetida* and *Eudrilus eugenae* live in the upper layer of soil surface. The burrowing types are pale, 20 to 30 cm long and live for 15 years. The non-burrowing types are red or purple and 10 to 15 cm long but their life span is only 28 months. The non-burrowing earthworms eat 10% soil and 90% organic waste materials; these convert the organic waste into vermicompost faster than the burrowing earthworms. They can tolerate temperatures ranging from 0 to 40°C but the regeneration capacity is more at 25 to 30°C and 40–45% moisture level in the pile. The burrowing types of earthworms come onto the soil surface only at night. These make holes in the soil up to a depth of 3.5 m and produce 5.6 kg casts by ingesting 90% soil and 10% organic waste.

Types of composting earthworms and their species description

- **Red wigglers (*Eisenia fetida*)**

The red wiggler is the most common type of composting worm. It can process large amounts of organic matter and, under ideal conditions, can consume feed proportional to its body weight each day. It also reproduces rapidly, and is very tolerant against variation in growing conditions. Other names for red wigglers include Tiger worms, Garlic worms, Manure worms, and Brandling worms.

- **Redworms (*Lumbricus rubellus*)**

The redworm is a very good composting worm. In bright light, it is a very wriggly and is not at all suitable for fish bait. Redworms are very effective in aerating and mixing the soil, and consume a large amount of organic material, although less than red wigglers. These worms are commonly found in decomposing animal manures and compost piles. Other names include bloodworms and red wriggler (but not the same as the worm described above).

- **Red tiger (*Eisenia andrei*)**

The Red Tiger worm makes an excellent composting worm. It is a close relative of the Red Wiggler (*E. fetida*) and is commonly used as a bait worm because they exude coelomic fluid, which attracts fish. These worms are very active wigglers in sunlight. Other common name for this is Tiger worm

- **Blue worms (*Perionyx excavatus*)**

Blue worms have become more popular in recent years as a composting worm. These worms do very well in warm climates, but their degradative potential drops in the cold. Although these worms eat fairly large amounts of organic materials and are fairly prolific breeders, they are also very sensitive to changes in their growing environment. In such case, these worms will migrate to other suitable area or may move outside the bin. Other common names include Indian Blue and Malaysian Blue worms.

- **African night crawlers (*Eudrilus eugeniae*)**

The African Night crawler can be a good composting worm, but is very sensitive to changes in its environment. The entire contents of some worm bins have been known to move out in less than a day if the prevailing conditions are not suitable for them. These worms also perform much better in warmer climates and are not recommended for areas where the temperature falls below 10° C. The other common name for this worm is the giant night crawler.

Table1: Ecological categories and niches of earthworms and their characteristic features and beneficial traits

Species	Ecological category	Ecological niche	Characteristic features	Beneficial trait
Eisenia foetida, Lumbricus rubellus L. castaneus, L. festivus Eiseniella tetraedra Bimastus minusculus B. eiseni, Dendrodrilus rubidus, Dendrobaena veneta, D. octaedra	Epigeics	Superficial soil layers, leaf litter, compost	Smaller in size, body uniformly pigmented, active gizzard, short life cycle, high reproduction rate and regeneration, tolerant to disturbance, phytophagous	Efficient bio-degraders and nutrient releasers, efficient compost producers, aids in litter combination and early decomposition
Aporrectodea caliginosa A. trapezoides, A. rosea Millsonia anomala	Endogeics	Topsoil or subsoil	Small to large sized worms, weakly pigmented, life cycle of medium duration, moderately tolerant to isturbance, geophagous	Brings about pronounced changes in soil physical structure, can efficiently utilize energy from poor soils hence can be used for soil improvements
Octolasion cyaneum O. lacteum	Polyhumic endogeic	Top soil (A1)	Small size, unpigmented, forms horizontal burrows, rich soil feeder	
Pontoscolex corethrurus Allolobophora chlorotica	Mesohumic endogeic	A and B horizon	Medium size npigmented, forms extensive horizontal burrows, bulk (A1) soil feeder	

Species	Ecological category	Ecological niche	Characteristic features	Beneficial trait
Aminthas sp.	Oligohumic endogeic	B and C horizon	Very large in size, unpigmented, forms extensive horizontal burrows, feeds on poor, deep soils	
L. terrestris *L. polyphemus* *A. longa*	Anecics		Permanent deep burrows in soil Large in size, dorsally pigmented, forms extensive, deep, vertical permanent burrows, low reproductive rate, sensitive to disturbance, phytogeophagous, nocturnal	Forms vertical burrows affecting air-water relationship and movement from deep layers to surface helps in efficient mixing of nutrients

Adapted from Pathma and Sakthivel, 2012

Table2: Comparison of Some Aspects of the Biology of the vermicomposting Species

Characters	Eisenia fetida	Eisenia andrei	Dendrobaena rubida	Dendrobaena veneta	Drawida nepalensis	Eudrilus eugeniae	Perionyx excavatus	Lumbricus rubellus
Color	Brown and buff bands	Red	Reddish purple	Reddish and purple bands	—	Reddish brown	Reddish brown	Reddish brown
Size of adult earth-worms	4–8 mm × 50–100 mm	4–8 mm × 50–100 mm	3–4 mm × 35–60 mm (0.11–	5–7 mm × 50–80 mm (0.2–	—	5–7 mm × 80–190 mm	4–5 mm × 45–70 mm	4 mm × 70–150 mm
Mean weight of adults	0.55 g	0.55 g	0.25 g	0.92 g	0.82 g	2.7–3.5 g	0.5–0.6 g	0.80 g
Time to maturity (days)	28–30	21–28	54	65	34–42	40–49	28–42	74–91
Number of cocoons day–1	0.35–0.5	0.35–0.5	0.20	0.28	0.15	0.42–0.51	1.1–1.4	0.07–0.25
Mean size of cocoons	4.85 mm × 2.82 mm	4.8 mm × 2.82 mm	3.19 mm × 1.97 mm	3.14 mm × 1.93 mm	—	—	—	3.50 mm × 2.46 mm
	18–26	18–26	15–40	42.1	24	12–16	18	35–40

Characters	Eisenia fetida	Eisenia andrei	Dendrobaena rubida	Dendrobaena veneta	Drawida nepalensis	Eudrilus eugeniae	Perionyx excavatus	Lumbricus rubellus
Hatching viability (%)	73–80	72	85	20	75–88	75–84	90	60–70
Number of worms cocoon–1	2.5–3.8	2.5–3.8	1.67	1.10	1.93	2–2.7	1–1.1	1
Self-	+	+	+	—	+	—	—	—
Life cycle (days)	45–51	45–51	75	100–150	100–120	50–70	40–50	120–170
Limits and optimal Temperature	25°C (0°C–35°C)	25°C (0°C–35°C)	—	25°C	—	25°C		—
Limits and optimal moisture	80-85% (70-90%)	80-85% (70-90%)	-	75% (65-85%)	--	80% (70-85%)		

Source: Dominguez and Edwards, 2011.

Variation in different types of earthworms in soil

The effects of EWs on soil biological processes and fertility level differ in ecological categories. In the majority of habitats and ecosystems (Table 3), it is usually a combination of these ecological categories which together or individually are responsible for maintaining the fertility of soils.

Table3: Effect of land conversion and management practices on changes in functional catagories of earthworms in the Indo-Gangetic plains,(±SE, n = 10)

Sites	Density (Anecic) (Individuals m^{-2} $year^{-1}$)	Biomass (Anecic) (gm^{-2} $year^{-1}$)	Density (Endogeics) (Individuals m^{-2} $year^{-1}$)	Biomass (Endogeics) (gm^{-2} $year^{-1}$)
Primary forest	141 (±3.2)a	123 (±11.6)a	2127 (±13.8)a	2255.8 (±20.6)a
Productive groeco-system	1141 (±11.6)b	1323 (±23.5)b	275 (±6.3)b	2157.5 (±13.3)b
Low productive agro-ecosystem	1106 (±7.9)c	1318 (±27.8)b	245 (±3.2)c	294.5 (±6.8)c
Agriculture fallow	164 (±3.8)d	142 (±2.9)c	2274 (±14.6)d	2518.7 (±42.6)d
Sodic ecosystems	0	0	0	0
5-year-old reclaimed agroecosystem	0	0	143 (±12.7)e	114.4 (±5.8)c

(Contd.)

Sites	Density (Anecic) (Individuals m^{-2} year^{-1})	Biomass (Anecic) (gm^{-2} year^{-1})	Density (Endogeics) (Individuals m^{-2} year^{-1})	Biomass (Endogeics) (gm^{-2} year^{-1})
10-year-old reclaimed agroecosystem	0	0	282 (±24.7)d	160.6 (±15.3)b
Acacia plantation in reclaimed soils	144(±5.3)a	1132 (±5.9)a	2133 (±9.6)a	2279.3(±21.5)e

Values followed by the different superscript letters are significantly different in different sampling sites. Values followed by different subscript numbers are significantly different in same sampling sites [Bhadauria, 2008]. Adapted from Bhadauria and Saxena, 2010.

What is vermicomposting?

Vermicomposting is a simple biotechnological process of composting, in which certain species of earthworms are used to enhance the process of waste conversion and produce a better end product. Vermicomposting differs from composting in several ways (Gandhi *et al.*, 1997). It is a mesophilic process, utilizing microorganisms and earthworms that are active at 10–32°C (not ambient temperature but temperature within the pile of moist organic material). The process is faster than composting; because the material passes through the earthworm gut, a significant but not yet fully understood transformation takes place, whereby the resulting earthworm castings (worm manure) are rich in microbial activity and plant growth regulators, and fortified with pest repellence attributes as well! In short, earthworms, through a type of biological alchemy, are capable of transforming garbage into 'gold' (Vermi Co 2001, Tara Crescent 2003).

Materials not suitable for vermicomposting

Epigeic earthworms reject certain food materials like all types of citrus products (orange, lemon and lime), grape fruits, onion, garlic, cloves, extremely hot and heavily spiced food and high acid food. They also avoid oleanders and other poisonous plants, meat, chicken, dairy foods, dog and cat manure. It is also important to note that earthworms do not feed on plastics, foils, metals, chemicals, oils, solvents, insecticides, soaps, paint etc.

Fresh organic matter that have been treated with herbicides or pesticides should be avoided for vermicomposting, however, small amount of pesticide treated plants (*e.g.,* grass clippings) may be mixed in the pile as long as one is careful to allow thorough decomposition prior to addition into vermibed.

Some materials may pose a health hazard or create a nuisance and therefore should not be used to make vermicompost. Adding human or pet faces cannot be recommended because they may transmit diseases. Meat, bones, grease, whole eggs and dairy products should not be added because they can attract rodents to the site.

Criteria for Selection of suitable earthworm species

Epigeic or manure worms are found on the surface and are reddish brown; they do not process the soils but are efficient in composting of organic wastes. Some basic criteria for selection of earthworm species for composting are given below:

i. The worm should have wide adaptability to environmental factors, especially to varying temperature conditions.
ii. The worm should have feeding preference and adaptability for wide range of organic materials.
iii. It should have high consumption, digestion and assimilation rate.
iv. Worm should be efficient converter of plant and animal bio-mass to body proteins so that its growth rates are high.
v. The worm should produce large number of cocoons and hatching of cocoons and the maturity from young to adult stage should be fast.
vi. Worms should have adaptability with other worms in the composting tank.
vii. Earthworm species should be resistant/tolerant to disease attack.

Biomass and waste materials for Vermicomposting

All most all types of biodegradable waste materials can be used as worm feeds. Earth worm feeding habit and multiplication rate depend on the quality of food materials. Easily degradable biomass and waste materials are preferred by the compost worms. Some organic materials, which are not easily degradable, are pre-treated with nitrogenous materials, cow dung or mineral amendments and partially decomposed prior to application in vermi-pit.

Table 4: List of some waste materials, their C:N ratios and suitability for vermicompost production

Organic matters	Carbon: nitrogen ratio	Suitability for vermicompost
a. Biomass from dhaincha, subabul, water hyacinth, azolla; leaves of leguminous plants; vegetable wastes of cabbage, cauliflower, potato etc.	15-25	Very good substrates for Vermicomposting
b. Cattle dung, wastes from goattery, poultry, duckery, vegetables and fruits wastes, grass clippings, wastes from pulse and oil seed crops	25-50	These are also suitable for Vermicomposting
c. Paddy, wheat and maize straw, sugarcane trash etc.	50-100	Suitable
d. Saw dust, coconut coir, paddy husk etc.	More than 100	less suitable

Feed mixture for vermicompost production

Dried dung of cattle, sheep, horses, pigs or droppings of poultry birds and small shredded pieces of vegetable waste form are the ideal feed for the worms. Cattle dung can be fed as such, but other dung material or vegetable waste can be mixed in equal quantity with cattle dung for the feed acceptability. Mixing strategy should aim to increase the worm population as well as to improve the quality of vermicompost. Some of the tested pre-treatment procedure and feed mixture are mentioned below:

1. Cattle dung or mixed dung along with agriculture waste/kitchen waste/ vegetable waste/sewage sludge/semi crushed leaf litter/water hyacinth/ wheat bran/gram bran in the ratio of 10:3.
2. Cattle dung, sheep dung and horse dung mixed in equal quantities.
3. Cattle dung of at least seven days old (alone can serve as a good substrate for compost worms.
4. Only agricultural waste or sewage sludge or kitchen waste or leaf litter or their mixture.
5. Poultry droppings mixed with biogas sludge or cattle dung in equal quantities.
6. Standard diet given by Prof. R.D. Kale includes cow dung or mixed dung along with gram bran, wheat bran and vegetable waste in the ratio of 10:1:1:1 + some powered egg shell.
7. Rock phosphate can be used to increase phosphorus content of vermicompost. As per the report of ICRIT (Report No.9 : 2004) up to 2 kg Rock phosphate can be used in feed mixture along with 50 kg dry

organic wastes and 15 kg dung slurry.

8. Biogas slurry

Advantages of vermicompost

As the wastes are pulverised as they pass through the worm, the surface area of the material increases which in turn helps as base for nutrients. Vermicompost, apart from supplying nutrients and growth enhancing harmones to plants, improves the soil structure leading to increase in water and nutrient holding capacities of soil. Chemical fertilizer in moderate doses can go along with vermicoposting. Vermicompost has several other advantages-

- Vermicompost is rich in nutrients (N, P, and K) and increases organic carbon content of soil and productivity of soil. Earthworms (EWs) are a major component of soil fauna communities in most ecosystems and comprise a large proportion of macrofauna biomass. Their activity is beneficial because it can enhance soil nutrient cycling through the rapid incorporation of detritus into mineral soils. In addition to this mixing effect, mucus production associated with water excretion in earthworm guts also enhances the activity of other beneficial soil microorganisms. This is followed by the production of organic matter. So, in the short term, a more significant effect is the concentration of large quantities of nutrients (N, P, K, and Ca) that are easily assimilable by plants in fresh cast depositions. In addition, earthworms seem to accelerate the mineralization as well as the turnover of soil organic matter. Earthworms are known also to increase nitrogen mineralization, through direct and indirect effects on the microbial community. The increased transfer of organic C and N into soil aggregates indicates the potential for earthworms to facilitate soil organic matter stabilization and accumulation in agricultural systems, and that their influence depends greatly on differences in land management practices (Bhadauria and Saxena, 2010). Vermicompost being a rich source of microorganisms, growth promoting substances and various enzymes exhibits beneficial effects on soils and crops. Several researchers have documented the presence of plant growth regulators such as auxins, gibberellins, cytokinins of microbial origin (Krishnamoorthy and Vajranabhiah, 1986; Grappelli *et al.*, 1987; Tomati *et al.*, 1988; Muscolo *et al.*, 1999) and humic acids (Senesi *et al.*, 1992; Masciandaro *et al.*, 1997; Atiyeh *et al.*, 2002) in vermicompost in appreciable quantities. Cytokinins produced by Bacillus and Arthrobacter spp. in soils increase the vigour of seedlings (Inbal and Feldman, 1982; Jagnow, 1987). Microbially produced gibberellins influence plant growth and development (Mahmoud *et al.*, 1984; Arshad and Frankenberger, 1993) and auxins produced by Azospirillum brasilense affects the growth of plants belonging to paoceae (Barbieri *et al.*, 1988). Extensive investigations on the biological activities of humic substances showed that they also posses plant growth stimulating property (Chen and Aviad, 1990).

- Vermicompost encourages mycorrhizal growth in rooting zone of plant. It also serves as excellent rooting media for artificial propagation.
- Vermicompost increases the soil texture, fertility and soil moisture and reduces the water requirement in long run.
- The optimal carbon/nitrogen (C/N) ratio is available in vermicompost which determines the quality of compost. Vermicompost suppress the development of plant diseases due to higher number of bacteria and fungi. Earthworm feeding reduces the survival of plant pathogens such as *Fusarium* sp. and *Verticillium dahlia* (Yeates 1981; Moody *et al*., 1996) and increases the densities of antagonistic fluorescent *pseudomonads* and filamentous *actinomycetes* while population densities of *Bacilli* and *Trichoderma* spp. remains unaltered (Elmer, 2009). Earthworm activities reduce root diseases of cereals caused by Rhizoctonia (Doube *et al*., 1994). L. terrestris showed a significant reduction of disease caused by *Fusarium oxysporum F. asparagi* and *F. proliferatum* on susceptible cultivars of asparagus (*Asparagus officinalis*), *Verticillium dahliae* on eggplant (*Solanum melongena*) and *F. oxysporum, F. lycopersicirace* on tomato. Two possible mechanisms of pathogen suppression have been described, one depends on systemic plant resistance and the other is mediated by microbial competition, antibiosis and hyper parasitism (Hoitink and Grebus 1997). The microbially mediated suppression is again classified into two mechanisms viz., 'general suppression' where a wide range of microbes suppress the pathogens such as Pythium and Phytopthora (Chen *et al*., 1987) and 'specific suppression' where a narrow range of organisms facilitates suppression, for instance disease caused by Rhizoctonia (Hoitink *et al*., 1997). Addition of organic amendments helped in suppression of various insect pests such as *Spodoptera litura, Helicoverpa armigera,* leaf miner (Apoaerema modicella), jassids (*Empoasca kerri*), aphids (*Aphis craccivora*) and spider mites on groundnuts (Rao *et al*., 2001; Rao 2002, 2003) and psyllids (*Heteropsylla cubana*) on a tropical leguminous tree (*Leucaena leucocephala*) (Biradar *et al*., 1998). Vermicompost amendment decreased the incidence of sucking pests under field conditions (Ramesh 2000) and suppressed the damage caused by of two-spotted spider mite (*Tetranychus spp*.), aphid (*Myzus persicae*) (Edwards *et al*., 2007) and mealy bug (*Pseudococcus* spp.) under green house conditions (Arancon *et al*., 2007). Vermicompost are rich in humic acid and phenolic compounds. Phenolic compounds act as feeding deterrents and hence significantly affect pest attacks (Kurowska *et al*., 1990; Summers and Felton 1994; QiTian 2004; Hawida *et al*., 2007; Koul 2008; Mahanil *et*

al., 2008; Bhonwong *et al.*, 2009). Vermicompost amendments appreciably suppress plant parasitic nematodes under field conditions (Arancon *et al.*, 2003b). Vermicompost also suppressed the attack of Meloidogyne incognita on tobacco, pepper, strawberry and tomato (Swathi *et al.*, 1998; Edwards *et al.*, 2007; Arancon *et al.*, 2002; Morra *et al.*,1998) and decreased the numbers of galls and eggmasses of Meloidogyne javanica Organic matter addition to the soil stimulates the population of bacterial and fungal antagonists of nematodes (*e.g., Pasteuria penetrans, Pseudomonas* spp. and chitinolytic bacteria, *Trichoderma* spp.), and other typical nematode predators including nematophagous mites viz., *Hypoaspis calcuttaensis* (Bilgrami 1996), Collembola and other arthropods which selectively feeds on plant parasitic nematodes. (Thoden *et al.*, 2011). Vermicompost amendment promoted fungi capable of trapping nematode and destroying nematode cysts (Kerry 1988) and increased the population of plant growth-promoting rhizobacteria which produce enzymes toxic to plant parasitic nematodes (Siddiqui and Mahmood 1999). Provide safe & clean environment. Vermicomposting does not involves a thermophilic phase which might increase the risk of using this technology for management of infectious wastes, but surprisingly Vermicomposting resulted into a noticeable reduction in the pathogen indicators such as fecal coliform, Salmonella sp., enteric virus and helminth ova in the biosolids (Eastman 1999; Sidhu *et al.*, 2001).

- Vermicompost is non-phytotoxic and does not pollute soil.

Mechanism of vermicomposting

The transformation of organic wastes into vermicompost is of double interest: on the one hand, a waste is converted into value added product, *i.e.*,, vermicompost and, on the other; it controls solid waste pollution that is a consequence of increasing population, industrialization, urbanization and intensive agriculture. Another positive aspect associated with vermicomposting is that it can be done at any scale from household vermicomposting of food waste to community or city scale vermicomposting (Edwards and Lofty 1972). Various steps of waste degradation by earthworms are given below.

- Ingestion of organic waste material.
- Softening of organic waste material by the saliva in the mouth of the earthworms.
- Softening of organic waste and neutralization by calcium (excreted by the inner walls of oesophagus) and passed on to the gizzard for further action in oesophagus region of worm body.
- Waste is finely ground into small particles in the muscular gizzard.
- Digestion of organic waste by a proteolytic enzyme in stomach.

- Decomposition of pulped waste material components by various enzymes including proteases, lipases, amylases, cellulases, chitinases etc. secreted in intestine and then the digested material is absorbed in the epithelium of intestine.
- Excretion of undigested food material from worm castings.

The success of vermicomposting process depends on a number of abiotic and biotic factors. Some of these factors are given below:

Importance of vermicomposting

Vermicomposting is a biotechnological process in which earthworms are employed to convert the organic wastes into humus like material known as vermicompost. Certain earthworm species are capable of consuming a wide range of organic wastes from sewage sludge, animal wastes, agricultural residues, domestic wastes, to industrial wastes. Under favorable conditions of temperature and moisture, earthworms maintain the aerobic conditions in the vermicomposting process ingest organic waste materials and egest a humus-like substance which is more homogeneous than the organic wastes or raw materials used (Arancon *et al.*, 2003; Edwards and Burrows 1988). The actions of the earthworms in this process are both physical and biochemical. The physical actions include fragmentation, turnover and aeration. Whereas biochemical actions include enzymatic digestion, nitrogen enrichment, transport of inorganic and organic materials (Edwards and Lofty 1972). During this process, important plant nutrients such as nitrogen, potassium, phosphorus and calcium present in the waste materials are converted through microbial action into such chemical forms which are much more soluble and available to the plants than those in the parent substrate (Ndegwa and Thompson 2001).

Source of plant nutrients

Earthworms consume various organic wastes and reduce the volume by 40–60%. Each earthworm weighs about 0.5 to 0.6 g, eats waste equivalent to its body weight and produces cast equivalent to about 50% of the waste it consumes in a day. These worm castings have been analyzed for chemical and biological properties. The moisture content of castings ranges between 32 and 66% and the pH is around 7.0. The worm castings contain higher percentage (nearly twofold) of both macro and micronutrients than the garden compost (Table 5).

Table5: Nutrient composition of vermicompost and garden compost

Nutrient element	Vermicompost (%)	Garden compost (%)
Organic carbon	9.8–13.4	12.2
Nitrogen	0.51–1.61	0.8
Phosphorus	0.19–1.02	0.35
Potassium	0.15–0.73	0.48
Calcium	1.18–7.61	2.27
Magnesium	0.093–0.568	0.57
Sodium	0.058–0.158	<0.01
Zinc	0.0042–0.110	0.0012
Copper	0.0026–0.0048	0.0017
Iron	0.2050–1.3313	1.1690
Manganese	0.0105–0.2038	0.0414

From earlier studies also it is evident that vermicompost provides all nutrients in readily available form and also enhances uptake of nutrients by plants. Sreenivas *et al.* (2000) studied the integrated effect of application of fertilizer and vermicompost on soil available nitrozen (N) and uptake of ridge gourd (*Luffaacutangula*) at Rajendranagar, Andhra Pradesh, India. Soil available N increased significantly with increasing levels of vermicompost and highest N uptake was obtained at 50% of the recommended fertilizer rate plus 10 t ha^{-1} vermicompost. Similarly, the uptake of N, phosphorus (P), potassium (K) and magnesium (Mg) by rice (Oryza sativa) land was highest when fertilizer was applied in combination with vermicompost (Jadhav *et al.*, 1997).

Vermicompost Production

- Processing involving collection of wastes, shredding, mechanical separation of the metal, glass and ceramics and storage of organic wastes.
- Pre digestion of organic waste for twenty days by heaping the material along with cattle dung slurry. This process partially digests the material and fit for earthworm consumption. Cattle dung and biogas slurry may be used after drying. Wet dung should not be used for Vermi-compost production.
- Preparation of earthworm bed. A concrete base is required to put the waste for Vermicompost preparation. Loose soil will allow the worms to go into soil and also while watering, all the dissolvable nutrients go into the soil along with water.
- Collection of earthworm after Vermi-compost collection. Sieving the composted material to separate fully composted material. The partially composted material will be again put into Vermicompost bed.

- Storing the Vermi-compost in proper place to maintain moisture and allow the beneficial microorganisms to grow.

Pit method

Pit method is commonly used for small scale production of vermicompost. These steps can be followed:

- Construct a pit of 3 – 2 – 1 m size (L – W – D) over ground surface using bricks. Size of pit may vary as per availability of raw materials.
- Fill the pit with following four layers:
    - 1st layer – sand or sandy soil of 5-6 cm. This layer helps to drain excess water from the pit.
    - 2nd layer-paddy straw or other crop residue of 30 cm above 1st layer which will be used for providing aeration to the pit.
    - 3rd layer-15 to 30 days old dung over paddy straw layer at a thickness of 20-30 cm. This helps in initiating microbial activity.
    - 4th layer-pre-digested material about 50 cm
    - Inoculate earthworm @ 1000 worms per square meter area or 10 k g earthworm in 100 kg of organic matter.
- Spray water on the bed and gunny bag. Maintain 50-60% moisture of the pit by periodical water spraying.

Windrows method

This method is widely used for large scale production of vermi-compost. Involve following steps:

- Load the organic wastes in the form of bed (preferably 10 feet L – 3 feet W – 1.5 feet H). Size of bed may vary as per availability of organic waste.
- After loading, the fresh bed should be covered with jute mate or dry agriculture wastes such as rice-bran, banana-leaf, maize residue etc.
- Sprinkle water over the covered vermin-bed to maintain 40% moisture in bed.
- Moisture percent can be checked by forming lump of organic waste using hand. it should easily form lump.
- Bed should be mixed thoroughly to prevent it from becoming compact.
- The first lot of vermicompost is ready for harvesting after $2\text{-}2^{1/2}$ months and the subsequent lots can be harvested after every six weeks of loading.
- Watering of bed should be stopped for at least 2-3 days before harvesting. Earthworms go down in the moist soil and the compost is collected from the top

without disturbing the lower layers of vermi bed having earthworm. Vermicompost harvested will be of dark brown colour and free flowing.

- The harvested compost should be stored in dark and cold place.

Vermicomposting in detail

Temperature

Temperature values and changes in temperature will affect the success of all of the activities of vermicomposting process, which is a bio-oxidation and stabilisation process of organic material that, in contrast to composting, involves the joint action of earthworms and micro-organisms (Dominguez *et al.*, 1997). Compost worms are the agents of turning, fragmentation and aeration of waste, resulting in the production of a homogeneous and stabilized humus-like product. They cannot tolerate temperatures above 35°C (Edwards, 1995) and as a consequence, the microbial processes are not the same as conventional composting in which the optimal temperature range is 45-65°C. Vermicomposting does not achieve high temperatures but sometimes, prior to adding the worms, a thermophilic stage is used to kill insects and pathogens (Frederickson *et al.*, 1997).

Vermicomposting systems operating at lower ambient temperatures of 6.3 ± 2.3 °C showed lower earthworm reproduction rates, when compared to those operating at the moderate temperature of 13.7 ±0.8°C, where significantly increased worm reproduction rates have been identified by Frederickson and Howell (2003). They further added that vermicomposting is the joint action of earthworms and the aerobic micro-organisms that flourish and decompose the waste at these lower temperatures (mesophilic conditions). Hence, it is common with vermicomposting systems to apply waste frequently in form of few centimeters thick layer to reactors, containing earthworms in order to prevent thermophilic conditions and to help keep the waste aerobic.

Moisture content

Feedstock moisture content is one of the critical factors to consider during vermicomposting, especially when deciding what proportions of various materials (feedstock) are to be mixed. Appropriate moisture content is also necessary to support the metabolic processes of the microbes. Earthworms breathe through their skins, which must be moist for exchange of air and excretion of waste to take place. If necessary, one can add water to dry bedding and feed. Appelhof (1997) suggested that too much moisture present as stagnant water in the reactor can reduce available oxygen and cause worms to drown. According to some researchers (Edwards, 1995 and 1998, Neuhauser *et al*., 1988, Kaushik and Garg, 2003), in order to achieve optimum conditions, moisture content should be maintained within the range of 70-90% (w/w). However, studies conducted by Frederickson *et al.* (1997), and Bansal and Kapoor (2000) successfully performed vermicomposting by using lower moisture contents of 66±2% and 60%, respectively.

It seems likely that, in order to maintain optimum moisture conditions during vermicomposting of different materials, moisture content can be altered by adding either water or mixing certain types of materials (*e.g.*, sawdust, cardboard) to the feed material. Suthar (2007), Garg *et al.* (2006) and Maboeta and Rensburg (2003) vermicomposted biodegradable waste, partially composted waste, woodchips and sewage sludge in different combinations and maintained moisture contents of 65-70%, 55-60% and 70% respectively, by periodically sprinkling the mixture with an adequate amount of water. In contrast to this, Marsh *et al.* (2005) used a mixed feedstock of aquaculture sludge and cardboard for regulating the moisture content from 95% to 80%. However, Alidadi *et al.* (2005) adjusted the moisture content of the sewage sludge (80%, w/w) to 60% by adding sawdust prior to vermicomposting.

Feedstocks

Earthworms will consume animal manures, compost, food scraps, shredded or chopped cardboard or paper, almost any decaying organic matter or waste product However, excellent feeds are horse, rabbit, swine, dairy, or steer manures, whereas, poultry manure is not recommended as it is too high in nitrogen and mineral matter (Edwards, 1998 and Slocum, 2002). Many researchers (Loehr *et al.*, 1985, Maboeta and Rensburg 2003 and Alidadi *et al.*, 2005) have studied the use of earthworms to digest sewage sludge. Recent studies (Sharma *et al.*, 2005) showed that a variety of organic solid wastes, including domestic, animal, agro-industrial and human wastes can also be vermicomposted. It was suggested that the first step in vermiculture is to select suitable feed materials for earthworms, which can be nitrogen-rich material such as cattle dung, pig manure, poultry manure, or some other organic material such as leguminous agrowaste with a C:N ratio of less than 40. Similarly, Marsh *et al.* (2005) used mixed aquaculture effluent and cardboard. They reported that maximum growth of *E. fetida* was observed by increasing the percentage of aquaculture sludge up to 50%, equating to a C:N ratio of 98:1. In contrast to this, Ndewga and Thompson (2000) claimed that an optimum C:N ratio of 25:1 is more suitable for the growth of *E. fetida*.

Suthar (2007) utilized different biodegradable waste categories (crop residue, farm-yard manure consisting of cattle excreta, discarded cattle feedstock, vegetable market waste and household waste) amended with animal manure in different ratios to evaluate the efficiency of P. sansibaricus for vermicomposting. They reported an increase in organic C, total N and available phosphorus and potassium after vermicomposting of all used substrates. Moreover, their study indicated that die end product had a lower C:N ratio, when compared to the initial ratio. Similarly, Garg *et al.* (2006) mixed kitchen waste, agro-residue and institutional waste separately with the cow dung and soil in the ratio of 6:3:1 (dry weight basis). Furthermore, the mixture was partially decomposed for 15 days, followed by vermicomposting using *E. fetida* for 85 days. Their results showed lower organic carbon and pH in the product and suggested that this was due to the production of CO_2. However, an increase in the electrical conductivity, along with total N, P and K, was also reported and it was concluded that the worm casts obtained from the textile industrial waste had the highest N and P content, when compared to the institutional waste, agro-residue and kitchen waste.

Loh *et al.* (2005) studied the growth and reproduction of E. fetida using cattle and goat manures. They were of the view that cattle manure provided a better environment for the earthworm to grow and produced a higher value of vermicast than goat manure because the C:N ratio in cattle manure was higher than in goat manure. Bansal and Kapoor (2000) analysed vermicomposting of mustard residue and sugarcane trash mixed with cattle dung and reported a significant reduction in C:N ratio and an increase in mineral N after three months. In light of all the above discussion, The Composting Association, concluded that earthworms will process more waste and reproduce more quickly when fed with fresh, finely-shredded organic materials containing a carbon-nitrogen (C:N) ratio in the range of 15:1 to 35:1.

Stocking density

Stocking density is another parameter that has a significant influence on vermicomposting. It refers to the initial weight of worm biomass per unit area of reactor/bedding. In the literature, various researchers have used a wide range of stock densities for their studies. Frederickson *et al.* (1997) observed a significant reduction in growth and reproduction of *E. andrei* with an increase in stocking density. During further studies Frederickson (2001) also estimated that the working density should range between 1 to 4 kg m^{-2}. Short *et al.* (1999) employed a density of 4 kg m^{-2} and Elvira *et al.* (1998) used 0.56 kg m^{-2} of stock density. Another research undertaken by Wright (2002) on Vermicomposting of a combination of substrates (fruits, vegetables, shredded newspapers, green waste and composted biodegradables) used 3 kg m^{-2}, whereas Roberts (2005) vermicompost in-vessel pre-composted green waste and biosolids using D. veneta at a stock density of 5 kg m^{-2}

Loh *et al.* (2005) suggested that in vigorously growing worm beds, feeding is required every 3-5 days with an optimal stocking density of 1.6 kg m^2 along with a feeding rate of 0.75 kg feed per day (Ndegwa *et al.* 2000). They further added that overfeeding must be avoided as it can lead to excessive fermentation in the bed and cause the worms to shrink and eventually die. Appelhof (1997) recommended the use of a worm to garbage (biodegradables) ratio of 2:1, based on the initial weight and on the average daily amount of garbage fed to the worms (*e.g.*, 225g of garbage daily for 500 g of worms). However, Bansal and Kapoor (2000) employed 100 mature earthworms to vermicompost 10 kg (dry weight) of crop residue with 7:3 ratio of cattle dung in cemented pits (60 – 60 – 90 cm) at a moisture content of 60%.

Physical, chemical and biological properties of vermicompost

Physical qualities of vermicompost

Vermicompost is one of the best forms of organic manure and is practically 100 percent like humus. Vermicompost has some unique physical quality parameters by which we can feel and smell the differences from other types of compost and manures.

- Vermicast is dark brown/black humus like coarse material, soft in feel and free from any live weed seeds and other contaminations.
- Mucus type substance coated on each particle that provides excellent water retention properties and improves the drainage in heavy soils.
- Contains sufficient moisture (20-35%) at the time of harvest to ensure high level of biological activities.
- Vermicompost are not lumpy, it is friable and free flowing.
- It is devoid of any foul smell.
- Free from inert materials.
- Minimum 90 per cent of the materials should pass through 4mm IS sieve.
- Bulk density of the materials should be in the range of 0.7 to 0.9 g/cm^3

Chemical properties of vermicompost

The available nutrients are more in vermicompost as compared to regular farmyard manure or compost. Moreover, secretions of earthworms present in vermicompost serve as growth stimulatory factors for plants.

The high variability of some important compost and vermicompost parameters suggests an urgent need for establishing quality assurance procedures in order to classify the available materials. The organic amendments prepared from different organic wastes (raw material), with different kind (composting or vermicomposting) and time of process, produce a final product which differs in its quality.

Biological properties of vermicompost

- Beneficial bacteria and fungi in vermicast introduces very high level of biological activities in the soil.
- Micro organisms present in vermicompost inactivate and suppress the growth of pathogens.
- Further modification in the organic waste to be converted to vermicompost also influences the microfloral population in the vermicompost. Addition of 2% neem cake to decomposing organic waste results in higher level of nitrogen-fixer population in vermicompost.
- Addition of 10% rock phosphate to organic waste containing 10% cow dung results in increase in phosphate solubilizing microorganisms.

Heavy metal content

Urban solid waste is known to contain heterogeneous materials both biodegradable and non-degradable with contamination of heavy metals originating from diverse sources. When the vermicompost, prepared from municipal solid waste, used for plant growth heavy metals become available for plant uptake. Some of these metals are beneficial to the plants when present in small quantity but when in high concentration adversely affects the plant growth as well as health of human and cattle.

Cadmium and chromium are not essential for the growth of plant. Cadmium accumulates in leafy vegetables, which affects human lever when consumed in high quantity. Similarly chromium is corrosive to membranes and may lead to lung cancer.

Copper, manganese and zinc are required in certain minimum concentrations by the plants but are injurious to them at high concentrations. Hence the ministry of Environment and Forest have prescribed limits for heavy metals in compost. Comparisons of the limits of heavy metals in different countries with India. Plant nutrients contain in urban solid wastes can be beneficially utilized provided the heavy metal content in the compost does not exceed the limits laid down in the standards.

Table 6: Fertilizer Association of India has also included standard specification of vermicompost in The Fertilizer (Control) Order

S. No.	Parameters of vermicompost	Quality Standard
1.	Moisture, per cent by weight	15.0-25.0
2.	Colour	Dark brown to black.
3.	Odour	Absence of foul dour.
4.	Particle Size	90% material should pass through 4.0 mm IS sieve
5.	Bulk density (g/cm^3)	0.7-0.9
6.	Total organic carbon, per cent by weight, Minimum	18.0
7.	Total nitrogen (as N), per cent by weight, Minimum	1.0
8.	Total phosphate (as P_2O_5), per cent by weight, Minimum	0.8
9.	Total potassium (as K_2O), per cent by weight, Minimum	0.8
10.	Heavy metal contents (mg/kg), Mximum	
	a. Arsenic (as As_2O_3)	10.00
	b. Cadminum (as Cd)	5.00
	c. Chromium (as Cr)	50.00
	d. Mercury (as Hg)	0.15
	e. Nickel (as Ni)	50.00
	f. Lead (as Pb)	100.00

Source: The Fertilizer (Control) Order 1985 (As amended upto June, 2017), The Fertilizer Association of India, New Delhi-110067

Nutrients transformation

According to Mulongoy and Bedoret (1989) and Ganeshamurthy *et al.* (1998), earthworms enhance the nutrient availability through casting. It is believed that the passage of the ingested material through the worm's guts may increase microbial activity in the casts and thus the availability of the nutrients. Many studies (Edwards and Burrows 1998, Atyieh *et al.* 2002a, Chaoui *et al.* 2003) have shown that the use of earthworm casts can serve as an alternative to fertiliser and results in the slow release of nutrients for the plant's uptake.

Buchanan *et al.* (1988) stated that vermicomposts have more available nutrients per weight than the organic waste from which they are produced. Furthermore, Pramanik *et al.* (2007) argued that vermicomposting allows obtaining an organic source of nutrients for the crop, which are physically, nutritionally and bio-chemically improved over compost in relatively less time. They also suggested that it is important to determine the characteristics and extent of organic C, N, P, K, humic acids, micro-organism and enzymatic activities in order to determine the dynamics of vermicomposting. Similarly,

Albanell *et al.* (1988) indicated that in comparison to the feed material, vermicompost possesses a reduced salt content, greater cation exchange capacity and increased content of total humic substances.

It has also been reported that the increase in nutrient content is attributed to accelerated mineralisation of organic matter after vermicomposting (Atiyeh, *et al.*, 2000c and Kaushik and Garg 2003, Loh *et al.*, 2005). Tripathi and Bhardwaj (2004) reported that E. fetida showed faster and higher decomposition rates, mineralisation of nutrients and breeding than did Lampitomauritii, when fed on kitchen waste plus cow dung. They also claimed that worm castings had an increased concentration of nitrogen, phosphorus and potassium and they observed reductions in C:N and C:P ratios after 150 days of vermicomposting as compared to the control Bansal and Kapoor (2000) studied vermicomposting of mustard residue and sugarcane trash mixed with cattle dung using E. fetida in a 90-day experiment. They reported a significant increase in Zn and observed changes in the total P, K and Cu contents in the final product.

Microbial activities

The digestive system of earthworms consists of a pharynx, an oesophagus and a gizzard, followed by an anterior intestine that secretes enzymes and a posterior intestine that absorbs nutrients (Sharma *et al.*, 2005). During progress through this digestive system, there is a dramatic increase in the number of microorganisms by as much as a factor of 1000 (Edwards, 1999). The casts excreted by earthworms contained significant quantities of polysaccharides, which stimulate microbial activity and induce changes in the structure of the intrinsic microbial community (Edwards and Bohlen 1996).

The burrowing and casting activities of earthworms contribute to the activity of soil micro-organisms (Edwards and Bohlen 1996) and nutrient enriched earthworm casts serve as a good source that supports microbial growth (Lee, 1985). However, Edwards (1998) suggested that vermicompost is rich in diverse microbial populations, particularly fungi, bacteria and actinomycetes. It is believed that bacteria are of minor importance in the diet, whereas algae are of moderate importance, while both protozoa and fungi are major sources of nutrients. Worms produced under sterile conditions could live on individual cultures of certain bacteria, fungi and protozoa but they grow best on various mixtures of micro-organisms (Sharma *et al.*, 2005). The total microbial load in the different regions of the guts of worms has also shown more intense colonisation of microbes in the anterior part of the intestine than the other parts.

Kale (1993) and Edwards (1999) argued that microbes, which transform nutrients into forms readily taken up by plants, are diverse due to the mesophilic transformation of organic matter. Vermicompost contains larger microbial populations than thermophilic compost, leading to greater potential for odour reduction and nutrient mineralisation.

Groups of soil microorganisms that live in very intimate contact with roots are the arbuscularmycorrhizal (AM) fungi. These fungi are known to assist the plant in uptake of nutrients and to improve plant growth (Douds *et al.*, 2005). It has also been demonstrated that vermicomposts can stimulate root elongation and lateral root emergence and mycorrhizal association (Kale *et al.*, 1987 and Cavender *et al.*, 2003).

Table 7: Biodiversity of vermicompost bacteria and their beneficial traits

Vermicompost earthworm	Names of bacteria	Beneficial traits	Reference
Pheretima sp.	*Pseudomonas oxalaticus*	Oxalate degradation	Khambata and Bhat, 1953
Unspecified	*Rhizobium trifolii*	Nitrogen fixation and growth of leguminous plants	Buckalew *et al.*, 1982
Lumbricus rubellus	*R. japonicum, P. putida*	Plant growth promotion	Madsen and Alexander 1982
L. terrestris	*Bradyrhizobium japonicum*	Improved distribution of nodules on soybean roots	Rouelle, 1983
Aporrectodea trapezoids	*P. corrugata, 214OR*	Suppress Gaeumannomyces graminis var. Tritd in wheat	Doube *et al.*, 1994
A. rosea			
A. trapezoids Microscolex dubius	*R. meliloti L5-30R*	Increased root nodulation and nitrogen fixation in legumes	Stephens *et al.*, 1994b
Eisenia foetida	*Bacillus spp., B. megaterium, B. pumilus, B. subtilis*	Antimicrobial activity against *Enterococcus faecalis* DSM 2570, *Staphylococcus aureus* DSM 1104	Vaz-Moreira *et al.* 2008
L. terrestris	*Fluorescent pseudomonads Filamentous actinomycetes*	Suppress *Fusarium oxysporum* f. sp. *asparagi* and *F. proliferatum* in asparagus, *Verticillium dahlia* in eggplant and *F. oxysporum* f. sp. *lycopersici* Race 1 in tomato	Elmer, 2009
Eudrilus sp.	Free-living N2 fixers *Azospirillum,*	Plant growth promotion by nitrification, phosphate solubilisation and plant disease suppression	Gopal *et al.*, 2009

Vermicompost earthworm	Names of bacteria	Beneficial traits	Reference
	Azotobacter, Autotrophic Nitrosomonas, Nitrobacter, Ammonifying bacteria, Phosphate solubilizers, Fluorescent pseudomonads		
E. foetida	*Proteobacteria, Bacteroidetes Verrucomicrobia, Actinobacteria, Firmicutes*	Antifungal activity against *Colletotrichum coccodes, R. solani, P. ultimum, P. capsici and F. moliniforme*	Yasir *et al.*, 2009a
Unspecified	*Eiseniicola composti YC06271*^{T}	Antagonistic activity against F. moniliforme	Yasir *et al.*, 2009b

Adapted from Pathma and Sakthivel, 2012

Process of vermicomposting

Vermicompost is prepared by heap method to reduce the cost of production. The size of the bed is 20 ft long – 3 ft width and 1.5 ft height. The feed mix of about 1.5 ft thick is poured on the vermibeds. Thereafter healthy young earthworms of the identified species is released in this bedding followed by feeding material. The pre-decomposed material need not be added in layers but, its depth should not be more than 2 ft. The heap should be well aerated and kept moist by periodic sprinkling of water. Moisture should be about 50% and temperature between 25 to 35°C. During the process of multiplication, the earthworms produce manure one third of their body weight daily.

The decomposition in the 1st cycle may take about 60-90 days depending on various factors and efficiency of the system and thereafter the period will get reduced to 4 to 6 weeks. Therefore one bed may be used for 4-6 times a year for preparation of vermi compost and rearing up worms.

A few days before the harvest, watering of the tank is discontinued to allow migrating of worms towards bottom of the bed where moisture still exists. The top layer of compost is then transferred outside without disturbing the bed and heaped on a plain open surface. The colour of this material is black and light in weight. The compost is sieved through a 3 mm mesh and then packed in bags.

The worms are harvested by hand. During the first harvest, the number of adults obtained can be same as that of the initially introduced worms. But the culture now contains the worms of the other stages of life cycle like cocoons, juveniles and sub-adults. Sub-adults may be collected and along with the adults may be released to new beds for expansion of units or can be sold in the market. Remaining ones *i.e.,* juveniles and cocoon can be left in the bed and pre-decomposed material may be added to it and the process continues. At times after removing the top layer of compost during the expansion phase instead of harvesting the adults and sub-adults, they are allowed to stay in the same bed and pre-decomposed material is released to it. However, bedding may be discarded after 6 months and fresh bedding should be prepared to maintain culture.

Environmental benefits of vermicomposting in large scale

The importance of earthworms to ecosystem functioning has led to many studies on the impacts of metals on earthworms. Fly-ash is a useful organic amendment. Huge amounts of fly-ash originate from different coal fired power station throughout the world, presenting several problems, the most important ones being environment pollution. Fly-ash has great potentiality in agriculture due to its efficacy in modification of soil health and crop performance. Vermicomposting is an excellent technique for reducing the toxic heavy metals in fly-ash using earthworm. Earthworms convert organic matter both physically and chemically to increase the soil porosity and therefore, they are also called as 'ecosystem engineers' (Usmani and Kumar), Earthworms can remediate heavy elements from fly-ash through two different pathways: by direct dermal contact with the elements dissolved in the fly-ash solution and/ or by ingestion and ensuing gut uptake from bulk fly-ash or specific fly-ash components. Earthworms are potential bio accumulators of heavy metals and therefore they have been successfully demonstrated in mitigating the toxicity of industrial and municipal waste by vermicomposting technology (Sinha *et al.*, 2010) (Table 8). Earthworms have different capacities to uptake and bio accumulate both essential (phosphorous and nitrogen and non-essential (heavy metals) elements. It has been hypothesized that this behavior is the result of inter-specific differences in the dietary intake of elements and in physiological and morphological characteristics, such as structure of skin, chemical species requirements, excretion and detoxification (Morgan and Morgan, 1992; Morgan and Morgan, 1999, Suthar *et al.*, 2008) Earthworms can tolerate and bio-accumulate high concentrations of heavy metals like cadmium (Cd), mercury (Hg), lead (Pb) copper (Cu), manganese (Mn), calcium (Ca), iron (Fe) and zinc (Zn) in their tissues without affecting their physiology and this particularly when the metals are mostly non bio available. The species *Lumbricus terrestris, L. rubellus* and *D. rubidawas* found to bio accumulate very high levels of lead (Pb) and Cadmium (Cd) in their tissues. Earthworms are capable of reducing possible toxic effects of superfluous heavy metals by utilizing them for physiological metabolism.

Table 8: Remediation of heavy metals from fly-ash by earthworms

S. No.	Methodology	Species	Accumulation of heavy metals by earthworms	Reference
1.	Fly-ash was mixed with cow-dung in 20, 40, 60 and 80%. Heavy metals conc. before and after vermicomposting were determined.	*Eisenia fetida*	Cu, Cr, Pb, Ni, Zn	Gupta *et al.*, 2005
2.	Seven different combinations of cow-dung and fly-ash-1:1, 1:2, 1:3,2:1, 3:1 were prepared for vermicomposting.	*Eisenia fetida* *Eudrilus eugeniae*	Zn, Mn	Jeyanthi *et al.*, 2010
3.	Fly-ash (FA) was composted with and without epigeic earthworm along with cow-dung (OM) in ratios – FA alone, FA+OM (1:1), FA+OM (1:3) and FA+OM (3:1).	*Eisenia fetida*	Cr, Pb, Cd	Bhattacharya *et al.*, 2012
4.	Remediation of heavy metals by three earthworm species by three different urban wastes –Flower waste (FW), Municipal solid waste (MSW) and Market waste (MW)	*Eudrilus eugeniae, Eisenia fetida, Perionyx excavatus*	Cu, Pb, Zn, Mn, Cd	Pattnaik and Reddy, 2009
5.	Treatment includes factorial combination of 3 rates of fly-ash and two different species of earthworms.	*Megascolecid* sp and exotic *Aporrectodea trapezoides*	Cu, Cr, Zn	Muir *et al.*, 2007

Adapted from Usmani and Kumar,

Precautions for vermiculture and vermicomposting

Vermiculturists should also be aware of the several precautions in doing such process to ensure that the project would turn out successful and fruitful. From our hands-on experiences, vermicompost pit should be protected from direct sun light so that the vermi worm would survive. Direct heat possibly causes the worms to die. Spray water on the pit as when required to maintain moisture level because vermi worms are fond of it. We should also protect the worms from ant, rat, bird and excessive rain.

How to use vermicompost?

- **Vermicompost can be used for all crops:** ***agricultural, horticultural, ornamental and vegetables at any stage of the crop.***
- **For general field crops**: Around 2–3 t ha^{-1} vermicompost is used by mixing with seed at the time of sowing or by row application when the seedlings are 12–15 cm in height. Normal irrigation is followed.
- **For fruit trees:** The amount of vermicompost ranges from 5 to 10 kg per tree depending on the age of the plant. For efficient application, a ring (15–18 cm deep) is made around the plant. A thin layer of dry cow dung and bone meal is spread along with 2–5 kg of vermicompost and water is sprayed on the surface after covering with soil.
- **For vegetables:** For raising seedlings to be transplanted, vermicompost at 1 t ha^{-1} is applied in the nursery bed. This results in healthy and vigorous seedlings. But for transplants, vermicompost at the rate of 400–500 g per plant is applied initially at the time of planting and 45 days after planting (before irrigation).
- **For flowers:** Vermicompost is applied at 750–1000 kg ha^{-1}.
- For vegetable and flower crops vermicompost is applied around the base of the plant. It is then covered with soil and watered regularly.

Vermiwash and its use

What is Vermiwash?

Vermiwash is a liquid leachate collected by allowing excess water to saturate the actively vermicomposting substrate in which the water washes the substrate in such a way that the vermicast excreted by the earthworms feeding on the substrate as well as the earthworm's body surface (Varghese and Prabha, 2014**)**. It is well known, vermiwash is a rich source of vitamins, hormones, enzymes, macronutrients and micronutrients which when applied to plants help in efficient growth.

Use of vermiwash

(a) Plant Protecting Chemicals: It contains several enzyme, plant growth hormones, vitamins along with micro and macronutrients (Shield Earl, 1982) which increases the resistance power of crops against various diseases and enhance the growth and productivity of crops (Anand *et al.,* 1995; Pathak, 2004; Suthar *et al.,* 2005; Umamaheswari *et al.,* 2003; Yadav *et al.,*2005). Karuna *et al.,* 1999. The treatment of vermiwash of vermicompost has been shown to reduced disease by necrotrophs as well as biotrophs (Fokkema, 1993; Al-Dahmani *et al.,* 2003) and significantly decreases in soil borne pathogens and various pests (Rodriguez *et al.,* 2000; Szczech *et al.,* 1993; Nakasone *et al.,* 1999; Orlikowski, 1999). Generally, foliar spray of vermiwash of vermicomposting would offer a method of supplying nutrients to higher plants more rapidly than methods involving soil and root application (Marschner, 1995). In dry condition with a lack of water in the top soil and corresponding decline in nutrients available foliar application of nutrients is much more effective than soil application (Grundon, 1980). Foliar spray containing nutrients can also compensate for the decline in nutrients uptake by roots with the onset of the reproductive stage a result increase in number of fruit (Buckerfield *et al.,* 1999). It was demonstrated that growth of ornamental plant after adding vermiwash showed similar growth pattern as with addition of auxines, gibberellins and cytokinines through the soil (Tomati and Gallii, 1995; Grappelli *et al.,* 1987).

Vermiwash was used in three different dilutions viz., 1:1, 1:2 and 1:4 by mixing with water, both as seedling dip treatment and foliar spray to evaluate its efficacy against thrips and mites along with organic amendments. Recommended dose of fertilizer along with dimethoate and dicofol sprays was maintained as standard check. Six rounds of vermiwash sprays were taken up at 15 days interval commencing at two weeks after transplanting. Vermiwash was obtained by culturing earthworms (*Eudrilus eugeniae* Kinberg) on organic substrates (65% pre-composted crop wastes and 35% animal manure). Application of vermicompost @ 2.5 t/ha along with vermiwash 1:1 sprays registered significantly maximum dry chilli yield where as significantly lower yields were recorded in other treatments where vermiwash dilution varied from 1 : 2 to 1 : 4 proportions. The positive effect of vermiwash on crop growth and yield in the present study is in conformity with the studies of Bucker field *et al.* (1999) who reported that weekly applications of vermiwash increased radish yield by 7.3% and Thangavel (2003) who observed that both growth and yield of paddy increased with the application of vermiwash and vermicast extracts (George *et al.,* 2006).

(b) Liquid Nutrient: Vermicompost leachate has also been reported to be suitable as formulation for liquid fertilizer (Gutirréz-Miceli *et al.,* 2008) and other agricultural uses (Tejada *et al.,* 2008), (Zaller, 2006) had reported that foliar applications of vermicompost leachate improved certain quality parameters of tomatoes besides suppressing *Phytophthora* disease. However, most of the studies mentioned here report the beneficial effects of vermiwash or vermicompost leachate produced from animal manure + earthworms or from the earthworms alone.

Different methods of vermiwash preparation or extraction

(a) From the vermicompost bed–The vermicompost bed is made 1% slope so that the drained liquid may be collected at the down tank. In this case the pipe is fitted just above the sand or gravell layer from each pit and all the leachestes were collected from the pit to one tank.

(b) Extracted through drums-Vermiwash extracted through vermiwash collecting device which is made up of Mud drum having capacity of 5 liter and a tap at the bottom the drum filled with broken bricks, about 10cm thickened which is followed by sand layer of 2-3cm thickness lastly with filled with vermicompost with heavy population of earth worms simultaneously added fresh water in to drum and a container kept bellow the tap of drum. The watery yellowish to black extract of vermicompost, vermiwash drainage out off drum. After 1to 2 days the process of extraction has been completed.

Chemical and Microbiological composition of Vermiwash Ph	7.48 – 0.03
Electro conductivity dS/m	0.25 – 0.03
Organic Carbon%	0.008 – 0.001
Total Kjeldhal Nitrogen%	0.01 – 0.005
Available Phosphate%	1.69 – 0.05
Potassium (ppm)	25 – 2
Sodium (ppm)	8 – 1
Calcium (ppm)	3 – 1
Copper (ppm)	0.01 – 0.001
Ferrous (ppm)	0.06 – 0.001
Magnesium (ppm)	158.44 – 23.42
Manganese (ppm)	0.58 – 0.040
Zinc (ppm)	0.02 – 0.001
Total Heterotrophs (CFU/ml)	1.79 – 10^3
Nitrosomonas (CFU/ml)	1.01 – 10^3
Nitrobacter (CFU/ml)	1.12 – 10^3
Total Fungi (CFU/ml)	1.46 – 10^3

Precautions of vermiwash

- Water should be poured slowly.
- Do not mix un-decomposed material while, watering.

- Do not add any green material.
- Do not allow to compact the contents.

To farmers

- Less reliance on purchased inputs of nutrients leading to lower cost of production
- Increased soil productivity through improved soil quality
- Better quantity and quality of crops
- For landless people provides additional source of income generation

To industries

- Cost-effective pollution abatement technology

To environment

- Wastes create no pollution, as they become valuable raw materials for enhancing soil fertility

To national economy

- Boost to rural economy
- Savings in purchased inputs
- Less wasteland formation

Conclusions

The vermiculture and vermicomposting activity is such a worthwhile and exciting venture. We have learned a lot specifically in the methodologies, benefits and significance of this activity. After almost three months, project delivery and execution, we can therefore conclude that:

- Vermiculture is a substantial way of reducing wastes, producing fertilizers and maintaining the balance of the ecological environment;
- Vermicomposting can produce high-quality fertilizers which are better compared to other commercial fertilizers in the market;
- Vermiculture converts farm wastes into organic fertilizer, making it an environment-friendly technology;
- Vermiculture increases crop yield and lessens dependence on chemical fertilizers thus mitigating climate change;
- Vermiculture can be made into a livelihood program and become a source of extra income through selling the vermicast and also the vermi worms;
- Taking worms out of their natural environment and placing them in the vermi

beds creates a human responsibility. They are living creatures with their own unique needs, so it is important to create and maintain a healthy habitat for them to do their work. If you supply the right ingredients and care, your worms will thrive and make compost for you.

Reference

Aiyyar, M. E. and Ananthakrishnan, T. N. (1993). A manual of Zoology, Volume-I, S. Viswanathan Publishers and Printers, Chennai, India

Albanell, E., Plaixats, J. and Cabrero, T. (1988). Chemical changes during vermicomposting (*Eisenia fetida*) of sheep manure mixed with cotton industrial wastes. *Biol.Fertil Soils.*, 6:266-269.

Al-Dahmani, J.H., Abbasi, P.A., Miller, S.A. and Hoitink, H.A.J. (2003). Suppression of bacterial spot of tomato with foliar sprays of compost extracts under greenhouse and field conditions. *Plant Dis*., 87: 913-919.

Alidadi, H., Parvaresh, A.R., Shahmansouri, M.R. and Pourmoghadas, H. (2005). Combined compost and Vermicomposting process in the treatment and bioconversion of sludge. *Iranian Journal of Environmental Health Science and Engineering*. 2 (4): 251-254.

Anand, J.A., Wilson, M.D.P. and Kale, R.D. (1995). Effect of vermiwash on seed germination and seedling growth. *J Soil Biol Ecol*., 15:90-95.

Appelhof, M. (1997). Worms eat my garbage. Second edition. Flower press, Kalamazoo, Michigan.

Arancon, N.Q, Edwards, C.A., Yardim, E.N., Oliver, T.J., Byrne, R.J. and Keeney, G. (2007). Suppression of two-spotted spider mite (*Tetranychus urticae*), mealy bug (*Pseudococcus* sp) and aphid (*Myzus persicae*) populations and damage by vermicomposts. *Crop Prot*., 26:29-39.

Arancon, N.Q., Edwards, C.A. and Lee, S. (2002). Management of plant parasitic nematode populations by use of vermicompost. In: Proceedings Brighton Crop Protection Conference – *Pests and Diseases*., 8(B2):705-716.

Arancon, N.Q., Edwards, C.A., Bierman, P., Metzger, J.D., Lee, S., and Welch, C., (2003). Effects of vermicomposts on die growth and marketable fruits of field-grown tomatoes, peppers and strawberries. *Pedobiologia*. 47: 731-735.

Arancon, N.Q., Edwards, C.A., Bierman, P., Welch, C. and Metzger, J.D. (2003b). Effects of vermicomposts on growth and marketable fruits of field-grown tomatoes, peppers, and strawberries. *Pedobiologia*, 47:731-735.

Arshad, M. and Frankenberger Jr. W.T. (1993). Microbial production of plant growth regulators. In: Metting, Jr., F.B. (Ed.), Soil Microbial Ecology: Applications in Agricultural and Environmental Management. Marcell Dekker, New York, pp. 307-347.

Atiyeh, R, M., Arancon, N.Q., Edwards, C.A, and Metzger, J.D. (2002a). Influence of earthworms processed pig manure on the growth and productivity of marigolds. *Bioresource Technology*. 81:103-108.

Atiyeh, R.M., Arancon, N., Edwards, C.A. and Metzger, J.D. (2000b). Influence of earthworms-processed pig manure on the growth and yield of greenhouse tomatoes. *Bioresource Technology*. 75: 175-180.

Atiyeh, R.M., Dominguez, J., Subler, S. and Edwards, C.A. (2000c). Changes in biochemical properties of cow manure during processing by earthworms (*Eisenia andrei*, Bouché) and the effects on seedling growth. *Pedobiologia*. 44: 709-724.

Bansal, S. and Kapoor, K.K. (2000). Vermicomposting of crop residues and cattle dung with *Eisenia foetida*. *Bioresour Technol*. 73: 95-98.

Barbieri, P., Bernardi, A., Galli, E. and Zanetti, G. (1988). Effects of inoculation with different strains of Azospirillum brasilense on wheat roots development. In: Klingmüller W (ed) Azospirillum IV, Genetics, Physiology, Ecology. Springer Verlag, Berlin, pp 81–188

Bhadauria, T. (2008). "Impact of intensive agricultural practices on diversity of earthworms in Raebareli District of Indogangetic plains, India," Final Report, Department of Science And Technology, Govt of India, New Delhi, India.

Bhadauria, T. and Saxena, K.G. (2010). Role of Earthworms in Soil Fertility Maintenance through the Production of Biogenic Structures. *Applied and Environmental Soil Science*. 1-7.

Bhattacharya, S. S., Iftikar, W., Sahariaha, B. and Chattopadhyay, G. N. (2012). Vermicomposting converts fly-ash to enrich soil fertility and sustain crop growth in red and lateritic soils. *Resour Conserv and Recy*., 65:100-106.

Bhonwong, A., Stout, M.J., Attajarusit, J. and Tantasawat, P. (2009). Defensive role of tomato polyphenol oxidases against cotton bollworm (*Helicoverpa armigera*) and beet army worm (*Spodoptera exigua*). *J. Chem. Ecol*., 35:28-38.

Bilgrami, A.L. (1996). Evaluation of the predation abilities of the mite Hypoaspis calcuttaensis, predaceous on plant and soil nematodes. *Fund. Appl. Nematol*., **20**:96-98.

Biradar, A. P., Sunita, N. D., Teggel, R. G. and Devaradavadgi, S. B. (1998). Effect of vermicompost on the incidence of subabul psyllid. *Insect-Environ*., 4:55-56.

Buchanan, M.A, Russel, G., and Block, S.D. (1988). Chemical characterization and nitrogen mineralization potentials of vermicomposts derived from differing organic wastes. In C.A Edwards and E.F. Neuhauser (eds.) Earthworms in waste and environmental management. SPB Academic Publishing, *The Hague*. Pp. 231-239.

Buckalew, D. W., Riley, R. K., Yoder, W. A. and Vail, W. J. (1982). Invertebrates as vectors of endomycorrhizal fungi and Rhizobium upon surface mine soils. *West Virginia Acad. Sci. Proc.,* 54:1.

Buckerfield, J. C., Flavel, T. C., Lee, K. E. and Webster, K. A. (1999). Vermicompost in Solid and Liquid Forms as a Plant-Growth Promoter. *Pedobiologia,* 43: 753-759.

Buckerfield, J. C., Flavel, T. C., Lee, K. E., Webster, K.A., Diazcozin, D. J., Jesus, J. B., Trigo, D. and Garvin, M. H. (1999). Vermicompost in solid and liquid forms as a plant growht promoter. Sixth International Symposium on Earthworm Ecology, Vigo, Sain, *Pedobiologia,* 43:753-759.

Buckerfield, J.C., Flavel, T.C., Lee, K.E. and Webster, K.A. (1999). Vermicompost in solid and liquid forms as a plant-growth promoter. *Pedobiologia.* 43: 753-759.

Cavender, N. C. D., Atiyeh, R. M and Keen, M. (2003). Vermicompost stimulates mycorrhizal colination of roots of sorghum bicolour at the expense of plant growth. *Pedobiologia,* 47: 85-89.

Chaoui, H. I., Zibilske, L. M., and Ohno, T. (2003). Effect of earthworm casts and compost on soil microbial activity and plant nutrient availability. *Soil Biol-Biochem.* 35: 295-302.

Chen, W., Hoitink, H. A., Schmitthenner, A. F. and Touvinen, O. (1987). The role of microbial activity in suppression of damping off caused by *Pythium ultimum. Phytopathology.* 78:314-322.

Chen, Y, and Aviad, T. (1990). Effects of humic substances on plant growth. In: Mac Carthy P, Clapp CE, Malcolm RL, Bloom PR (eds) Humic Substances in Soil and Crop Sciences. Selected Reading ASA and SSSA, Madison, pp 161-186

Darwin, F. and Seward, A. C. (1903). More letters of Charles Darwin. In: John M (ed) A record of his work in series of hitherto unpublished letters, vol 2. London, p 508.

Dominguez, J. and Edwards, C.A. (2011). Biology and Ecology of Earthworm Species Used for Vermicomposting. by Taylor & Francis Group, LLC

Dominguez, J., Edwards, C. A. and Subler, S. (1997). A comparison of Vermicomposting and composting. *Biocycle. ABI/INFORM Global.* 38 (4): 57.

Doube, B. M., Stephens, P. M., Davorena, C. W. and Ryderb, M. H. (1994). Interactions between earthworms, beneficial soil microorganisms and root pathogens. *Appl. Soil Ecol.,* 1:3-10.

Douds, D., Nagahashi, G., Pfeffer, P., Kayser, W. and Reider, C. (2005). On-farm production and utilization of arbuscular mycorrhizal fungus inoculum. *Canadian Journal of Plant Science.* 85(1): 15-21.

Edwards, C.A., Arancon, N. Q., Emerson, E. and Pulliam, R. (2007). Supressing plant parasitic nematodes and arthropod pests with vermicompost teas. *Biocycle.,* 38–39

Edwards, C. A. and Bohlen, P. J. (1996). Biology and Ecology of Earthworms (3rd Ed.), Chapman and Hall, London, U.K.

Edwards, C. A. and Lofty, J. R. (1972). Biology of Earthworms; Chapman and Hall, London, pp: 283.

Edwards, C.A. (1995). Histroical overview of vermicomposting. *Biocycle.*, 36: 56-58.

Edwards, C.A. (1998). The use of earthworms in the breakdown and management of organic wastes. In C.A Edwards (ed.).Earthworm Ecology, CRC Press, Boca Raton, FL. Pp. 327-354.

Edwards, C.A. (1999). Soil invertebrate controls and microbial interactions in nutrient and organic matter dynamics in natural and agroecosystems. In: Webmaster Functions in Agroecosystems: Invertebrates, Climate and Substrate Quality. Ed. D.C. Coleman and P. Hendrix, CABI Publ., Wallingford, U.K., Chapter 7, Pp. 141-159.

Edwards, C.A. and Burrows, I. (1988). The Potential of Earthworms Composts as Plant Growth Media. In Edward, C.A. and E.F. Neuhauser (Eds.). Earthworms in Waste and Environmental Management. SPB Academic Publishing, The Hague, The Netherlands; ISBN 90-5103-017-7, pp: 21-32.

Elmer, W. H. (2009). Influence of earthworm activity on soil microbes and soilborne diseases of vegetables. *Plant Dis.*, 93:175-179.

Elvira, C., Sampedro, L., Benítez, E. and Nogales, R. (1998). Vermicomposting of sludges from paper mill and dairy industries with Eisenia Andrei: a pilot-scale study. *Bioresour Technol.*, 63: 205-211.

Fokkema, N. J. (1993). Opportunities and problems of control of foliar pathogens with micro-organisms. *Pestic. Sci.*, 37:411-416.

Frederickson, J. (2001). Timely research at worm research centre. Composting News, The Composting Association, UK. 5(3): 4-5.

Frederickson, J. and Howell, G. (2003). Large-scale vermicomposting; emission of nitrous oxide and the effect of temperature on earthworms population, *Pedobiologia*, 47: 724-730.

Frederickson, J., Butt, K. R., Morris, R. M. and Daniel, C. (1997). Combining Vermiculture with Traditional Green Waste Composting Systems. *J. of Soil Biology and Biochemistry*, 29: 725-730.

Gandhi, M., Sangwan, V., Kapoor, K. K. and Dilbaghi, N. (1997). Composting of household wastes with and without earthworms. *Environment and Ecology*.15(2):432-434.

Ganeshamurthy, A. N., Manjaiah, K. M. and Rao, A. S. (1998). Mobilisation of nutrients in tropical soil through worm casting; availability of macronutrients. *Soil Biol-Biochem*. 30(13): 1671-1676.

Garg, P., Gupta, A. and Satya, S. (2006). Vermicomposting of different types of waste using Eisenia foetida: A comparative study. *Bioresource Technology*. 97: 391-395.

Gavrilov, K. (1962). Role of earthworms in the enrichment of soil by biologically active substances. *Voprosy Ekologii Vysshaya Shkola (Moscow)*.7: 34.

George, S., Giraddi, R. S. and Patil, R. H. (2006). Utility of Vermiwash for the Management of Thrips and Mites on Chilli (*Capsicum annuum* L.) Amended with Soil Organics. *Karnataka J. Agric. Sci.*, 20(3): 657-659.

Gopal, M., Gupta, A., Sunil, E. and Thomas, V. G. (2009). Amplification of plant beneficial microbial communities during conversion of coconut leaf substrate to vermicompost by Eudrilus sp. *Curr Microbiol.* 59:15-20.

Grappelli, A., Galli, E. and Tomati, U. (1987). Earthworm casting effect on Agaricus bisporus fructification. *Agrochimica,* 31(4-5): 457-461.

Grundon, N.J. (1980). Effectiveness of soil dressing and foliar spray of copper sulphate in correcting copper deficiency of wheat (*Triticum aestivum*) in Queensland. *Australian Journal of Experimental Agriculture and Animal Husbandry*, 20, Pp.717-723.

Gupta, S. K., Tewari, A., Srivastava, R., Murthy, R. C. and Chandra, S. (2005). Potential of Eisenia fetida for sustainable and efficient vermicomposting of fly-ash. *Water and air pollut.*, 163:293-302.

Gutiérrez-Miceli, F. A., Moguel-Zamudio, B., Abud-Archila, M., Gutiérrez-Oliva, V. F. and Dendooven, L. (2008). Sheep manure vermicompost supplemented with a native diazotrophic bacteria and mycorrhizas for maize cultivation. *Bioresource Technology*. 99(15): 7020-7026.

Hawida, S., Kapari, L., Ossipov, V., Ramtala, M. J., Ruuhola, T., and Haukioja, E. (2007). Foliar phenolics are differently associated with Epirrita autmnata growth and immune competence. *J. Chem. Ecol.*, 33:1013-1023.

Hoitink, H. A. and Grebus, M. E. (1997). Composts and Control of Plant Diseases. In: Hayes MHB, Wilson WS (eds) Humic Substances Peats and Sludges Health and Environmental Aspects. Royal Society of Chemistry, Cambridge, pp 359 –366

Hoitink, H. A., Stone, A. G. and Han, D.Y. (1997). Suppression of plant diseases by compost. *Hort. Sci.*, 32:184-187.

Inbal, E. and Feldman, M. (1982). The response of a hormonal mutant of common wheat to bacteria of the Azospirillium. *Israel J. Bot.*, 31:257-263.

Jadhav, A. D., Talashilkar, S. C. and Pawar, A. G. (1997). Influence of the conjunctive use of FYM, vermicompost and urea on growth and nutrient uptake in rice. *Journal of Maharashtra Agricultural Universities*. 22(2):249-250.

Jagnow, G. (1987). Inoculation of cereal crops and forage grasses with nitrogen-fixing rhizosphere bacteria: a possible cause of success and failure with regard to yield response-a review. *Z. Pflanzenernaehr Dueng Bodenkde.* 150:361-368.

Jeyanthi, A., Balakumar, S. and Mahalakshmi, T. (2010). Increasing bioavailability of nutrients in fly-ash through vermicomposting. *J. Biosci. Tech.*, 2: 100-113.

Kale, R. D. (1993). Vermiculture: scope for new biotechnology. In A.K. Ghosh (ed.) Earthworm Resources and Vermiculture. Zoological Survey of India, Calcutta, 105-108.

Kale, R. D., Bano, K., Sreenivasa, M. N. and Bagyaraja, D. J. (1987). Influence of worm cast (Vee Comp. E. UAS'83) on the growth and mycorrhizal colonisation of two ornamental plant. *South India Horticulture.* 35: 433-437.

Karuna, K., Patil, C. R. and Narayanswamy, P. (1999). Kale stimulatory effects of earthworm body fluid (vermiwash) on Crinkle red Varity of Vnthurium andreanum Lind. *Crop Res.,* 17(2):253-257.

Kaushik, P. and Garg, V. K. (2003). Vermicompost of mixed solid textile mill sludge and cow dung with epigeic earthworm (Eisenia Foetida). *Bio resource Technology.* 90: 311-316.

Kerry, B. (1988). Fungal parasites of cyst nematodes. *Agriculture, Ecosystems & Environment.* 24(1-3): 293-305.

Khambata, S. R. and Bhat, J. V. (1953). Studies on a new oxalate-decomposing bacterium, Pseudomonas oxalaticus. *J. Bacteriol.* 66:505-507.

Koul, O. (2008). Phytochemicals and insect control: an antifeedant approach. *Crit. Rev. Plant Sci.,* 27:1-24

Krishnamoorthy, R.V. and Vajranabhaiah, S.N. (1986). Biological activity of earthworm casts: an assessment of plant growth promoter levels in the casts. *Proceedings of Indian Academic Science (Animal Science).* 95:341-351.

Kurowska, A., Gora, J. and Kalemba, D. (1990). Effects of plant phenols on insects. *Pol. Wiad. Chem.,* 44:399-409.

Lee, K. (1985). Earthworms: Their Ecology and Relationships with Soils and Land Use; Academic Press: New York.

Loehr, R. C., Neuhauser, E. F. and Malecki, M. R. (1985). Factors affecting the vermistabilisation process. *Water Res.,* 19(10): 1311-1317.

Loh, T. C., Lee, Y. C., Liang, J. B. and Tan, D. (2005). Vermicomposting o f cattle and goat manure by Eisenia Foetida and their growth and reproduction performance. *Bioresource Technology.* 96: 111-114.

Maboeta, M. S. and Rensburg, L. V. (2003). Vermicomposting of industrially produced woodchips and sewage sludge utilising Eisenia Fetida. *Ecotoxicology and Environmental Safety.* 56: 265-270.

Madsen, E. L. and Alexander, M. (1982). Transport of Rhizobium and Pseudomonas through soil. *Soil Sci. Soc. Am. J.,* 46:557-560.

Mahanil, S., Attajarusit, J., Stout, M. J. and Thipayong, P. (2008). Over expression of tomato phenol oxidase increases resistance to the common cutworm. *Plant Sci.,* 174:456-466.

Mahmoud, S. A., Ramadan, Z., Thabet, E. M. and Khater, T. (1984). Production of plant growth promoting substance rhizosphere organisms. *Zentrbl Mikrobiol.,* 139:227-232.

Marschner, H. (1995). Mineral nutrition of higher plants, Academic Press, London.

Marsh, L., Subler, S., Mishra, S., and Marini, M. L. (2005). Suitability of aquaculture effluent solid mixed with cardboard as a feedstock for vermicomposting. *Bioresource Technology*. 96: 413-418.

Martin, J. P. (1976). Darwin on earthworms: the formation of vegetable moulds. Bookworm Publishing, Ontario.

Masciandaro, G., Ceccanti, B. and Gracia, C. (1997). Soil agro-ecological management: fertigation and vermicompost treatments. *Bioresour Technol*. 59:199-206.

Moody, S. A., Piearce, T.G. and Dighton, J. (1996). Fate of some fungal spores associated with wheat straw decomposition on passage through the guts of Lumbricus terrestris and Aporrectodea longa. *Soil Biol. Biochem.*, 28:533-537.

Morgan, J. E. and Morgan, A. J. (1992). Heavy metal concentrations in the tissues, ingesta and faeces of ecophysiologically different earthworm species. *Soil Biol. Biochem.,* 12: 1691-1697.

Morgan, J. E. and Morgan, A. J. (1999). The accumulation of metals (Cd, Cu, Pb, Zn and Ca) by two ecologically contrasting earthworm species (*Lumbricus rubellus* and *Aporrectodea caliginosa*): implications for ecotoxicological testing. *Appl Soil Ecol.*, 13:9-20.

Morra, L., Palumbo, A. D., Bilotto, M., Ovieno, P. and Ptcascia, S. (1998). Soil solarization: organic fertilization grafting contribute to build an integrated production system in a tomato-zucchini sequence. *Colture-Protte.*, 27:63-70.

Muir, M. A., Yunusaa, I. A. M., Burchett, M. D., Lawrieb, R., Chanb, K.Y. and Manoharan, V. (2007). Short-term responses of two contrasting species of earthworms in an agricultural soil amended with coal fly-ash. *Soil Biol. Biochem.*, 39:987-992.

Mulongoy, K. and Bedoret, A. (1989). Properties of worm cast and surface soil under various plant cover in the humid tropics. *Soil Biol-Biochem.*, 21(2): 197-203.

Muscolo, A., Bovalo, F., Gionfriddo, F. and Nardi, S. (1999). Earthworm humic matter produces auxin-like effect on Daucus carota cell growth and nitrate metabolism. *Soil Biol. Biochem.,* 31: 1303-1311.

Nakasone, A. K., Bettiol, W., de Souza, R. M. (1999). The effect of water extracts of organic matter on plant pathogens. *Summa Phytopathology.*, 25: 330-335.

Ndegwa, P, M., Thompson, S. A. and Das, K.C. (2000). Effects of stocking density and feeding rate on vermicompost of Biosolids. *Bioresource Technology.*, 71: 5-12.

Ndegwa, P. M. and Thompson, S. A. (2000). Effects of C-to-N ratio on Vermicomposting of Biosolids. *Bioresource Technology* 75: 7-12.

Ndegwa, P. M. and Thompson, S. A. (2001). Integrating composting and vermicomposting in the treatment and bioconversion of biosolids. 76: 107-112.

Neuhauser, E. F., Loehr, R. C. and Malecki, M. R. (1988). The potential of earthworms for managing sewage sludge, hi Edwards, C.A., and Neuhauser, E.F., (eds.) Earthworms in Waste and Environmental Management, SPB Academic Publishing, The Hague, 9-20.

Nielson, R. L. (1965). Presence of plant growth substances in earthworms demonstrated by paper chromatography and the Went pea test. *Nature*. 208:1113-1114.

Orlikowski, L. B. (1999). Vermicompost extract in the control of some soil borne pathogens. *International Symposium on Crop Protection*. 64: 405-410.

Parle, J. N. (1963a). Microorganisms in the intestines of earthworms. *J. Gen. Microbial.* 31:1-13.

Parle, J. N. (1963b). A microbilogical study of earthworm casts. *J. Gen. Microbial.* 31: 13-23.

Parle, J. N. (1963). A Microbiological Study of Earthworm Casts. *J. Gen. Microbiol.* 31:13-22.

Pathak, R. K. and Ram, R. A. (2004). Manual on Jai Octochaetidae: Oligochaeta vic Krishi, Central Institute for Subtropical Horticulture, Rehmankhera, P.O. Kokari, Lucknow-227107, 24: 31-32.

Pathma, J. and Sakthivel, N. (2012). Microbial diversity of vermicompost bacteria that exhibit useful agricultural traits and waste management potential. *Springer Plus.*, 1(1): 1-19.

Pathma, J. and Sakthivel, N. (2012). Microbial diversity of vermicompost bacteria that exhibit useful agricultural traits and waste management potential. *Springer Plus.*, 1:26.

Pattnaik, S. Reddy, M. V. (2009). Assessment of municipal solid waste management in Puducherry (Pondicherry), India. Resour. *Conserv. Recy.*, 54:512-520.

Pramanik, P., Ghosh, G.K., Ghosal, P.K. and Banik, P. (2007). Changes in organic-C, N, P and K and enzyme activities in vermicompost of biodegradable organic wastes under liming and microbial inoculants. *Bioresource Technology*. 98: 2485-2494.

QiTian, S. (2004). Research on prevention and elimination of agricultural pests by ginkgo phenols phenolic acids. *Chem. Ind. For Prod.*, 24: 83.

Ramesh, P. (2000). Effects of vermicomposts and vermicornposting on damage by sucking pests to ground nut (Arachis hypogea). *Indian J. Agri. Sci.*, 70:334.

Rao, K. R. (2002). Induce host plant resistance in the management sucking pests of groundnut. *Ann. Plant Protect. Sci.,* 10:45-50.

Rao, K. R. (2003). Influence of host plant nutrition on the incidence of Spodoptera litura and Helicoverpa armigera on groundnuts. *Indian J. Entomol.*, 65:386-392.

Rao, K. R., Rao, P. A. and Rao, K. T. (2001). Influence of fertilizers and manures on the population of coccinellid beetles and spiders in groundnut ecosystem. *Ann. Plant Protect Sci.*, 9:43-46.

Roberts, P. (2005). Private communications, School of Agriculture and Forestry Science, University of Wales, Bangor.

Rodríguez, J. A., Zavaleta, E., Sanchez, P. and Gonzalez, H. (2000). The effect of vermicompost on plant nutrition, yield and incidence of root and crown rot of Gerbera (Gerbera jamesonii H Bolus). *Fitopathologia.*, 35: 66-79.

Rouelle, J. (1983). Introduction of an amoeba and Rhizobium Japonicum into the gut of Eisenia fetida (Sav.) and Lumbricus terrestris L. In: Satchell JE (ed) Earthworm Ecology: From Darwin to Vermiculture. Chapman and Hall, New York, pp 375 – 381

Senesi, N., Saiz-Jiminez, C., and Miano, T. M. (1992). Spectroscopic characterization of metal-humic acid-like complexes of earthworm-composted organic wastes. *Science of the total Environment.*, 117: 111-120.

Sharma, S., Pradhan, K., Satya, S. and Vasudevan, P. (2005). Potentiality of earthworms for waste management and in other uses-A review. *The Journal of American Science.*, 1(l): 4-16.

Shield, E. B. (1982) Raising earthworms for profit. Shields Publication. PO Box 669 Eagle River Wisconsin, 128 p

Short, J. C. P., Frederickson, J. and Morris, R M. (1999). Evaluation of traditional windrow composting and vermicomposting for stabilization of waste paper sludge. *Pedobiologia.*, 43: 1-9.

Siddiqui, Z. A. and Mahmood, I. (1999). Role of bacteria in the management of plant parasitic nematodes: a review. *Bioresour. Technol.,* 69:167-179.

Sidhu, J., Gibbs, R. A., Ho, G. E. Unkovich, I. (2001). The role of indigenous microorganisms in suppression of Salmonella re-growth in composted biosolids. *Water Res.* 35:913-920.

Sinha, R. K., Heart, S., Valani, D. and Chauhan, K. (2010). Earthworms – the environmental engineers: review of vermiculture technologies for environmental management and resource development. *Int. J. Global Environ.*, 10(3-4):265-292.

Slocum, K. (2002). Proc Start Your Own Worm Farm, North Carolina State University, Raleigh, NC.

Sreenivash, C. H., Muralidhar, S. and Rao, M. S. (2000). Vermicompost: A viable component of IPNS in nitrogen nutrition of ridge gourd. *Annals of Agricultural Research.* 21: 108-113.

Stephens, P. M., Davoren, C. W., Ryder, M. H. and Doube, B. M. (1994b). Influence of the earthworm Aporrectodea trapezoids (Lumbricidae) on the colonization of alfalfa (*Medicago sativa* L.) roots by Rhizobium melilotti strain LS-30R and the survival of L5-30R in soil. *Biol. Fertil. Soils.*, 18:63-70.

Summers, G. and Felton, G.W. (1994). Prooxidation effects of phenolic acids on the generalist herbivore Helicoverpa zea: potential mode of action of phenolic compounds on plant anti-herbivory chemistry. *Insect Biochem. Mol. Biol.*, 24:943-953.

Suthar, S. (2007). Vermicomposting potential of Perionyx sansibaricus (Perrier) in different waste materials. *Bioresource Technology*. 98: 1231-1237

Suthar, S., Choyal, R. R., Singh, S. and Sudesh, R. (2005). Stimulatory effect of earthworm body fluid on seed germination and seedlings growth of two legumes. *J. Phytol. Res.*, 1(2): 219-222.

Suthar, S., Singh, S. and Dhawan, S. (2008). Earthworms as bioindicator of metals (Zn, Fe, Mn, Cu, Pb and Cd) in soils: Is metal bioaccumulation affected by their ecological category? *Ecol. Eng.*, 32: 99-107.

Swathi, P., Rao, K. T. and Rao, P. A. (1998). Studies on control of root-knot nematode Meloidogyne incognita in tobacco miniseries. *Tobacco Res.*, 1:26-30.

Szcech, M., Rondomanski, W., Brzeski, M. W., Smolinska, U. and Kotowski, J. F. (1993). Suppressive effect of a commercial earthworm compost on some root infecting pathogens of cabbage and tomato. *Biol. Agric. and Hortic.*, 10:47-52.

Tara, Crescent. (2003). Vermicomposting. Development Alternatives (DA) Sustainable Livelihoods. (http://www.dainet.org/livelihoods/default.htm)

Tejada, M., Gonzalez, J. L., García-Martínez, A.M. and Parrado, J. (2008). Application of a green manure and green manure composted with beet vinasse on soil restoration: Effects on soil properties, *Bioresource Technol.* 99: 4949-4957.

The Fertilizer (Control) Order 1985 (As amended upto June, 2017), The Fertilizer Association of India, New Delhi-110067

Thoden, T. C., Korthals G. W. and Termorshuizen, A. J. (2011). Organic amendments and their influences on plant-parasitic and free-living nematodes: a promising method for nematode management? *Nematology*. 13: 133-153.

Tomati, U., Grapppelli, A. and Galli, E. (1988). The hormone like effect of earthworm casts on plant growth. *Biol. Fertil. Soils.*, 5:288-294.

Tomati, V. and Galli, E. (1995). Earthworms, Soil Fertility and Plant Productivity. *Acta Zoologica Fennica*. 196: 11-14.

Tripathi, G. and Bhardwaj, P. (2004). Decompostion of kitchen waste amended with cow manure using an epigeic species (Eisenia Fetida) and an anecic species (Lampito mauritti). *Bioresource Technology*. 92: 215-218.

Umamaheswari, S., Viveka, S. and Vijaylakshmi, G. S. (2003). Indigenous vermiwash collecting device, The Hindu, Jul. 17.

Usmani, Z. and Kumar V. Remediation of Heavy Metals from Fly-Ash with Aid of Ecosystem Engineers: Earthworms – A Review. Strategic Technologies of Complex Environmental Issues-A Sustainable Approach.

Varghese, S. M. and Prabha, M. L. (2014). Biochemical Characterization of Vermiwash and its Effect on Growth of Capsicum frutescens. *Malaya Journal of Biosciences*. 1(2): 86-91.

Vaz-Moreira, I., Silva, M. E., Manaia, C. M. and Nunes, O. C. (2008). Diversity of bacterial isolates from commercial and homemade composts. *Microb. Ecol.*, 55: 714-722.

Vermi, Co. (2001). Vermicomposting technology for waste management and agriculture: an executive summary. (http://www.vermico.com/summary.htm) PO Box 2334, Grants Pass, OR 97528, USA: Vermi Co.

Waksman, S. A. and Martin, J. P. (1939). The role of micro-organisms in the conservation of the soil. *Science*. 90: 304-305.

Wright, L, M., 2002. Sustainable waste management and vermicompost of biodegradable municipal waste, Phd thesis. Cardiff School of Engineering. Cardiff University.

Yadav, A. K., Kumar, K., Singh, S. and Sharma, M. (2005). Vermiwash-A liquid biofertilizer. *UP. J. Zool.*, 25(1):97-99.

Yasir, M., Aslam, Z., Kim, S. W., Lee, S. W., Jeon, C. O. Chung, Y. R. (2009a). Bacterial community composition and chitinase gene diversity of vermicompost with antifungal activity. *Bioresour. Technol.*, 100:4396-4403.

Yasir, M., Aslam, Z., Song, G. C., Jeon, C. O. and Chung, Y. R. (2009b). Eiseniicola compostigen. nov., sp. nov., with antifungal activity against plant pathogenic fungi. *Int. J. Sys. Evol. Microbiol.*, 60:268

Yeates, G. W. (1981). Soil nematode populations depressed in the presence of earthworms. *Pedobiologiaogia.*, 22:191–202

Zaller, J. G. (2006). Foliar spraying of vermicompost extracts: Effects on fruit quality and indications of late-blight suppression of field-grown tomatoes. *Biological Agriculture and Horticulture.*, 24(2): 165-180.

4

A Low Cost Aquatic Green Manure for Agricultural Crops (*Azolla*)

Bappa Paramanik, Parimal Panda, Partha Sarathi Patra, Indranil Das Naba Kishore Mahato, Ashok Choudhury* and *Asit Mandal

Introduction

Cyanobacteria are often called "blue-green algae", this name is convenient for talking about organisms in water that make their own food, but does not reflect any relationship between the cyanobacteria and other organisms called algae. Cyanobacteria are relatives to bacteria, not eukaryotes, and it is only the chloroplast in eukaryotic algae to which cyanobacteria are related. Some cyanobacteria are aquatic and photosynthetic, that is, they live in water, and can manufacture their own food. They are quite small and usually unicellular, though they often grow in colonies large enough to see. In fact, it may surprise you then to know that the cyanobacteria are still around; they are one of the largest and most important groups of bacteria on earth (Berry *et al.,* 2008). The great contribution of cyanobacteria is the origin of plants chloroplast with which plants make food for themselves is actually a cyanobacterium living within the plant's cells. Sometime in the late Proterozoic or in the early Cambrian, cyanobacteria began to take up residence within certain eukaryote cells, making food for the eukaryote host in return for a home. This event is known as endosymbiosis, and is also the origin of eukaryotic mitochondrion (Issa *et al*., 2002).

Azolla is a free floating water fern that floats in the water and fixes atmospheric nitrogen because of its association with the nitrogen fixing cyanobacterium Anabaena (Moore, 1969). It is widely distributed both in temperate and tropical regions. An *Azolla-Anabena* system is ideal for the cultivation of rice under tropical conditions because of its ability to fix atmospheric nitrogen and capacity to multiply at faster rates. The relatively quick decomposition of the biomass and rapid availability of its nitrogen to the standing crop makes it agronomically outstanding. *Azolla* is eco-friendly as it never leads to contamination of the environment. It also does not compete with the rice plants for photosynthesis or nutrition. Such biological systems are able to provide 1.5-2.0 million tonnes of nitrogen for crop production in India whereas at least 3.3-4.4 million tonnes of urea will be required to give similar amounts of nitrogen. The use of *Azolla* is quite popular among the farmers of Vietnam and China. In fact the initial reports of using *Azolla* as a cheap source of nitrogen was realized by the Chinese farmers and there were several records on the use of *Azolla* as manure in the beginning of 17th century in China and Vietnam. They described this plant as miniature nitrogen fertilizer factories. The book written by Jia Ssu Hsieh in 540 A.D describes the cultivation and use of Azolla in rice fields. In India, Singh and co-workers popularized the use of Azolla in rice cultivation (Singh, 1977; Singh, 1979; Singh and Srivastava, 1984; Singh, 1988; Singh and Singh, 1995). At present many progressive farmers and non-governmental organizations are employing Azolla as an invaluable input in agriculture and some of them are actively engaged in its popularization as biofertilizer, green manure, poultry feed and fodder. However, certain key scientific issues need to be addressed to make the system more compatible with the present day need and demand.

Species and Distribution

The genus Azolla was placed in the family Salviniaceae in the order Salviniales. However, Azolla has been placed in the monotypic family Azollaceae (Konar and Kapoor, 1972). There are seven extinct and twenty five fossil species of Azolla (Hills and Gopal, 1967). The most commonly found forms of Azolla are A. pinnata, A. filiculoides, A. rubra, A. microphylla, A. imbricata and A. caroliniana. Azolla nilotica is a rare and found in nile river, because of its size, an especially conspicuous Azolla species which is in strong need of light and warmth. Azolla Mexicana is found in Mexico stream side mud (Kleinman

and Zimmerman,). Azolla pinnata is the most widely distributed species in India and throughout the world in both tropical and temperate regions. It is generally found to inhabit in paddy fields, canals, ditches and rivers. In India A. pinnata has been found to be distributed widely in stagnant and shallow waters. The growth is luxuriant during the months of August to March and it dies in summer due to increase in the temperature.

Characteristics

Azolla develops a symbiotic relationship with a blue-green algae, Anabaena *Azolla*e. An *Azolla* plant is a fern frond consisting of a main stem growing at the surface of the water, with alternate leaves and adventitious roots at regular intervals along the stem. Secondary stems develop at the axil of certain leaves. *Azolla* fronds are triangular or polygonal and float on the water surface individually or in mats. Plant diameter ranges from 1/3 to 1 inch (1-2.5 cm) for small species, such as *Azolla pinnata,* to 6 inches (15 cm) or more for *A. nilotica. A. filiculoides* was introduced to Hawaii in the early 20th century. A shallow fresh-water pond, similar to the environment found in a taro is the ideal environment for Azolla· When taro reaches maturity, it shades out the Azolla below the canopy, gradually killing it and resulting in the release of nutrients into the soil-water system, where they become available for uptake by the taro plants.

Classification and identification of different Azolla species

Traditionally, seven species of Azolla are recognized and they are divided into three series (Yin-Qiu and Yu, 2003). In *Azolla,* two subgenera, *Azolla* and *Rhizosperma* (Meyen) Strasburger, to account for forms having fluked glochidia and trichome-like glochidia respectively (Ashton and Walmsley, 1984). Besides another series of *Azolla* is Tetrasporocarpia.

A. *Azolla* subgenus *Azolla* (Meyen) Strasburger SYNONYM

Azolla section *EuAzolla* Meyen (*sensu* Sadebeck) Three 'floats' on the megaspore; rigid, barbed glochidia on the microsporangial massulae. This subgenus is found in tropical areas of the Americas, New Zealand, and Australia.

(a) *Azolla filiculoides* Lam

Glochidia not septate, or rarely with 1 or 2 septa at apex; plants elongate (frequently 20-50 mm long), alternately branched and branching freely, with closely appressed, imbricate, papillose, oblong to ovate leaves (1-2 mm long); microsporangia 25-1; 20 in an indusium; massulae 4-10; megaspore with raised, irregularly hexagonal markings (Ashton and Walmsley, 1984). It is a floating aquatic fern, with very fast growth, capable of spreading over lake surfaces to give complete coverage of the water in only a few months. Has green tinged pink, orange or red at the edges and breaking into smaller sections as it. It is not tolerant of cold temperatures, and in temperate regions it largely dies back in winter, surviving by means of submerged buds.

(b) *Azolla caroliniana* Willd

Glochidia not septate; plants small (5-15 mm diameter), dichotomously branched; the leaves not closely imbricate, small (0.5 mm long), nearly smooth, suborbicular and divaricate; microsporangia 8-40 in an indusium; megaspore occasionally with small raised markings (Ashton and Walmsley, 1984).

(c) *Azolla mexicana* Presl

Glochidia multi-septate; plants dichotomously branched (10-1 5 mm diameter), with upper leaf lobes (green or red photosynthetic lobe that has a colorless margin) 0.7 mm long, the lower lobes (a translucent cup shaped lobe) much larger; microsporangia usually with 4 massulae; megaspore pitted (Ashton and Walmsley, 1984). *Azolla* mexicana is a small aquatic fern. It bears two kinds of spore bearing structures, called sporocarps. The male sporocarp is much larger and contains many sporangia, which in turn contain many spores. The smaller female sporocarp contains only one spore. *Azolla* mexicana can also grow in streamside mud. The upper lobe has a chamber within which *Anabaena Azollae,* a nitrogen fixing cyanobacterium (blue-green alga) grows (Kleinman and Zimmerman).

(d) *Azolla microphylla* Kaulf

Glochidia multi-septate; plants small (10-20 mm long), pinnately branched, with nearly orbicular leaves 1 mm long; megaspore smooth (Ashton and Walmsley, 1984).

(e) *Azolla rubra*

Aquatic, floating, dark-red (when exposed) or glaucous-green plants, usually forming ovate to ovoid patches on the surface of water bodies. Branching irregular; roots peg-like, simple (not branched). Leaves triangular; apex rounded; margins membranous, translucent; surface smooth, bright red or glaucous-green. Microsporangiate massulae spherical, surface conspicuously barbed. Megasporocarps brown, partially obscured by leaf lobe. *Azolla pinnata* is best distinguished from *A. rubra* by its pinnate rather than simple roots.

B. *Azolla* subgenus *Rhizosperma* (Meyen) Strasburger SYNONYM:

Azolla section *Rhizosperma* Meyen (sensu Sadebeck) Nine 'floats' on the megaspore; glochidia on the microsporangial massulae (microspores of one sporangium released as one mass) thread-like (trichome-like) or absent (Ashton and Walmsley, 1984). This subgenus is occurring in Africa, Australia, and Asia.

(a) *Azolla pinnata* R.Br

Glochidia thread-like, present only on the inner surface of massulae; sporocarps in pairs; stems pinnate or bipinnate, plants appearing deltoid in outline, 10-20 mm in length; leaves papillose (each papilla 2-celled), ovate, obtuse, loosely imbricate, 0.8-1.2 mm long; numerous root hairs on each root. *Azolla pinnata* has very regular branching. Plants can be red or green depending on whether they are in the open or shade, respectively (Ashton and Walmsley, 1984).

Shen (1961) considered the species *Azolla imbricata* (Roxb.) Nakai and *Azolla africana* Desv. to be synonyms for *Azolla pinnata* R.Br. Sweet & Hills (1971) recognized two varieties on the basis of vegetative morphology. These are *Azolla pinnata* var. *pinnata* occurring in Africa, Madagascar, Australia and New Guinea, and *Azolla pinnata* R. Br. var. *imbricata* (Roxb.) Bonap., occurring in India, Nepal, Burma,SE Asia, China, Japan and the Phillipines. The dorsal leaf lobes of *Azolla pinnata* var. *pinnata* are thick and curled, with acute apices and irregularly distributed papillae. In addition, the protruding tips of cells at the edge of the hyaline leaf margin confer a somewhat serrated appearance to the leaf. In contrast, the dorsal leaf lobes of *Azolla pinnata* var. *imbricata* are thin and flat, with rounded apices, regularly distributed papillae and smooth edges to the hyaline leaf margin (Ashton and Walmsley, 1984).

C. *Azolla nilotica* Decne. SYNONYM

Azolla section Tetrasporocarpia. This section is distributed only in Africa. Glochidia absent from massulae; sporocarps in tetrads; stems elongated (60-150 mm long), branches often arising in tufts from main axis; leaves nearly smooth, obtuse, not closely imbricate, 3-5 mm long (Ashton and Walmsley, 1984).

A molecular phylogenetic study has now more or less confirmed this taxonomic scheme as Rhizosperma, *Azolla* and Tetrasporocarpia (Reid, 2002), with the only change that *A. nilotica* and *A. pinnata* form a monophyletic instead of a paraphyletic group as in the morphological cladistic analysis (Sanders and Fowler,1993).

Environmental requirements

Azolla is found in ponds, ditches, and wetlands of warm-temperate and tropical regions throughout the world. It must grow in water or wet mud, and it dies within a few hours under dry conditions. *Azolla* can survive a water pH range of 3.5-10, but optimum growth occurs when the water is between pH 4.5 and

7. The optimum temperature for *Azolla* is between 64 and 82°F (18-28°C). The growth rate gradually declines as salinity increases. Wilting occurs where salt concentrations surpass 1.5 g/l during the summer. *Azolla* grows in full to partial shade (100-50% sunlight), with growth decreasing quickly under heavy shade. In the tropics direct sunlight at midday suppresses the growth of the fern, while light exposure on cloudy days is more favourable. Wind tends to push all the leaves together so they collect on the same part of the water surface (*uses.plantnet-project.org/en/Azolla pinnata (PROSEA).*

Distribution, Habit and Habitat and Morphology

Azolla is found in both temperate and tropical regions. It grows luxuriantly in ditches, fresh water ponds and paddy fields. The *Azolla* plants are delicate, small and triangular or polygonal in shape. It is free floating and aquatic but can grow on moist soils as long as the moisture persists in the soil. The sporophytic plant has a horizontal rhizome of 0.5 to 7 cm in diameter with branches having densely arranged and overlapping leaves. A leaf consists of a thick dorsal lobe and a thin ventral lobe. The symbiotic Blue Green Alga (Anabaena) is confined to the dorsallobe (Peters and Mayne, 1974). An epidermis covers the surface of the dorsal lobe and the epidermis has vertical rows of single celled stomata and trichomes of one or more cells. The ventral lobe which helps in floating due to its convex surface touching water has a few stomata and trichomes (Eames, 1936).

Biology of the symbionts of *Azolla* and their interaction mechanism

Anabaena *Azolla*e, although it can exist on it's own, is usually found within ovoid cavities inside the leaves of the water fern *Azolla*. *A. azollae* and *Azolla* have formed a symbiotic relationship in which the cyanobacterium receives carbon and nitrogen sources from the plant in exchange for fixed nitrogen. This relationship has proven to be useful to humans in the production of food; specifically in the fertilization of rice paddies. *A. azollae* is a small filamentous phototrophic cyanobacteria generally seen as a multicellular organism with two distinct, interdependent cell types. The first is a small, circular, photoautotrophic "vegetative" cell that performs oxygenic photosynthesis and is typically blue green in color. The second is a heterocyst; a larger, paler, more homogenous cell produced by Anabaena to fix atmospheric nitrogen. *Azolla* is used as an organic biofertilizer to increase the nitrogen levels in the rice paddies. *A. azollae* is a model organism commonly used in labs for the study of gene differentiation in the formation of heterocysts. *A. azollae* typically undergoes a type of asexual reproduction called fragmentation; a process in which, a section of the cyanobacteria chain will split off from the parent chain and float away. These new sections called hormogonia are small motile segments of the parent unit. Over time hormogonia will begin to elongate via the separation of adjacent cell walls and form their own chains. Fragmentation can also occur when a live portion of a chain separates from a section of dead cells. Interestingly, how *A. azollae* passes on to new generations of *Azolla* is not completely understood. It is believed that hormogonia may become entrapped by the embryo of the *Azolla* plant during differentiation of its shoot apex and dorsal lobe primordia of the first leaves. (*microbewiki.kenyon.edu/ index.php/Anabaena azollae).*

Like all phototrophic bacteria, *A. Azollae* uses solar energy along with bacteriochlorophyll and CO_2 to carry out photosynthesis in the cytoplasm of its cells. Photosynthetic electron transport in Anabaena proceeds from oxidation of H_2O in Photosystem II in which ATP and NADPH are formed through the movement of excited electrons down a series of proteins to Photosystem I. In Photosystem I, electrons are separated from an organic electron donor and are ultimately transferred to form NADH or NADPH. This electron transport system is represented in Z-scheme photosynthesis (*microbewiki.kenyon.edu/ index.php/Anabaena azollae*).

Heterocysts function as the sites for nitrogen fixation under aerobic conditions and are formed in response to a lack of fixed nitrogen. Terminal and intercalary heterocysts are created from Anabaena vegetative cells, which have undergone morphological differentiation as well as by biochemical changes. Although they vary in size, heterocysts are usually larger, paler, and more homogeneous than the cells from which they are formed. Some of the major changes that occur include the production of three additional cell walls (including one composed of glycolopid which forms a hydrophobic barrier to oxygen), production of nitrogenase and other proteins for nitrogenfixation, production of enzymes that eliminate oxygen from the cell, rearrangement of the thylakoid, and the up-regulation of glycolytic enzymes. The mature heterocysts contain no functional photosystem II and cannot produce oxygen. Instead, they only contain photosystem I, which enables them to carry out cyclic photophosphorylation and ATP regeneration. Loss of photosystem II means that heterocysts are no longer capable of producing necessary reductants required for nitrogen fixation. These need to be supplied by neighboring vegetative cells largely in the form of sucrose, produced during photosynthesis. These changes provide appropriate conditions for the functioning of oxygen-sensitive nitrogenase. Under environmental stress, such as light limitation, cold temperatures, or phosphate starvation, filamentous cyanobacteria such as *Anabaena* form specialized spores called akinetes. An akinete forms a long, oval cell adjacent to a heterocyst, where it stores nitrogen and developes a thickened envelope. Like other types of spores, akinetes are able to resist desiccation and remain viable for long periods of time. Akinetes can lie dormant for decades until favorable conditions allow them to germinate and grow in to new vegetative filaments.

The basic biology of the *Azolla·Anabaena* symbiosis has been extensively investigated in the last two decades by different methodological approaches with special attention to the nitrogen fixation processes (Albertano and Caiola, 1991). The nitrogen fixed as ammonium by *A. azollae*, and released in the leaf cavities, is assimilated as glutamate by the fern (Uheda and Maejima, 2009), and then translocated apo-or symplasticaUyto the other parts of *Azolla* and probably from the apexes to the vegetative cells of *Anabaena* (Albertano and Caiola, 1991). While the transfer cell ultrastructure in the terminal cells of the primary branched hairs of *Azolla* could be functional in supplying nitrogen to the apical colonies, it has been hypothesized that the ammonium assimilation could be due to the body cells of the secondary branched hairs (Albertano and Caiola, 1991). Nevertheless, at present the role of the epidermal hairs of the fern in the nitrogen exchanges among the symbionts has to be confirmed, and the different nitrogen compounds arising from the nitrogenase activity have to be localized in the symbiotic system (Albertano and Caiola, 1991).

Growth and Multiplication

Azolla is found in both temperate and tropical regions and it grows luxuriantly in ditches, fresh water ponds and paddy fields. The pH requirement of the soil is 7.2 for an ideal growth coupled with a temperature of 32°C. However, studies conducted elsewhere in India have shown that the region specific choice of the species for their use as biofertilizer. A study was conducted at Centre for Conservation and Utilization of Blue Green Algae, Indian Agricultural Research Institute, New Delhi to compare the biomass production and nitrogen fixation potential of different species of *Azolla* (Arora and Singh, 2003). Based on the experiments *A. microphylla* was found to perform better and hence selected for mass multiplication. This strain is maintained round the year as it is able to withstand both high as well as low temperature conditions

Production Technology for *Azolla* Biofertilizer and Modes of Application

The propagation of *Azolla* is generally carried out in soil-based nurseries. However, to maintain the germplasm medium based cultures is advisable. Effects of nutrient status of the medium on the productivity and nitrogen fixation in *Azolla* are reported (Kushari and Taheruzzaman, 1990). A comparative study on the growth, pigments, nitrogen fixation and nutrient status of *Azolla* maintained as soil and nutrient based cultures showed that the soil based cultures are as good as those raised on nutrient media (Dawar and Singh, 2002). *Azolla* production can be carried out in nursery plots, ponds, ditches, canals, concrete tanks and polythene lined ditches. The field selected for *Azolla* cultivation needs to be thoroughly prepared and leveled uniformly. Generally 20 m – 2 m size plots are made in the field with suitable bunds and irrigation channels with a water depth of at least 10 cm and in each plot water (20 liters) is added and inoculated with *Azolla* (8-10 Kg). Single super phosphate (100 g) in 2-3 split doses is applied at an interval of 4 days to each plot. Furadon or carbofuran (3% active granules) can be applied in the plots (100 g plot^{-1}) with or after a week of inoculation. 100-150 Kg fresh *Azolla* can be harvested from each plot after 15 days from each plot. The same method can be used to produce *Azolla* in bigger plots. The quantities of in-puts need to be varied as per requirement. Depending on the availability and need of the inocula, *Azolla* can also be maintained in nursery in trays or earthen or cemented pots of any dimension. If the production is carried out in a pond or canal fertilizers and insecticides are not applied. Cattle slurry and animal dung for the production and utilization of *Azolla* as biofertilizer for rice has been reported (Singh *et al.*, 1993). The cattle slurry and animal dung are effective as phosphorous fertilizers and application of *Azolla* is found to enhance the C, N and available P content of the soil.

Precaution

- Plant should not be allowed to enter maturity stage or sporulation stage by periodic application of cow dung slurry, super phosphate and other macro and micronutrients except nitrogen.
- Temperature should be retained below 30°C centigrade incase the temperature goes up, the light intensity should be maintained by providing shade net or other devices.
- Bio-mass should be removed every day or alternative days to avoid overcrowding. pH should be tested periodically to see that it never goes below 5.5 and above 7.
- Seed stock is maintained separately treated with pesticides and fungicides.
- *Azolla* should be well washed with water before feeding to livestock to get rid of smell of cow dung.
- Biomass collected from the field applied with the pesticide should not be used as a feed for livestock.

Methods of Application of *Azolla*

The most common mode of application of *Azolla* in the field is as green manure or as a dual crop along with rice. In case of application as green manure *Azolla* collected directly from ponds/ditches is applied in the field. It may be grown in nurseries as specified earlier and can also be applied in the field. A thick mat of *Azolla* will be formed after application in about 2-3 weeks' time and can be incorporated in the soil. Rice can also be transplanted in the field subsequently. Single super phosphate (25-50 kg ha^{-1}) is applied in split doses. After analyzing the soil P-status the dosage of the same can be reduced. Cattle dung or slurry may also be used instead of single super phosphate. In case of pest infestation or attack, pest control measures have to be undertaken. *Azolla* application by this mode contributes around 20-40 kg N ha^{-1}.

Biological Nitrogen Fixation and *Azolla* in Relation to Soil Fertility

The ability to fix atmospheric nitrogen at substantially higher rates has led to the exploitation of the organism as biofertilizer. Application of *Azolla* in rice paddy fields has a positive role in improving the soil fertility index. The ability of nitrogen fixation is due to the presence of the symbiotic cyanobacterium *Anabaena* that occurs in the dorsal leaf cavities of the fronds (Peters and Meeks, 1989). The symbiont is able to meet the entire nitrogen requirement of the association. Calvin cycle operates in both the partners and the primary end product of photosynthesis is sucrose (Van Hove, 1989). A strong interaction exists between nitrogen fixation and the photosynthesis and the source of ATP and NADPH is photosynthesis. The capacity of *Azolla* to fix nitrogen in the field has been estimated to be 1.1 kg N ha^{-1} day^{-1} and this fixed nitrogen is sufficient to meet the entire nitrogen requirement of rice crop within a few weeks (Lumpkin and Plucknett, 1980). The rapid and substantially higher rates of nitrogen fixation coupled with production of high biomass have made the organism an outstanding agronomic choice.

Azolla strains have been successfully exploited as an efficient biofertilizer for rice paddy fields because of the nitrogen fixing potential. *A. pinnata* fixes 75 mg N g^{-1} dry weight day^{-1}and produces a biomass of 347 ton fresh weight ha^{-1} in a year. This biomass contains 868 kg N which is equivalent to 1900 Kg urea. A wide variability regarding growth and nitrogen fixation among different strains of *Azolla* is observed by a study conducted at Central rice Research Institute, Cuttack (Singh, 1988). Among the several factors that influence the growth and nitrogen fixing potential of *Azolla* are nutrient availability, rate and the time of inoculation etc (Kannaiyan, 1993; Singh and Singh, 1995). In addition to sustaining rice yields, inoculation of *Azolla* has been reported to enhance the soil biological health. It is important to optimize use of organic, inorganic and biological inputs in an integrated manner taking into consideration the ecological and soil conditions to sustain crop productivity.

Soil enzyme activity is considered as an index of microbial activity and fertility of the soil. *Azolla* decomposes rapidly in soil and supply nitrogen to the crop plants. It contributes significant amounts of phosphorus, potassium, sulfur, zinc, iron and molybdenum in addition to other micronutrients besides addition of nitrogen. The biological health of the soil due to application of *Azolla* has resulted in improving mineralization and consequent increase in the microbial status of the soil. In low land rice cultivation mineralization of organic nitrogen to ammonia is an important process (Sahrawat, 1983). The rate of mineralization is influenced by factors such as C: N. *Azolla* species with a low C: N mineralized in 2 days while the species with high C:N mineralized in 5 days (Wang *et al.*, 1987). The decomposed organic matter plays an active role in the development of microbial population irrespective of the time taken for mineralization. Soil fertility is also influenced by the humic substances formed during the decomposition of *Azolla* (Bhardwaj and Gaur, 1970).

Improvement in Crop Productivity Due to Application of *Azolla*

Productivity of rice is related to the availability of nitrogenous fertilizers. Nitrogen can also be supplemented to the rice crops through biofertilizers such as *Azolla* and cyanobacteria. Application of *Azolla* to rice plants is ideal as the growth requirement for both the organisms are similar. The organism is able to multiply fast which is a remarkable aspect in its exploitation as biofertilizer. Successful exploitation of this organism in the Asian context has also been highlighted (Singh, 1989). Increase in yield due to application of *Azolla* was demonstrated by several studies conducted at several locations. A single crop of *Azolla* provides 20-40 kg N ha^{-1} (Singh, 1977b). Similar results have been obtained with integrated application of *Azolla* and chemical nitrogen fertilizers. When a comparison of *Azolla* is made with other biofertilizers highest grain yield in rice is observed (Singh *et al.*, 1992). After the decomposition of the organism nitrogen content is made available to the rice plants. It was found that the plant decomposed 8-10 days after incorporation into the soil and rice plants are benefitted after 20-30 days (Singh, 1977d). The release of N by *Azolla* is slow and its availability to the first crop of rice is about 70% to that of ammonium sulphate. Moreover, the release of nitrogen is faster as compared to nitrogen fixing cyanobacteria (Saha *et al.*, 1982). Fresh *Azolla* releases its N faster as compared to dried *Azolla* due to rapid mineralization and mineralization of N is faster at room temperatures (Singh, 1979a; Singh, 1979b).

Bioremediation Potential of *Azolla*

Another interesting and applied aspect of utilization of *Azolla* which is gaining popularity in the recent times is its application in the process of bioremediation. Increase in population and rapid rates of industrialization has resulted in the release of huge quantities of pollutants into water resources. The pollutants released are toxic in nature and pose great concern due to their adverse impact on plant and animal health. Therefore phytoremediation offer an excellent option to ward off the pollutants from aquatic ecosystems. *Azolla* may be used to clean up the polluted and contaminated waters (Table 1). Efficient removal of gold from waste water solutions using the fern *Azolla filiculoides* is reported (Antunes *et al.*, 2001). Similarly Elmachliy *et al.*, (2010) demonstrated removal of Silver and Lead ions from waste waters using *Azolla filiculoides*. Successful cultivation of *A. microphylla* biomass in secondary treated Municipal waste water of Delhi is attempted by Arora and Saxena (2005). *Azolla* biomass is also employed in the phytoremediation of toxic heavy metals. Recently the phytoremediation potential of three different species of *Azolla* such as *A.microphylla*, *A. pinnata* and *A. filiculoides* is studied and it is found that *A. microphylla* accumulated more metal as compared to the other species (Arora *et al.*, 2006). The biomass of *Azolla* (both live and dead) has also been used in the bio-sorption of several heavy metals (Umali *et al.*, 2006; Mashkani and Ghazvini, 2009). Extensive work has been done using *Azolla* in the removal of heavy metals from aquatic environments (Rai 2008; Rai and Tripathi, 2009; Rai 2010a; Rai, 2010b).

Table 1: Bioaccumulation potential of heavy metals by various *Azolla* spp.

Azolla spp.	Heavy metal	Concentration of heavy metal accumulated (μg metal g^{-1})
A. pinnata	Cadmium	2759
	Mercury	450
A. caroliniana	Lead	416
	Chromium	964
A. filiculoides	Nickel	28443
	Chromium	12383
A. microphylla	Nickel	21785
	Cadmium	1805

Source: Sood *et al.*, (2011)

Nutritional value of *Azolla*

For both pigs and poultry, there is lack of specific information on the levels to which *Azolla* can be incorporated in rations. This information is needed before any practical recommendations on nutritive value can be made. Because *Azolla* is a heavy metal accumulator, toxicity to animals may be a problem. Alternatively *Azolla* may be an important source of essential trace minerals for pig and poultry rations, when incorporated at appropriate levels. To mix dried *Azolla* as supplement to commercial layer or broiler rations, the freshly harvested *Azolla* should be washed thoroughly to remove soil and pesticide residues. Then it is dried in the sun for 3-7 days, until it crumbles when squeezed. Subsequently, it is ground, either using a commercial grinder, or by crushing the dried plants in bags with the feet. The *Azolla* meal is mixed with the layer/broiler rations 1:10 by weight or 1:5 by volume. To harvest the spores, the harvested plants are placed in a 50 cm deep, well-drained pit. The plants are left to decay for two weeks during summer or up to one month in the rainy season. The decayed plants are collected and dried in the sun and open air for 24 hours. The dried mass is sieved through a coarse screen (1 mm mesh). The particles that pass through are immersed in water and left to soak overnight. The floating spores are finally scooped from the water (*uses.plantnet-project.org/en/Azolla_pinnata_(PROSEA)*. (Write about *Azolla* as fish feed from Bishu)

Chemical content of *Azolla*

Per 100 g (dry weight basis) *A. pinnata* contains approximately: protein 13-30 g, fat 4-6 g, carbohydrates 41-45 g, fibre 39 g, cellulose 5-15 g, hemicellulose 9-18 g, lignin 9-35 g, ash 10-24 g (ranges of the elements N 2-5%, P 0.1-1.6%, K 0.3-6%, Ca 0.4-1.7%, Mg 0.2-0.6%, S 0.2-0.7%, Si 0.1-3.5%, Na 0.1-1.3%, Cl 0.6-0.9%, Al 0.04-0.6%, Fe 0.04-0.6%, Mn 0.06-0.3%, Cu 0-0.03% and Zn 0.002-0.1%). The N, K and Si contents are reasonably high. Deficiencies in methionine, histidine and fibre make *A. pinnata* unfit to be the sole feed source for animals, but the amino acid composition compares well with reference protein sources and although the methionine content is low, the lysine content is more than twice that of maize. In the tropics 60-75% of the N will be released as ammonia within 6 weeks after the onset of mineralization in the soil in a flooded field without rice. *Azolla pinnata* and *Azolla rubra* also posses antioxidant efficacy due to the presence of high phenolic (Total phenolic content – 95.25 and 92.16 μg Gallic Acid Equivalents/mg extract, respectively) and flavonoid content (41.13 and 39.66 μg Catechin Equivalents/mg extract (*uses.plantnet-project.org/en/Azolla_pinnata_(PROSEA)*. Antioxidant efficacies of these phenolic compounds are due to radical scavenging effect, inhibition of lipid peroxidation and chelation of metal ions (Noor Nawaz *et al.* 2014).

Storage techniques of *Azolla*

Azolla is most commonly fed in the fresh form harvested daily. It was proposed that if suitable storage techniques such as sun-drying or ensilaging were available, the problems of maintaining supply when growth rates are low could be overcome. Dried *Azolla* has been found less palatable than fresh *Azolla,* and different strains of fresh *Azolla* vary in palatability. *Azolla* has been successfully ensiled with salt in China. The group decided there was a need to investigate further various techniques of storing *Azolla* as feed for animals and fish.

Disease and pests of *Azolla*

A serious disease of *A. pinnata* in rice fields is caused by the fungus *Myrothecium verrucaria*, characterized by white spots of mycelium and a rapid rot and death of *Azolla. A. pinnata* is also highly susceptible to rice sheath blight, caused by *Sclerotinia sclerotiorum* and to black rot disease caused by *Rhizoctonia solani.* Disease severity is higher when the leaves are also attacked by snails. Numerous other fungal genera have been isolated from diseased *Azolla.* Crude garlic extract was found to be very effective against fungal pathogens. *(uses.plantnet-project.org/en/Azolla_pinnata_(PROSEA).*

In South-East Asia pests are considered the major limiting factor in *Azolla* cultivation. In the Philippines, the most important pests are the webworm (causing yield losses of 6-74 %), the lepidopteran spinningworm (*Ephestiopsis vishnu*) and the caseworm complex (*Elophila Nymphula*) *enixalis*, *E. nigralbalis*, *E. responsalis*). In Thailand, *Chironomus glauciventris, Cryptoblabes* sp., *Polypedilum johannseni* and *Elophila enixalis* are the main pests. In Thailand, *Azolla* is also damaged by high temperatures in April and by competition from blue-green algae. The best insect control and the highest yield was obtained by spraying 0.5 kg/ha monocrotophos, applied 1 and 10 days after inoculation. Foliar sprays of carbosulfan or chlorpyrifos applied after field inoculation also gave good results. Treatment with these or with monocrotophos resulted in 3-fold increases in yield. A preparation (Bactospeine) containing *Bacillus thuringiensis*, applied after inoculation, provided some insect control and significantly increased the yield. To reduce application costs, it is recommended that the *Azolla* stock culture be sprayed 2-3 days before it is transferred to the propagation field. Another application is necessary 7 days later. Because *Azolla* cultivation should begin after transplanting rice, insecticide applications also have some effect on rice pests. Destruction by snails is an obstacle to *Azolla*-breeding in India. Snails primarily responsible are *Lymnaea luteola*, *Bellamya bengalensis,* and *Pila globosa.* The golden apple snail, *Pomacea canaliculata*, originating from South America, was introduced as a promising export industry in South-East Asia but it has also escaped to rice fields in Taiwan, Japan and the Philippines (here *Pomacea cuprina* and *Pomacea gigas* have also been introduced) and *Azolla* and young rice seedlings are preferred hosts. Control is possible by manual removal, draining of rice fields, releasing carp (prey on young snails), application of pesticides such as metaldehyde ("namekil", effective with low fish toxicity). The *Azolla* midge (*Polypedilum ivinoense*), which is aquatic at larval instar stage, can devastate an *Azolla* stand in 3-5 days. It is light red, 2-3 mm long, and makes its nest on the underside of *Azolla*, eating roots and young leaves. Methods to prevent damage by *Azolla* midge include protecting the *Dytiscidae* which are natural enemies, applying maceration extract of cake of tea oil and insecticides such as deltamethrin, carbofuran, carbaryl, and temephos. Pesticides and herbicides can have effects on the growth and nitrogen assimilation of the *Azolla-Anabaena* symbiosis. Most rice herbicides (*e.g.,* bipyridylium and phenolic herbicides, chloramben, dicamba, simazine, benzoic, triazine, dinitroaniline and urea herbicides) are deleterious to *Azolla* but delaying inoculation with *Azolla* until after herbicide application reduces or eliminates the problem. (*uses.plantnet-project.org/en/Azolla pinnata (PROSEA)*.

The benefits of *Azolla* application

- Basal application of green manure @ 10-12 tones/hectare increases soil nitrogen by 50-60 kg/ha and reduces 30-35 kg of nitrogenous fertilizer requirement of rice crop.
- Under low land conditions a thick mat does not allow weeds to grow in rice field thus *Azolla* suppress the weed growth and creates congenial condition for rice production. Additionally, *Azolla* reduces evaporation of surface water thus increases water use efficiency in rice.
- *Azolla* may be used for the production of hydrogen fuel and biogas, control of weeds and mosquitoes, and the reduction of ammonia volatilization that accompanies the application of chemical nitrogen fertilizer.
- Hosts a symbiotic blue green algae *Anabaena azolle*, which is responsible for the fixation and assimilation of atmospheric nitrogen.
- Can double in size every several days under optimal nutrient and water temperate conditions.
- Green manure plant for paddy agriculture for weed supression and fertility (used extensively in organic *duck and rice* systems).
- Nitrogen and mineral source for compost making.
- 25-30% Protein, and low in lignin with makes it digestible to many animals, as well as nutritious.
- Rich in essential amino acids, vitamins (vitamin A, vitamin B12 and Beta-Carotene), growth promoter intermediaries and minerals like calcium, phosphorous, potassium, ferrous, copper, magnesium etc.

- On a dry weight basis, it contains 25-35 percent protein, 10-15 percent minerals and 7-10 percent of amino acids, bio-active substances and bio-polymers.
- Palatable to: ducks, chickens, pigs, cows, goats, sheep and rabbits.
- Can increase milk production in cows by 15-20%.
- *Azolla* is a nutrient feeder and prevents Algal blooms in farm dams as a result, keeping water more usable for stock.
- Can be easily harvested with a scoop net, or grown in enclosed, floating rings which can be pulled to the edge for easy harvest.
- After initial collection/purchase, you get a lifetime supply.

Other Uses of *Azolla*

Azolla has also multidimensional uses and recently it has also gained importance as poultry feed and fodder. *Azolla* has many uses such as use as human food, animal feed, medicine, production of biogas, hydrogen fuel, water purifier, weed control, reduction of ammonia volatilization and because of the multifaceted uses it has been aptly referred to as "green gold mine" (G. M. Wagner, 1997). The potential of *Azolla* as meal for chicken was also reported (Alalade and Iyayi, 2006). They emphasized the importance of this unconventional feed as an effective alternative to feed chicken. *Azolla* is used to feed pigs and is a good source of minerals and essential amino acids (Leterme *et al.*, 2009). In the recent times *Azolla* is gaining tremendous importance in phytoremediation of polluted waters. *A.microphylla* biomass was successfully cultivated in secondary treated Municipal waters of Delhi (Arora and Saxena, 2005). Phytoremediation potential of three different species of *Azolla* to chromium has also been carried out (Arora *et al.* 2006). These results show that *Azolla* may be successfully employed in phytoremedation of polluted water bodies. The botanicals derived from *Azolla*-Anabaena system was shown to possess bioactivity. It is also an excellent composting material. Experiments conducted at IRRI, Philippines showed that conversion of compost from *Azolla* was rather quick but further research is required to evaluate the exact potential (Kar *et al.* 2000). Fresh *Azolla* collected from fields and ponds were used in compost preparation and since it has an excellent carbon nitrogen ratio it decomposes rapidly and accelerates the decomposition of other organic residues inside the compost pit (Pabby *et al.* 2000).

Used as a Green Manure

Azolla can be used as a green manure:

- by growing it as a monocrop and then incorporating it as a basal manure before the rice is transplanted; or transported to another site for use on upland crops;

- growing it as an intercrop and incorporating it as a top dressing manure after the rice is transplanted;
- or by growing it both as a monocrop and an intercrop.

All three systems can be successful but, as is common in agriculture, use of the green manure crop requires some adjustments in management of both the green manure and the main crop.

Intercropped *Azolla* is usually grown with the rice in places where there is no time available in the cropping system for the monocropping of *Azolla*. As an intercrop *Azolla* will be initially incorporated by hand or rotary rice weeder and then later killed by heavy shading and/or high temperatures-with subsequent decomposition and release of nitrogen to the Crop-at the stage of maximum rice tillering.

Monocrop *Azolla* is used in China and Vietnam during winter and spring to produce nitrogenfor the spring rice crop. The same technique is used to produce nitrogen for the early summer rice crop, but this is less common since the growth of *Azolla pinnata* is affected by high temperature and heavy pest attack during mid to late summer.

Growing both monocrop and intercropped *Azolla* is a technique that is designed to use the growing period for *Azolla* before the planting of the rice crop, *plus* production of added nitrogen for the crop through cultivation of intercropped *Azolla*, In this system two different varieties of *Azolla* may be used in each of the different periods. Different temperature and light sensitivities of *Azolla* varieties make this possible.

Reference

Alalade, O. A. and Iyayi, E. A. (2006). Chemical composition and the feeding value of *Azolla* (*Azolla pinnata*) meal for egg type chicks. *Int. J. Poultry Sci.*, 5: 137.

Albertano, P. and Caiola, M. G. (1991). Localization of nitrogen in the *Azolla-anabaena* symbiosis byeftem. In: *Nitrogen Fixation,* M. Polsinelli *et al.* (eds.), Springer Science Business Media Dordrecht.

Antunes, A. P. M., Watkins, G. M. and Duncan, J. R. (2001). Batch studies on the removal of gold (III) from aqueous solution by *Azolla filiculoides. Biotechnol Lett.,* 23: 249-251.

Arora, A. and Saxena, S. (2005). Cultivation of *Azolla microphylla* biomass on secondary-treated Delhi municipal effluents.*Biomass and Bioenergy,* 29: 60-64.

Arora, A. and Singh, P. K. (2003). Comparison of biomass productivity and nitrogen fixing potential of *Azolla* spp. *Biomass and Bioenergy*, 24: 175-178.

Arora, A., Saxena, S. and Sharma, D. K. (2006). Tolerance and phytoaccumulation of chromium by three *Azolla* species. *World J. Microbiol & Biotechnol.* 22: 97-100.

Ashton, P. J. and Walmsley, R. D. (1984). The taxonomy and distribution of *Azolla* species in southern Africa. *Botanical Journal of the Linnean Sociey*. 89: 239-247.

Berry, J.P., Gantar, M., Perez, M.H., Berry, G. and Noriega, F.G. (2008). Cyanobacterial toxins as allelo chemicals with potential applications as algaecides, herbicides and insecticides. *Marine Drugs,* 6:117-146.

Bhardwaj, K. K. R. and Gaur, A. C. (1970). Effect of humic and fulvic acid on growth and efficiency of nitrogen fixation by *Azotobacter chroococcum*. *Folia Microbial.,* 15: 364-367

Dawar, S. and Singh, P. K. (2002). Comparison of soil-and nutrient-based medium for maintenance of *Azolla* culture. *J. Plant. Nutr.*, 25:2719-2729.

Eames, A. J. (1936). Morphology of Vascular Plants. McGraw Hill Book Co, New York.

Elmachliy, S., Chefetz, B., Tel-Or E., Vidal, L., Canals, A. and Gedanken, A. (2010). Removal of silver and lead ions from water wastes using *Azolla filiculoides*, an aquatic plant, which adsorbs and reduces the ions into the corresponding metallic nanoparticles under microwave radiation in 5 min. *Water Air Soil Pollut.* DOI 10.1007/s11270-010-0650-3

Hills, L.V. and Gopal, B. (1967). *Azolla* primaeva and its phylo-genetic significance. *Can. J. Bot.* 45: 1179.

Issa, A. A., El-Enany, A. E. and Abdel-Basset, R. (2002). Modulation of the photosynthetic source: sink relationship in cultures of the cyanobacterium *Nostoc rivulare*. *Biologia plant*. 45: 212-225.

Kannaiyan, S. (1993). Nitrogen contribution by *Azolla* to rice crop *Proc Ind. Natl. Sci. Acad. Part B. Biol Sci.* 59: 309-314.

Kar, P. P., Mishra S. and Singh, D. P. (2000). Variability in *Azolla* sporulation in response to phosphorus application. *Biol. Fertil. Soils.*, 31: 458.

Konar, R. N. and Kapoor, R. K. (1972). Anatomical studies on *Azolla pinnata*. *Phytomorphology*. 22: 211.

Kushari, D. P. and Taheruzzaman, Q. (1990). Multiplication of *Azolla* pinnata in primary production units. National Symposium on Cyanobacterial Nitrogen Fixation, (Ed: Kaushik B D), pp. 353-358, NFBGAC, IARI, New Delhi, India.

Leterme, P., Londono, A. M., Munoz, J. E., Suarez, J., Bedoya, C. A., Souffrant, W. B. and Buldgen, A. (2009). Nutritional value of aquatic ferns (*Azolla filiculoides* Lam. and *Salvinia molesta* Mitchell) in pigs. *Anim. Feed Sci. & Technol.*, 149:135.

Lumpkin, T. A. and Plucknett, D. L. (1980). *Azolla:* Botany, Physiology and use as a green manure. *Econ Bot.,* 34: 111-153.

Mashkani, S. G.and Ghazvini, P. T. M. (2009). Biotechnological potential of *Azolla filiculoides* for biosorption of Cs and Sr: Application of micro-PIXE for measurement of Biosorption. *Bioresource Technol.*, 100: 1915-1921.

Moore, A.W. (1969). *Azolla*: Biology and Agronomic significance. *Bot. Rev.*, 35:17.

Noor Nawaz A. S, Junaid, S., Dileep, N., Rakesh, K. N. and Prashith Kekuda T. R. (2014). Antioxidant Activity of *Azolla Pinnata* and *Azolla Rubra*-A Comparative Study. *Scholars Academic Journal of Biosciences*. 2(10): 719-723

Pabby, A., Dua, S. and Ahluwalia, A. S. (2000). Changes in nitrogen metabolism of *Azolla microphylla* and *Azolla pinnata* on supplementation of nitrogen fertilizer. *Phykos*., 39: 51.

Peters, G. A. and Mayne, B. C. (1974). The *Azolla*-Anabaena *Azollae* relationship I. Initial characterization of the association. *Pl. Physiol*., **53**: 813-819

Peters, G. A. and Meeks, J. C. (1989). The *Azolla-Anabaena* symbiosis: basic biology. *Ann Rev Plant Physiol Plant Mol. Biol*., 40: 193-210.

Rai, P. K. (2008). Phytoremediation of Hg and Cd from industrial effluent using an aquatic free floating macrophyte *Azolla pinnata*. *Int. J. Phytoremed.,* 10: 430-439.

Rai, P. K. (2010a). Heavy metal pollution in lentic ecosystem of subtropical industrial region and its phytoremediation *Int. J. Phytoremed.,* 12: 226-242.

Rai, P. K. (2010b). Microcosom investigation of phytoremediation of Cr using *Azolla pinnata*. *Int. J. Phytoremed.*, 12: 96-104.

Rai, P. K. and Tripathi, B. D. (2009). Comparative assessment of *Azolla pinnata* and *Vallisneria spiralis* in Hg removal from G.B. Pant Sagar of Singrauli Industrial region India. *Environ Monitor and Assessment.* 148:75-84.

Reid J. D. (2002). Systematics of the genus *Azolla* (*Azolla*ceae). *Amer. J. Bot*., 88

Saha, K. C., Panigrahi, B. C. and Singh, P. K. (1982). Blue green algae and *Azolla* additions on the nitrogen and phosphorus availability and redox potential of flooded rice soil. *Soil Biol. Biochem*., 14: 23-26.

Sahrawat, K. L. (1983). Mineralization of soil organic nitrogen under waterlogged conditions in relation to other properties of rice soil. *Plant Soil.* 42: 305-308.

Sanders R. M. K. and Fowler K. (1993). The supraspecific taxonomy and evolution of the fern genus *Azolla* (*Azolla*ceae). *Pl. Syst. Evol*., 184:175-193.

Shen, E. Y. F. (1961). Concerning *Azolla imbricata*. *American Fern Journal.* 51: 151-155.

Singh, A. and Srivastava, O. N (1984). Effect of different soil pH on the growth of *Azolla* pinnata R Brown. *Geobios*. 3: 123.

Singh, D. P. and Singh, P. K. (1995). Response of *Azolla caroliniana* and rice to phosphorus enrichment of the *Azolla* inoculums and phosphorus fertilization during intercropping. *Exp. Agric*., 31: 21-26.

Singh, K. (1977). Multiplication and utilization of fern *Azolla* containing nitrogen fixing Algal symbionts as a green manure in rice cultivation. *Riso*., 46: 642.

Singh, P. K. (1977d). *Azolla* fern plant-rice fertilizer and chicken feed. *Kerala Karshakan*. 26: 5-6

Singh, P. K. (1979 b). Symbiotic algal nitrogen fixation and crop productivity *In*: Annual Review Plant Sciences Vol I, (Ed: C P Malik) pp 37-60 Kaliyani Publishers, New Delhi.

Singh, P. K. (1979a). Use of *Azolla* in rice production in India *In*: Nitrogen and rice pp 407-418, International Rice Research Institute, Philippines.

Singh, P. K. (1988). Biofertilization of rice crop, Biofertizers: Potential and Problems. (Eds: S P Sen and P Palit), pp 109-114, Plant Physiology Forum, Calcutta.

Singh, P. K. (1989). Use of *Azolla* in Asian Agriculture. *Appl. Agric. Res.*, 4(3): 149-161.

Singh, P. K., Singh, D. P. and Satapathy, K. B (1993). Use of cattle slurry and other organic manures for *Azolla* production and its utilization as biofertilizer for rice. In : Utilization of biogas slurry. Consortium on Rural Technology, New Delhi, pp 135-142.

Singh, P. K., Singh, D. P. and Singh, R. P. (1992). Growth, acetylene reduction activity. nitrate uptake and nitrate reductase activity of *Azolla caroliniana* and *Azolla Pinnata* at varying nitrate levels. *Biochem physiol Pflanzen.* 118: 121-127.

Singh, P.K. (1977b). *Azolla* fern plant-rice fertilizer and chicken feed. *Kerala Karshakan.* 26: 5-6.

Singh, P.K. (1979). Use of *Azolla* in rice production in India; in nitrogen and Rice pp. 407 (Philippines; IRRI).

Sood, A., Uniyal, P. L., Prasanna, R. and Ahluwalia, A. S. (2011). Phytoremediation potential of Aquatic Macrophyte, *Azolla Ambio*, DOI 10.1007/s13280-011-0159-z.

Sweet, A. and Hills, L. V. (1976). Early Tertiary species of *Azolla* subg. *Azolla* sert. *Krematospora* from western and arctic Canada. *Canadian Journal of Botany.*, 54: 334-351.

Uhed, E. and Maejima, K. (2009). Inhibition of nitrogen-fixing activity of the cyanobiont affects the localization of glutamine synthetase in hair cells of *Azolla*. *J. Plant Physiol.*, 166(15):1705-9.

Umali, L. J., Duncan, J. R. and Burgess, J. E. (2006). Performance of dead *Azolla filiculoides* biomass in biosorption of Au from wastewater. *Biotechnol Lett.* 28: 45-49.

Van Hove, C. (1989). *Azolla* and its multiple uses with emphasis on Africa, Food and Agricultural Organization, Rome.

Wagner, G. M. (1997). *Azolla*: a review of its biology and utilization. *Bot. Rev.*, 63: 1-26

Wang De-Xian, Zhao Miao-Zheng and Chen-De-Fu (1987) Studies on the of the mineralization rate and nutrient releasing dynamics, *Azolla* utilization, Proceedings of Workshop on *Azolla* use, pp 275, Fuzhou,Fujian, China, International Rice Research Institute, LosBanos, Philippines.

Yin-Qiu, and Yu. J. (2003). *Azolla*-A Model Organism for Plant Genomic Studies. *Genomics Proteomics Bioinformatics.* 1(1): 15-25.

5

Plant Growth Promoting Rhizobacteria (PGPR) Prospects for New Inoculants

Parimal Panda, Bappa Paramanik, Partha Sarathi Patra, Bisweswar Mahato, Ranajit Panda, Indranil Das, Kausik Mandal* and *Ashok Choudhury

Introduction

Plant growth in agricultural soils is influenced by a myriad of abiotic and biotic factors. As a routine practice growers normally use physical and chemical approaches to manage the soil environment to improve crop yields while the application of microbial products for this purpose is less common. An exception to this is the use of rhizobial inoculants for legumes to ensure efficient nitrogen fixation; a practice that has been occurring in North America for over 100 years (Smith, 1997). The region around the root, the rhizosphere, is relatively rich in nutrients, in expense of as much as 40% of plant photosynthates from the roots (Lynch and Whipps, 1991). Consequently, the rhizosphere supports large and active microbial populations capable of exerting beneficial, neutral, or detrimental effects on plant growth. The importance of rhizosphere microbial populations for maintenance of root health, nutrient uptake, and tolerance of environmental stress is now recognized (Bowen and Rovira, 1999; Cook, 2002). These beneficial microorganisms can be a significant component of management practices to achieve the attainable yield for in other words a crop yield limited only by the natural physical environment of the crop and its innate genetic potential (Cook, 2002).

The prospect of manipulating crop rhizosphere microbial populations by inoculation of beneficial bacteria to increase plant growth has shown considerable promise in laboratory and greenhouse studies however, responses have been variable in the field (Bowen and Rovira, 1999). The potential environmental benefits of this approach, leading to a reduction in the use of agricultural chemicals are fitting the sustainable management practices. Realizing the biological interactions occurring in the rhizosphere, recent progress are in need to fulfill the practical requirements of inoculant formulation and timely delivery which would increase the technology's reliability in the field and facilitate its commercial development.

Several approaches to determine rhizosphere colonization

Hiltner (1904) coined the term 'Rhizosphere' to designate the zone of intense microbial activity that immediately surrounds the roots, the soil-root interface has come to be recognized as an important functional component of plants contributing significantly to their performance and production in any ecosystem.

The rhizosphere microorganisms which interacts with roots and can alter plant growth, principally belonging to broad groups and are categorized as disease causing pathogens, deleterious rhizobacteria, plant growth promoting rhizobacteria, symbiotic and asymbiotic (root associated) N_2-fixing bacteria. A host of other microorganisms through their independent and interactive effects may also significantly alter and influence other phenomena in the rhizosphere to aid nutrition (*e.g.,* PO_4 – solubilization), resistance and escape to pathogens (*e.g.,* parasitism and inter-microbial antagonism) etc.

Plant growth-promoting rhizobacteria (PGPR) were first defined by Kloepper and Schroth (Kloepper and Schroth, 1978) to describe soil bacteria that colonize the roots of plants following inoculation onto seed and that enhance plant growth. The following are implicit in the colonization process: ability to survive inoculation onto seed, to multiply in the spermosphere (region surrounding the seed) in response to seed exudates, to attach to the root surface, and to colonize the developing root system (Kloepper, 1993). The ineffectiveness of PGPR in the field has often been attributed to their inability to colonize plant roots (Benizri *et al.*, 2001; Bloemberg and Lugtenberg, 2001). A variety of bacterial traits and specific genes contribute to this process, but only a few have been identified as yet (Benizri *et al.*, 2001; Lugtenberg *et al.*, 2001). The specific characteristics of these traits includes motility, chemotaxis to seed and root exudates, production of pili or fimbriae, production of specific cell surface components, ability to use specific components of root exudates, protein secretion, and quorum sensing (Lugtenberg *et al.*, 2001). The generation of mutants altered in expression of these traits is aiding our understanding of the precise role each one plays in the colonization process (Lugtenberg *et al.*, 2001; Persello-Cartieaux *et al.*, 2003). Progress in the identification of new, previously uncharacterized genes is being made using nonbiased screening strategies that rely on gene fusion technologies. These strategies employ reporter transposons (Roberts *et al.*, 1998) and in vitro expression technology (IVET) (Rainey, 1999) to detect genes expressed during colonization.

Using molecular markers such as green fluorescent protein or fluorescent antibodies it is possible to monitor the location of individual rhizobacteria on the root using confocal laser scanning microscopy (Bloemberg *et al.*, 2000; Bloemberg and Lugtenberg, 2001; Sorensen *et al.*, 2001). This approach has also been combined with an rRNA-targeting probe to monitor the metabolic activity of a rhizobacterial strain in the rhizosphere and showed that bacteria located at the root tip were most active (Lubeck *et al.*, 2000; Sorensen *et al.*, 2001).

An important aspect of colonization is the ability to compete with indigenous microorganisms already present in the soil and rhizosphere of the developing plant. Understanding the factors involved in these interactions has been hindered by the inability to culture and characterize diverse members of the rhizosphere community and also to determine how that community varies with plant species, plant age, location on the root, and soil properties. Phenotypic and genotypic approaches are now available to characterize rhizobacterial community structure. Phenotypic methods that rely on the ability to culture microorganisms include standard plating methods on selective media, community level physiological profiles (CLPP) using the BIOLOG system (Garland, 1996), phospholipid fatty acid (PLFA) (Tunlid and White, 1992), and fatty acid methyl ester (FAME) profiling (Germida *et al.*, 1998). Culture-independent molecular techniques are based on direct extraction of DNA from soil and 16S-rRNA gene sequence analysis, bacterial artificial chromosome or expression cloning systems (Rondon *et al.*, 1999). These are providing new insight into the diversity of rhizosphere microbial communities, the heterogeneity of the root environment, and the importance of environmental and biological factors in determining community structure (Baudoin *et al.*, 2002; Berg *et al.*, 2002; Smalla *et al.*, 2001). These approaches can also be used to determine the impact of inoculation of plant growth-promoting rhizobacteria on the rhizosphere community (Ciccillo *et al.*, 2002; Steddom *et al.*, 2002).

Process of root colonization

Root colonization can be considered to involve four stages. The initial stage of root colonization is the movement of microbes to the plant root surface. Bacterial movement can be passive, via soil water fluxes, or active, via specific induction of flagellar activity by plant-released compounds (chemotaxis). The second step in colonization is adsorption to the root. This is a step required before anchoring and can be defined as nonspecific and based on electrostatic forces, whereas anchoring can be defined as the firm attachment of a bacterial cell to the root surface. Following adsorption and anchoring, specific and/ or complex interactions between the bacterium anther host plant may ensue, thickhead to induction of bacterial gene expression. Table 1 cites examples of root exudate components involved in these processes.

Table 1: Root colonization

S. No.	Process	Excdate component	Microbial species	Reference
1.	Movement of microbes to root surface a. Passive b. Active—induction of flagellar activity (chemotaxis)	*Lureolin/ Phenolics* Acetosyringone Benzoate/ aromaties Sugars Amino acids, nucleotides and sugars Serine Unidentified Unidentified Unidentified	*Rhizobium meliloti* *Agrobacterium tumefaciens* *Azaspirllum spp.* *Agrobacerium timefaciens* *Pseudomonas lachrymans* *Pseudomonas aeruginosa* *Pseudomonas fluorescens* *Pseudomonas punida* *Azaspirillum brasilense*	80 81 82 83 83 83 83 83 84.85
2.	Adsorption			
3.	Anchoring a. Bacterial appendages (pili. fimbriae) b. Agglutination c. Formation of cell aggregate	Lectins	*Pseudomonas fluorescens* *Pseudomonas putido* *Enterobacter agglomerans*	86 87.88 89
4.	Gene expression a. Nodulation genes—production of nod factors b. Virulence genes—production of virlence factors	Flavonoids	*Rhicobian* spp. *Bradying obium* spp. *Agrobacterium tunefaciens*	90, 91, 92 90

Source: Melissa J. Brimecombe, Frans A. De Leij, and James M. Lynch. 2001, the Effect of Root Exudates on Rhizosphere Microbial Populations. In: The Rhizosphere (Eds. Pinton, R., Varanini, Z., Nannipieri. P.) Pp.105-106.

Mechanisms of action

PGPR enhance plant growth by direct and indirect means, but the specific mechanisms involved have not all been well-characterized (Glick, 1995;

Kloepper, 1993). Direct mechanisms of plant growth promotion by PGPR can be demonstrated in the absence of plant pathogens or other rhizosphere microorganisms, while indirect mechanisms involve the ability of PGPR to reduce the deleterious effects of plant pathogens on crop yield. Direct mechanisms for enhancement of plant growth by PGPR was evidenced by a variety of modes: fixation of atmospheric nitrogen that is transferred to the plant, production of siderophores (are defined as relatively low molecular weight, ferric ion specific chelating agents elaborated by bacteria and fungi growing under low iron stress) that chelate iron and make it available to the plant root, solubilization of minerals such as phosphorus, and synthesis of phytohormones (Glick, 1995). Direct enhancement of mineral uptake due to increases in specific ion fluxes at the root surface in the presence of PGPR has also been reported (Bashan and Levanony, 1991; Bertrand *et al.*, 2000). PGPR strains may use one or more of these mechanisms in the rhizosphere. Molecular approaches using microbial and plant mutants altered in their ability to synthesize or respond to specific phytohormones gave better understanding to the role of phytohormone synthesis as a direct mechanism of plant growth enhancement by PGPR (Glick, 1995; Persello-Cartieaux *et al.*, 2003). PGPR that synthesize auxins and cytokinins or that interfere with plant ethylene synthesis have been identified (Garcia de Salamone *et al.*, 2001; Glick, 1995, Persello-Cartieaux *et al.*, 2003).

Biological nitrogen fixation

Biological N_2-fixation (BNF) by soil microorganisms is considered one of the major mechanisms by which plants benefit from the association of micro partners. One of the benefits that diazotrophic microorganisms provide to plants is fixed nitrogen in exchange for fixed carbon released as root exudates. The beneficial effects of the symbiotic association between rhizobia and legumes are well known and these have been intensively investigated. Moreover, (Malik *et al.*, 1997; Biswas *et al.*, 2000a and Ladha *et al.*, 1998) observed that free-living bacteria as well as rhizobial strains can promote the growth of cereal plants (lowland rice plants) by contributing to N-economy through their ability to fix N_2. (Using a 15N isotopic dilution technique, (Malik *et al.*, 1997) investigated the nitrogen fixing ability of Azospirillum strain N-4 in rice and found a significant contribution of nitrogen fixed by the this organisms belonging to the group of PGPR. Similarly, (Biswas *et al.*, 2000a) reported that biological nitrogen fixation by diazotrophic PGPR strains may be a contributing factor of rice growth promotion in addition to other mechanisms.

Production of plant growth regulators

Plant growth regulators (PGRs) are organic compounds that influence the physiological processes in plants at extremely low concentrations. Production of PGRs by inocula has been suggested as one of the most possible mechanisms of action affecting plant growth. Numerous studies have shown an improvement in plant growth and development in response to seed or root inoculation with various microbial inoculants capable of producing PGRs. There are five classes of well-known PGRs, namely, auxins, gibberellins, cytokinins, ethylene, and abscisic acid. Soil micro biota, particularly the rhizosphere microflora are potential sources of these PGRs (Frankenberger and Arshad, 1995; Patten and Glick, 1996). The type and amount of PGR production by microorganisms is variable. Mansour *et al.* (1994) tested 24 isolates of thallobacteria (A class of bacteria of the division Firmicutes, consisting of simple, gram-positive, asporogenous and sporogenous rods and cocci) belonging to the genus Streptomyces for their potential to produce PGRs. All the isolates were capable of producing auxins (ranging from 10.5 to 39.0 µg mL^{-1}), gibberellins (1.21 to 13.3 µg mL^{-1}) and cytokinins (2.5 to 14.9 µg mL^{-1}) in liquid medium.

It is now known that microbial 'production of PGRs is one of the major mechanisms in modifying the growth and yield of plants. Young *et al.* (1991) screened the PGPR strains, Pseudomonas and Serratia for PGRs and observed a good correlation between induction of rice root elongation and phytohormone production. Likewise, Asghar *et al* (2002) reported a highly significant correlation between auxin production by PGPR in vitro and grain yield (r = 0.77*), number of pods (r = 0.78*) and number of branches (r = 0.77*) per plant of *Brassica juncea* It was hypothesized that the 'PGPR influenced the growth and yield of inoculated plants by producing auxins in the rhizosphere upon the release of tryptophan in the root exudates, although other mechanisms of action might have also contributed. Most Rhizobium species have been shown to produce IAA (Ahemad and Khan, 2012b,d,f; Ahemad and Khan, 2011e, j). Since, IAA is involved in multiple processes including cell division, differentiation and vascular bundle formation, these three processes are also essential for nodule formation. Gutierrez-Manero *et al* (2001) isolated *Bacillus pumilus* and *Bacillus lichenoformis* from the rhizosphere of *Alnus glutinosa* L. which had strong growth promoting activity on alder.

Non-pathogenic rhizosphere microorganisms can be detrimental to plant growth. Suslow and Schroth (1982) first highlighted the importance of bacteria responsible for the inhibition of root growth and they proposed the term "deleterious rhizobacteria" or "DRB". Subsequently, the term "deleterious' rhizosphere microorganisms" or "DRMO" has been used to include nonpathogenic fungi detrimental to root growth (Cherrington and Elliott, 1987). The mechanisms by which DRMOs adversely affect plants include high levels of IAA, siderophore-mediated competition for iron, ethylene (C_2H_4), HCN, and unidentified phytotoxins. PGPR can modify plant growth by producing and releasing substances other than lAA. Baker and Schippers (1987) established that 4% of the total aerobic bacteria and 40% of the *pseudomonads* from the rhizosphere of potato grown in short rotation were able to produce HCN in vitro. They also demonstrated that the cytochrome respiratory pathway of potato roots was particularly sensitive to cyanide. Schippers (1988) provides a model in which the DRMO use glycine and proline in potato root exudate to synthesize HCN, which is taken up by the root (Bowen and Rovira, 1999).

Several other workers have reported that C_2H_4 and abscisic acid (ABA) production is a common characteristic of soil and rhizosphere microflora (Arsad and Frankenberger, 1989). Glick and his co-workers have suggested the involvement of an enzyme, 1-aminocyclopropane-l-carboxylic acid (ACC) deaminase produced by Pseudomonas putida GR12-2 in modifying the root growth of different plants (Glick *et al.*, 1998). They found that this bacterium hydrolyzes ACC, the immediate precursor of C_2H_4 in higher plants. ACC deaminase might act to stimulate plant growth by sequestering and then 'hydrolyzing ACC from germinating seeds, thereby lowering the endogenous levels of ACC which subsequently results in plant growth promotion.

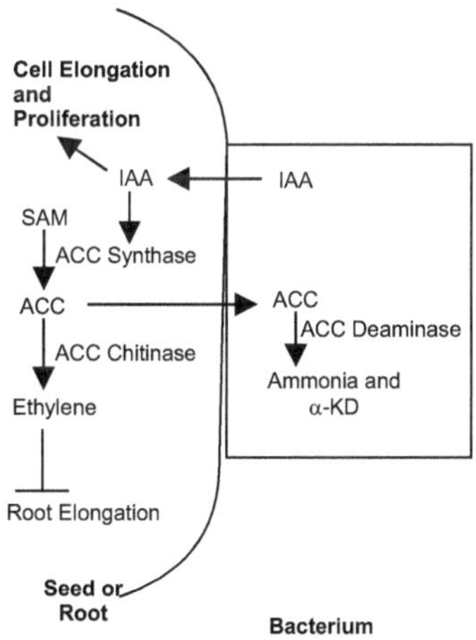

Solubilization and nutrient uptake

An adequate supply of mineral nutrients is necessary for optimum plant growth. However, when adequate amounts of essential nutrients are present in soil, plants may still show deficiencies due to the non-availability of these mineral nutrients. Solubilization of mineral nutrients such as phosphorus and iron by PGPR makes them more readily available for plant uptake, and this should be considered as a mechanism for enhanced plant growth (Glick, 1995). Several reports have suggested that PGPR can stimulate plant growth by increasing solubilization (via releasing siderophores or organic acids) and facilitate the uptake of mineral nutrients by the plant (Kloepper *et al.*, 1987; Dazzo *et al.*, 2000). Dazzo *et al.* (2000) reported that certain strains of rhizobia are able to solubilize phosphorus. Pandey and Maheshwari 2007; Badawi *et al.* 2011 observed few genera of rhizobia, for example, Bradyrhizobium and Rhizobium, have been capable of solubilizing P and secrete IAA Phosphorus uptake was increased (13-23%) significantly in response to rhizobial inoculation of rice compared to a non-inoculated control (Biswas *et al.*, 2000b). Toro *et al.* (1997) evaluated the interactive effect of phosphate-solubilizing bacteria (*Enterobacter* sp. and *Bacillus subtilis*) and arbuscular mycorrhizal (AM) fungi (*Glomus intraradices*) on onion (*Allium cepa* L.) with a soil of low P content. Several other researchers have reported a similar role of other bacteria in P and other mineral nutrient solubilization enhancing nutrient uptake and subsequently plant growth (Belimov *et al.*, 1995; Noel *et al.*, 1996; Glick *et al.*, 1998). PGPR may also alter the solubility of mineral nutrients by releasing organic acids, sugar acids and creating acidity via CO_2 (respiration). Pietr *et al.*, (1990) tested 748 bacterial strains isolated from the rhizoplane of different field crops and found that 25.6% strains were capable of dissolving calcium phosphate. They suggested that production of organic acids was the major mechanism of action by which insoluble phosphorus compounds were converted to more soluble forms. Many other scientists have also reported that PGPR can create an acidic environment to promote mineral nutrient solubilization. Further studies indicated that PGPR may increase the mobility and availability of micronutrients by the formation of high affinity siderophores.

Siderophores, as stated earlier, are low-molecular weight compounds that complex with Fe^{2+} and render it available to microorganisms (Leong, 1986). Kloepper *et al.* (1980) designated a siderophore as "pseudobactin" which is some fluorescent pseudomonads producing a yellow-green pigment. The role of pseudobactin in promoting the growth of potato was demonstrated when 10 μg of pseudobactin increased plant growth to the same extent as when the fluorescent pseudomonad was applied to potato seed pieces. Diversified organisms namely, *Bacillus, Rhizobium, Pseudomonas, Agrobacterium, Escherichia coli* and many fungi produce a wide range of iron-chelating compounds. Numerous plants are capable of using bacterial Fe siderophore complexes as a means of obtaining Fe from soil. The above statement is supported by the findings of Hughes *et al.* (1992) who reported enhanced Fe uptake in oat due to siderophore production. The quantity, quality, and optimal conditions for microbial production of siderophores and their influence on plant growth is yet to determine.

Biological Control

Soil-borne pathogens are well known for their devastating effects on plant health and yield. For successful disease management, it is important to find the most effective and economical ways to protect the plant from various pests or diseases. In recent years, the use of PGPR as an inducer of systemic resistance in crop plants against different pathogens has been demonstrated under field conditions (Vidhayasekaran and Muthamilan, 1999). The use of natural PGPR strains in plant defense may offer a practical way to render immunization. In recent years, biological control of disease has emerged as an important mode of action through which PGPR may benefit plant growth. PGPR have been reported to increase the resistance in plants against fungal, bacterial, and viral diseases (Maurhofer *et al.*, 1998), and nematodes (Sikora, 1988)

Studies on mode of action reveal that biological control by PGPR involves production of bacterial metabolites which decelerate the population or activities of pathogens or deleterious rhizosphere microflora (Glick, 1995). These metabolites may include siderophores that bind Fe making it less available to certain members of the native pathogenic micro flora (Kloepper *et al.*, 1987; Subba Rao, 1993). Berthelin *et al* (1991) reported that production of siderophores chelating iron in the plant rhizosphere makes it unavailable to harmful microflora in a sterilized medium. Certain fluorescent pseudomonads, particularly *Pseudomonas fluorescens* and *Pseudomonas putida* were adapted to colonize plant roots which promoted plant growth under field conditions most likely through siderophore formation (Kloepper *et al.*, 1987). Similarly, Subba Rao (1993) suggested that fluorescent pseudomonads producing siderophores which specifically recognize and sequester the limited supply of iron in the rhizosphere may reduce the availability of iron for the growth of pathogens (Subba Rao, 1993).

Antibiotic production is one of the most intensively studied aspects of biocontrol, but in many cases it is difficult to distinguish between antibiosis and competition. When antibiotic-negative mutants are inoculated in the rhizosphere and they do not affect the target organism as does the parent antibiotic producer, it is a good indication that antibiosis is a major factor. Several studies have demonstrated that production of antibiotics (*e.g.*, pyrrolnitrin, phycocyanin, 2,4-diacetyl phloroglucinol) by microbial inocula can cause suppression of pathogen (Kloepper, 1993, Subba Rao, 1993; Glick, 1995). Glick (1995) was of the view that the most effective mechanism that a

PGPR can employ to prevent proliferation of phytopathogens is the synthesis of antibiotics.

Other mechanisms for biological control of disease may include: competition for infection sites and nutrients, parasitism on pathogens *i.e.,* destruction of fungal pathogens by the action of lytic enzymes (*e.g.,* chitinase and β-1, 3-glucanase) that degrade fungal cell wall, and uncharacterized antifungal factors (Kloepper, 1993; Glick, 1995). Buchenauer (1998) reported various mechanisms for biological control such as competition for space and nutrients in the rhizosphere and spermosphere (spermosphere is the zone surrounding seeds where interactions between the soil, microbial communities and germinating seeds take place) lytic enzymes, HCN, and many other metabolites produced by rhizobacteria. Subba Rao (1993) believes in multifarious mechanisms for biocontrol including competition (competition has a negative effect on both of the species) for substrate and niche exclusion (niche is basically a species' ecological role, which is defined by the set of conditions, resources, and interactions it needs or can make use of. Competitive exclusion principle says that two species can't coexist if they occupy exactly the same niche or competing for identical resources). He further explains that the points of emergence of lateral roots are favorable sites for deleterious rhizobacteria and PGPR appear to compete for those sites very effectively. In addition, several microorganisms are able to hydrolyze the compound fusaric acid, which is a causative agent in the damage of plants infected with. *Fusarium* and, thus may prevent plant disease (Thomashow and Weller, 1995)

Renwick *et al* (1991) concluded that most of the naturally occurring biological controls result from a mixture of antagonists rather than from a high density of a single antagonist. The application of a mixture of introduced biocontrol agents would more closely mimic the natural environment and might broaden the spectrum of biological activity The work of various scientists have recently supported this broad spectrum of biocontrol activity by a mixture of PGPR strains However, in some cases, mixtures of different strains had no synergistic effect. Furthermore. a consortium that improves efficacy under one set of conditions or on one host may not perform under a different set of conditions and also on different hosts (Schisler *et al.*, 1997).

Recent work on the broad spectrum of PGPR-mediated induced systemic resistance (ISR) (is a state of increased defensive capacity developed by plants when appropriately stimulated, through activation of latent resistance induced by diverse agents including rhizobacteria) against different pathogens in different crops has gained importance. Induced systemic resistance by *Ps. putida strain 89B-27* and *S. marcescens strain 90-166* was effective against anthracnose of cucumber (Wei *et al.*, 1991), angular leaf spot caused by Ps. syringae pv. lachrymans, *Fusarium* wilt caused by *F. oxysporum*

f. sp. cucumerinum (Liu *et al.*, 1995b) and cucurbit wilt caused by *Erwinia tacheiphila* (Kloepper, 1993). Similarly, *Ps. fiuorescens strain* Pf1 induces resistance against different pathogens in different crops, viz. *Rhizoctonia solani* (Nandakumar, 1998) and *Colletotrichum falcatum* in sugarcane (Viswanathan and Samiyappan, 1999). The broad spectrum of biocontrol by PGPR has many applications in disease protection. Hence selecting suitable strains having the potential to induce systemic resistance against multiple pathogens and pests is an important task in the delivery of microbial inoculants to the field.

Selection and characterization of PGPR

One of the challenges in developing PGPR for commercial application is ensuring that an effective selection and screening procedure is in place, so that the most promising organisms are identified and brought forward. In the agricultural chemical industry, thousands of prospective compounds are screened annually in efficient high-throughput assays to select the best one or two compounds for further development. Similar approaches are not yet in place for PGPR. Effective strategies for initial selection and screening of rhizobacterial isolates are required. It may be important to consider host plant specificity or adaptation to a particular soil, climatic conditions or pathogen in selecting the isolation conditions, and screening assays (Bowen and Rovira, 1999; Chanway *et al.*, 1989). The spermosphere model, an enrichment technique that relies on seed exudates as the nutrient source, has been used for selection and isolation of promising N_2-fixing rhizosphere bacteria from rice (Thomas-Bauzon *et al.*, 1982). One approach for selection of organisms with the potential to control soil-borne phytopathogens is to isolate from soils that are suppressive to that pathogen (Weller *et al.*, 2002). Other approaches involve selection based on traits known to be associated with PGPR such as root colonization (Silva *et al.*, 2003), 1-aminocyclopropane-1-carboxylate (ACC) deaminase activity (Cattelan *et al.*, 1999; Glick, 1995), antibiotic property (Giacomodonato *et al.*, 2001) and siderophore production (Cattelan *et al.*, 1999). The development of high throughput assay systems and effective bioassays will facilitate selection of superior strains (Mathre *et al.*, 1999; McSpadden Gardener and Fravel, 2002).

Root and shoot growth under axenic condition

This is the most widely used approach for preliminary screening of rhizobacteria for selection of PGPR. In the past, when sensitive analytical techniques were not available, this approach was commonly employed for preliminary selection of PGPR from a large number of rhizobacterial isolates. In most cases, the mechanism(s) of action of PGPR isolates collected by this approach was not well understood. The rhizobacterial isolates are inoculated onto the seed and/or roots of the host plants under axenic conditions and the effects of inoculation on various growth parameters such as root/shoot growth are monitored. The rhizobacteria exhibiting positive effects on the growth parameters of inoculated host plants are then selected and sometimes subjected to further screening under repeated trials to confirm their growth promoting activities. Such selected rhizobacteria could be designated as PGPR and may be used as inocula under natural field conditions. Although this approach does not elucidate the mechanism of action of the PGPR, it does provide a good preliminary selection in testing of a high number of isolates under natural field conditions. However, this approach is extremely laborious and time consuming.

Production of growth regulators (*in vitro*)

Although the role of microbial produced biologically active substances in evoking a physiological response in inoculated plants was speculated long ago, it was not until the 1970s that these substances were unequivocally identified upon the availability of state-of-the-art analytical instruments. During last couple of decades, the approach of screening microflora for in vitro production of PGRs has become very popular. Different rhizobacteria vary greatly in their potential to produce PGRs (Lebuhon and Hartmann, 1993; Pal *et al.*, 1999; Padua *et al.*, 2000). Several studies have demonstrated a strong relationship between in vitro production of PGRs by PGPR and their plant growth-promoting activity in various plants. Glick (1995) reports that the primary mechanism most commonly invoked by PGPR on plants is the production of phytohormones, and IAA may play the most important role in plant growth promotion. Under gnotobiotic conditions, Noel *et al.* (1996) demonstrated the direct involvement of PGRs including, IAA and cytokinin in modifying the growth of canola and lettuce

ACC deaminase activity

Microbial inoculation has been shown to alter the endogenous levels of the plant hormone, C_2H_4, which subsequently leads to changes in growth and development of inoculated plants. Ethylene, which is believed to be produced by all plants, mediates a wide range of different plant responses and developmental processes. In some instances, the presence of C_2H_4 is stimulatory while in others it is inhibitory. Recently, it has been discovered that certain bacteria (such as *Ps. putida)* contain an enzyme, 1-aminocyclopropane-l-carboxylate (ACC)-deaminase which hydrolyses ACC (an immediate precursor of C_2H_4), into ammonia and a-ketobutyrate.

A model describing the role of bacterial ACC-deaminase in facilitating plant growth implies that bacteria inhabiting roots or seeds cause a decrease in the levels of endogenous ACC in plant roots. Decreased levels of ACC result in lower levels of endogenous C_2 and subsequently promote root growth (Glick *et al.*, 1998). Better root growth often results in better shoot growth.

Glick and his co-workers (Glick *et al.*, 1998) support this premise by providing the following evidence: (1) C_2H_4 has previously been found to be an inhibitor of plant root elongation in several different systems (Ables *et al.*, 1992). However, soil bacteria that contain the enzyme, ACC deaminase can lower C_2H_4 levels by decreasing the plant ACC content (Glick, 1995). Thus, the roots of C_2H_4sensitive plants are invariably longer when ACC deaminase-containing bacteria are present (Hall *et al.*, 1996); (2) the ACC level in extracts of roots from canola seedlings as measured by HPLC were lower when the seeds were treated with *Ps. putida* GR12-2 than in seedlings from the untreated seeds. In addition, the ACC levels in root extracts from the seedlings in which the seeds were treated with an ACC deaminase minus mutant of *Ps. putida* GR12-2 were similar to those of the untreated seedlings; (3) the increase in the length of the roots of young (five-to seven-day-old) seedlings of canol a, tomato, lettuce, wheat, oats, and barley following treatments of the seeds with the wild-type *Ps. putida* GR12-2 was similar to the response of these plants when their seeds were treated with the C_2H_4 inhibitor, A VG (Hall *et al.*, 1996). Similarly, plants that were stimulated to the greatest extent by treatment with either *Ps. putida* GR12-2 or AVG were those that were the most sensitive to root length inhibition by ethephon (Hall *et al.*, 1996). Thus, as far as root length is concerned, only C_2H_4 sensitive plants responded to the presence of PGPR that contained ACC deaminase.

Enrichment for soil bacteria that can utilize ACC as a sole N source is a means of isolating organisms with the presence of the enzyme, ACC deaminase. This trait appears to be limited to rhizobacteria that are capable of stimulating plant growth. Based on the interaction of ACC-deaminase carrying bacterial strains with plant roots, it is highly likely that manipulation of the ACC-deaminase gene could be used to improve agricultural production.

Field application of PGPR

The application of PGPR for control of fungal pathogens in greenhouse systems shows considerable promise (Paulitz and Belanger, 2001), due in part to the consistent environmental conditions and high incidence of fungal disease in greenhouses. Achieving consistent performance in the field where there is heterogeneity of abiotic and biotic factors and competition with indigenous organisms is more difficult. Knowledge of these factors can aid in determination of optimal concentration, timing and placement of inoculant, and of soil and crop management strategies to enhance survival and proliferation of the inoculant (Bowen and Rovira, 1999; McSpadden Gardener and Fravel, 2002). The concept of engineering or managing the rhizosphere to enhance PGPR function by manipulation of the host plant, substrates for PGPR, or through agronomic practices, is gaining increasing attention (Bowen and Rovira, 1999; Mansouri *et al.*, 2002). Development of better formulations to ensure survival and activity in the field and compatibility with chemical and biological seed treatments is another area of focus; approaches include optimization of growth conditions prior to formulation and development of improved carriers and application technology (Bashan, 1998; Bowen and Rovira, 1999; Date, 2001; Mathre *et al.*, 1999; Yardin *et al.*, 2000).

Challenges in commercialization of PGPR

Prior to registration and commercialization of PGPR products, a number of hurdles must be overcome (Fravel *et al.*, 1999; Mathre *et al.*, 1999; McSpadden Gardener and Fravel,2002). These include scale up and production of the organism under commercial fermentation conditions while maintaining quality, stability, and efficacy of the product. Formulation development must consider factors such as shelf life, compatibility with current application practices, cost, and ease of application. Health and safety testing may be required to address such issues as non-target effects on other organisms including toxigenicity, allergenicity and pathogenicity, persistence in the environment, and potential for horizontal gene transfer. The product claim, whether as a fertilizer supplement or for biological control, will determine to which federal agency applications for registration should be addressed in Canada and the USA. Capitalization costs and potential markets must be considered in the decision to commercialize. McSpadden Gardener and Fravel (McSpadden Gardener and Fravel, 2002) estimated that a minimum capitalization of $1 million US is required to register a biopesticide product in North America.

Future prospects

New PGPR products may becoming available in these day as our understanding of the complex environment of the rhizosphere, of the mechanisms of action of PGPR, and of the practical aspects of inoculant formulation and delivery increases.. The success of these products will depend on our ability to manage the rhizosphere to enhance survival and competitiveness of these beneficial microorganisms (Bowen and Rovira, 1999). Rhizosphere management will require consideration of soil and crop cultural practices as well as inoculant formulation and delivery (Bowen and Rovira, 1999; McSpadden Gardener and

Fravel, 2002). Genetic enhancement of PGPR strains to enhance colonization and effectiveness may involve addition of one or more traits associated with plant growth promotion (Bloemberg and Lugtenberg, 2001; Glick, 1995; Lubeck *et al.*, 2000). Genetic manipulation of host crops for root-associated traits to enhance establishment and proliferation of beneficial microorganisms (Mansouri *et al.*, 2002; Smith and Goodman, 1999) is being pursued. However, regulatory issues and public acceptance of genetically engineered organisms may delay their commercialization. The use of multi-strain inocula of PGPR with known functions is of interest as these formulations may increase consistency in the field (Jetiyanon and Kloepper, 2002; Siddiqui and Shaukat, 2002). They offer the potential to address multiple modes of action, multiple pathogens, and temporal or spatial variability.

PGPR offer an environmentally sustainable approach to increase crop production and health. The application of molecular tools is enhancing our ability to understand and manage the rhizosphere and will lead to new products with improved effectiveness

Reference

Ables, F. D., Morgan, P.W. and Salveit, M. E. Jr. (1992). "Ethelene in Plant Biology" (2nd ed.). *Academic Press, San Diego, CA.*

Ahemad, M. and Khan, M. S. (2011e). Ecotoxicological assessment of pesticides towards the plant growth promoting activities of Lentil (*Lens esculentus*)-specific Rhizobium sp. strain MRL3. *Ecotoxicology*. 20: 661-669.

Ahemad, M. and Khan, M. S. (2011j). Plant growth promoting fungicide tolerant Rhizobium improves growth and symbiotic characteristics of lentil (*Lens esculentus*) in fungicide-applied soil. *J. Plant Growth Regul.*, 30: 334-342.

Ahemad, M. and Khan, M. S. (2012b). Ecological assessment of biotoxicity of pesticides towards plant growth promoting activities of pea (*Pisum sativum*)-specific Rhizobium sp. strain MRP1. Emirates *J. Food Agric.*, 24: 334-343.

Ahemad, M. and Khan, M. S. (2012d). Effects of pesticides on plant growth promoting traits of Mesorhizobium strain MRC4. *J. Saudi Soc. Agric. Sci.*, 11: 63-71.

Ahemad, M. and Khan, M. S. (2012f). Productivity of green gram in tebuconazole-stressed soil, by using a tolerant and plant growth promoting Bradyrhizobium sp. MRM6 strain. *Acta Physiol. Plant.*, 34: 245-254.

Arsad, M. and Frankenberger, W. Jr. (1989). Biosynthesis of ethylene by Acremonium falciorme. *Soil Biol. Biochem.*, 21: 633-638.

Asghar, H. N., Zahir, Z. A., Arshad, M. and Khaliq, A. (2002). Relationship between in vitro production of rhizobacteria and their growth-promoting activities in *Brassica juncea L.Biol. Fertil. Soils.* 35: 231-237.

Badawi, F. S. F., Biomy, A. M. M. and Desoky, A. H. (2011). Peanut plant growth and yield as influenced by co-inoculation with Bradyrhizobium and some rhizo-microorganisms under sandy loam soil conditions. *Ann. Agric. Sci.* 56:17-25.

Baker, A. W. and Schippers, B. (1987). Microbial cyanide production in the rhizophere in relation potato yield reduction and Pseudomonas spp-mediated plant growth stimulation. *Soil Biol. Biochem.* 19:451-457.

Bashan, Y. (1998). Inoculants of plant growth-promoting bacteria for use in agriculture. *Biotechnol. Adv.*, 16: 729-770.

Bashan, Y. and Levanony, H. (1991). Alterations in membrane potential and in proton efflux in plant roots induced by *Azosprillum brasilense. Plant Soil.* 137: 99-103.

Baudoin, E., Benizri, E. and Guckert, A. (2002). Impact of growth stage on the bacterial community structure along maize roots, as determined by metabolic and genetic fingerprinting. *Appl. Soil Ecol.*, 19:135-145.

Belimov, A. A., Kojemiakov, A. P. and Chuvarliyeva, C. V. (1995). Interaction between barley and mixed cultures of nitrogen fixing and phosphate-solubilizing bacteria. *Plant Soil.* 173: 29-37.

Benizri, E., Baudoin, E. and Guckert, A. (2001). Root colonization by inoculated plant growth promoting rhizobacteria. *Biocontrol Sci. Technol.*, 11: 557-574.

Berg, G., Roskot, N., Steidle, A., Eberl, L., Zock, A. and Smalla, K. (2002). Plant-dependent genotypic and phenotypic diversity of antagonistic rhizobacteria isolated from different Verticillium host plants. *Appl. Environ. Microbiol.,* 68: 3328-3338.

Berthelin, J., Leyrol, C., Laheurte, F. and Degiudici, P. (1991). Some considerations on the relations between phosphate solubilizing rhizobacteria and their effect on seedling and plant growth related to phosphorus solubilization. In "Growth Promoting Rhizobacteria: Progress and Prospects" (C. Keel, B. Koller, and G. Defago, Eds.), pp. 359-364. IOBC, Switzerland.

Bertrand, H., Plassard, C., Pinochet, X., Toraine, B., Normand, P. and Cleyet-Marel, J. C. (2000). Stimulation of the ionic transport system in *Brassica napus* by a plant growth-promoting rhizobacterium (*Achromobacter* sp.). *Can. J. Microbiol.*, 46: 229-236.

Biswas, J. C., Ladha, J. K. and Dazzo, F. B. (2000a). Rhizobia inoculation improves nutrient uptake and growth of lowland rice. *Soil. Sci. Soc. Am. J.*, 64: 1644-1650.

Biswas, J. C., Ladha, J. K., Dazzo, F. B., Yanni, Y G. and Rolfe, B. G. (2000b). Rhizobial inoculation influences seedling vigor and yield of rice. *Agron. J.*, 92: 880-886.

Bloemberg, G. V. and Lugtenberg, B. J. J. (2001). Molecular basis of plant growth promotion and biocontrol by rhizobacteria. Curr. Opin. *Plant Biol.,* 4: 343-350.

Bloemberg, G. V., Wijfjes, A. H. M., Lamers, G. E. M., Stuurman, N. and Lugtenberg, B. J. J. (2000). Simultaneous imaging of Pseudomonas fluorescens WCS3 Roberts *et al.*,1998 populations expressing three different auto fluorescent proteins in the rhizosphere: New perspectives for studying microbial communities. Mol. *Plant-Microbe Interact.* 13:1170-1176.

Bowen, G. D. and Rovira, A. D. (1999). The rhizosphere and its management to improve plant growth. *Adv. Agron.*, 66: 1-102.

Buchenauer, H. (1998). Biological control of soil-borne diseases by rhizobacteria. *Journal of Plant Disease and Protection.* 105: 329-348.

Cattelan, A. J., Hartel, P. G. and Fuhrmann, J. J. (1999). Screening for plant growth-promoting rhizobacteria to promote early soybean growth. *Soil Sci. Soc. Am. J.*, 63: 1670-1680.

Chanway, C. P., Nelson, L. M. and Holl, F. B. (1989). Cultivar-specific growth promotion of spring wheat (*Triticum aestivum* L. by co-existent *Bacillus* species. *Can. J. Microbiol.*, 34: 925-929.

Cherrington, C.A. and Elliott, L. F. (1987). Induction of inhibitory *pseudomonads* in the pacific North-west. *Plant Soil.* 101: 159-165.

Ciccillo, F., Fiore, A., Bevivino, A., Dalmastri, C., Tabacchioni, S. and Chiarini, L. (2002). Effects of two different application methods of *Burkholderia ambifaria* MCI 7 on plant growth and rhizospheric bacterial diversity. *Environ. Microbiol.*, 4: 238-245.

Cook, R. J. (2002). Advances in plant health management in the twentieth century. *Ann. Rev. Phytopathol.* 38: 95-116.

Date, R. A. (2001). Advances in inoculant technology: a brief review. *Austral. J. Exp. Agric.*, 41: 321-325.

Dazzo, F. B., Yanni, Y. G., Rizk, R., deBruijn, F. J., Radermarker, J., Squartini, A., Corich, V., Mateos, P. and Martinez-Molina, E. (2000). Progress in multinational collaborative studies on the beneficial association between Rhizobium leguminosarum bv. trifolii and rice. *In* "The Quest for Nitrogen Fixation in Rice" (J. K. Ladha and P. M. Reddy *et al.*, Eds.), *pp.* 167-189. IRRI, Los Banos, Philippines.

Frankenberger, W. T. Jr. and Arshad, M. (1995)."Phytohormones in Soil: Microbial Production and Function". *Dekker, New York, USA, p.*503.

Fravel, D. R., Rhodes, D. J. and Larkin, R. P. (1999). Production and commercialization of biocontrol products. Pages 365-376 in: Integrated Pest and Disease Management in Greenhouse Crops. R. Albajes, M. L. Gullino, J. C. van Lenteren, and Y. Elad, eds. Kluwer Academic Publishers, Dordrecht.

Garcia, de Salamone, I. E., Hynes, R. K. and Nelson, L. M. (2001). Cytokinin production by plant growth promoting rhizobacteria and selected mutants. *Can. J. Microbiol.*, 47: 404-411.

Garland, J. L. (1996). Patterns of potential C source utilization by rhizosphere communities. *Soil Biol. Biochem.*, 28: 223-230.

Germida, J. J., Siciliano, S. D., de Freitas, J. R. and Seib, A. M. (1998). Diversity of root-associated bacteria associated with field-grown canola (*Brassica napus* L.) and wheat (Triticum aestivum). *FEMS Microbiol. Ecol.*, 26: 43-50.

Giacomodonato, M. N., Pettinari, M. J., Souto, G. I., Mendez, B. S. and Lopez, N. I. (2001). A PCR-based method for the screening of bacterial strains with antifungal activity in suppressive soybean rhizosphere. *World J. Microbiol. Biotechnol.* 17: 51-55.

Glick, B. R. (1995). The enhancement of plant growth by free-living bacteria. *Can. J. Microbiol.* 41:109-117.

Glick, B. R., Penrose, D. M. and Li, J. (1998). A model for lowering plant ethylene concentration by plant growth promoting rhizobacteria. *J. Theor. Biol.* 190: 63-68.

Gutierrez-Manero, F. G., Ramos- Solano, B., Probanza, A., Mehouachi, J., Tadeo, F. R. and Talon, M. (2001).The plant growth promoting rhizobacteria *Bacillus pumillus* and *Bacillus licheniformis* produce high amount of physiologically active gibberellins. *Physiol.Plant.*, 111: 206-211.

Hall, J. A., Peirson, D., Ghos, S. and Glick, B. R. (1996). Root elongation in virus agronomic crops by planth growth-promoting rhizobacterium *Pseudomonas putida* GR-12-2. *Israel J. Plant Sci.*, 44: 37-42.

Hiltner, L. (1904). Uber neuere Erfahrungen und Probleme auf dem Gebiete der Bodenbakteriologie unter besonderden berucksichtigung und Brache. *Arb. Dtsch. Landwirtsch. Gesellschaft.*, 98: 59-78.

Hughes, D. F., Jolley, V. D. and Brown, J. C. (1992). Roles for potassium in the iron-stress response mechanisms of strategy I and strategy II plants. *J. Plant Nutr.*, 15: 1821-1839.

Jetiyanon, J. and Kloepper, J. W. (2002). Mixtures of plant growth-promoting rhizobacteria for induction of systemic resistance against multiple plant diseases. *Biol. Control.* 24: 285-291.

Kloepper, J. W. (1993). Plant growth-promoting rhizobacteria as biological control agents. Pages 255-274 in: Soil Microbial Ecology: Applications in Agricultural and Environmental Management. F. B. Metting, Jr., ed. Marcel Dekker Inc., New York, USA.

Kloepper, J. W. and Schroth, M. N. (1978). Plant growth-promoting rhizobacteria on radishes. Pages 879-882 in: Proc. of the 4th Internat. Conf. on Plant Pathogenic Bacter. Vol. 2, Station de Pathologie Vegetale et Phytobacteriologie, INRA, Angers, France.

Kloepper, J. W., Hume, D. J., Scher, F. M., Singleton, C., Tipping B., Laliobert, E. M., Fraulay, K., Kutchaw, T., Simonson, C., Lifshitz, R., Zaleska, I. and Lee, L. (1987). Plant growth promoting bacteria on canola (rapeseed). *Phytopathology.* 71: 42-46.

Kloepper, J. W., Scroth, M. N. and Miller, T. D. (1980). Effects of rhizosphere colonization by growth promoting rhizobacteriaon potato plant development and yield. *Phytopathology*. 70: 1078-1082.

Ladha, J. K., Kirk, G. J. D., Bennet, J., Reddy, C. K., Reddy, P. M. and Sing, U. (1998). Opportunities for increased nitrogen use efficiency from improved low land rice germ plasm. *Field. Crops. Res.*, 56:41-71.

Lebuhon, M. and Hartmann, A. (1993). Method for determination of indole-3-acetic acid and related compounds of L-tryptophan catabolism in soils. *J.Chromatogr.*, 629: 255-266.

Leong, J. (1986). Siderophores: Their biochemistry, and possible role in the biocontrol of plant pathogens. *Ann. Rev. Phyto.pathol.*, 24: 187-209.

Liu, L., kloepper, J. W. and Tuzun, S. (1995b). Induction of systemic resistance in cucumber by plant growth promoting rhizobacteria: duration of protection and effects of host resistance on protection and root colonization. *Phytopathology*. 85: 1064-1068.

Lubeck, P. S., Hansen, M. and Sorensen, J. (2000). Simultaneous detection of the establishment of seed-inoculated *Pseudomonas fluorescens* strain DR54 and native soil bacteria on sugar beet root surfaces using fluorescence antibody and in situ hybridization techniques. *FEMS Microbiol. Ecol.*, 33: 11-19.

Lugtenberg, B. J. J., Dekkers, L. and Bloemberg, G. V. (2001). Molecular determinants of rhizosphere colonization by Pseudomonas. *Ann. Rev. Phytopathol.,* 38: 461-490.

Lynch, J. M. and Whipps, J. M. (1991). Substrate flow in the rhizosphere. Pages 15-24 in: The rhizosphere and plant growth. D. L. Keister and B. Cregan, eds. Beltsville Sympos. in Agric. Res. 14. Kluwer, Dordrecht, The Netherlands.

Malik, K. A., Bilal, R., Mehnaz, S., Rasul, G., Mirza,. S. and Ali, S. (1997). Association of nitrogen fixing, plant growth promoting rhizobacteria (PGPR)with callar grass and rice. *Plant Soil*. 194: 37-44.

Mansour, F. A., Ildesuguy, H. S. and Hamedo, H. A. (1994).Studies on plant growth regulators and enzyme production by some bacteria. *Qatar Univ.Sci. J.*, 14: 281-288.

Mansouri, H., Petit, A., Oger, P. and Dessaux, Y. (2002). Engineered rhizosphere: the trophic bias generated by opine-producing plants is independent of the opine type, the soil origin, and the plant species. *Appl. Environ. Microbiol.*, 68: 2562-2566.

Mathre, D. E., Cook, R. J. and Callan, N. W. (1999). From discovery to use. Traversing the world of commercializing biocontrol agents for plant disease control. *Plant Dis.*, 83: 972-983.

Maurhofer, M., Reimann, C., Sacherer, S. P., Heebs, S., Haas, D. and Defago, G. (1998). Salicylic acid biosynthetic genes expressed in *Pseudomonas fluorescens* strain P3 improve the induction of systemic resistance in tobacco against necrosis virus. *Phytopathol*. 88: 678-684.

McSpadden Gardener, B. B. and Fravel, D. R. (2002). Biological control of plant pathogens: Research, commercialization, and application in the USA. Online. Plant Health Progress doi:10.1094/PHP-2002-0510-01-RV.

Nandakumar, R. (1998). Induction of systematic resistance in rice with *fluorescens Pseudomona* for the management of sheath blight disease. M.Sc. (Agric.) Thesis TNAU, Coimbatore, India.

Noel, T., Sheng, C., Yost, C. K., Pharis, R. P. and Haynes, M. F. (1996). Rhizobium leguminosarum as a plant growth promoting rhizobacterium: Direct growth promotion of canola and lettuce. *Can. J. Microbiol.*, 42 : 279-283.

Padua, V. L. M., Masuda, H. P., Alves, H. M., Schwarez, K. D., Baldani, V. L. D., Ferreira, P.C.G. and Hemrly, A. S. (2000). Effect of endophytic bacterial indole-acetic acid (IAA) on rice development. Auburn University Web Site, available: http://www.ag.auburn.edu/argentina/pdfmanuscripts/padua.pdfAccessed 07/01/2001.

Pal, K. K., Dey, R., Bhatta, D. M. and Chauhan, S. (1999).Enhancement of groundnut growth and yied by plant growth promoting rhizobacteria. *Inter. Arachis Newsletter.* 19: 51-53.

Pandey P, and Maheshwari, D. K. (2007). Two-species microbial consortium for growth promotion of *Cajanus cajan. Curr. Sci.*, 92:1137-1142.

Patten, C. L. and Glick, B. R. (1996). Bacterial biosynthesis of indole-3-acetic acid. *Can. J. Microbiol.*, 42: 81-84.

Paulitz, T. C. and Belanger, R. B. (2001). Biological control in greenhouse systems. *Ann. Rev. Phytopathol.* 39:103-133.

Persello-Cartieaux, F., Nussaume, L. and Robaglia, C. (2003). Tales from the underground: Molecular plant-rhizobacteria interactions. *Plant Cell Environ.*, 26:189-199.

Pietr, S. J., Koran, B. and Stankiewez, M. (1990). "Influence of rock phosphate-dissolving rhizobacteria on the growth and the phosphorus uptake by oat-preliminary results". "Abstract of the second international workshop on plant growth promoting rhizobacteria. 26.Oct. 14-19, 1990 Interlaken, Switzerland.

Rainey, P. B. (1999). Adaptation of *Pseudomonas fluorescens* to the plant rhizosphere. *Environ. Microbiol.*, 1: 243-257.

Renwick, A., Campbell, R. and Coe, S. (1991). Assessment of in vivo screening systems for potential biocontrol agents of *Gaeumannomyceds graminis*. *Plant pathol.*, 40: 524-532.

Roberts, D. P., Yucel, I. and Larkin, R. P. (1998). Genetic approaches for analysis and manipulation of rhizosphere colonization by bacterial biocontrol agents. Pages 415-431 in: Plant-Microbe interactions and Biological Control. G. J. Boland and L. D. Kuykendall, eds. Books in Soils, Plants, and the Environment, vol. 63. Marcel Dekker Inc., New York, USA.

Rondon, M. R., Goodman, R. M. and Handelsman, J. (1999). The earth's bounty: Assessing and accessing soil microbial diversity. *Trends Biotechnol.* 17:403-409.

Schippers, B. (1988). Biological control of pathogens with rhizobacteria. *Phil. Trans. R. Soc. Lond. B.*, 15: 283-293.

Schisler, D. A., Slinger, P. J. and Bothast, R. J. (1997). Effects of antagonist cell concentration and two-strain mixtures on biological control of *Fusarium* dry root of potatoes. *Phytopathology*. 87: 177-183.

Siddiqui, I. A. and Shaukat, S. S. (2002). Resistance against damping-off fungus *Rhizoctonia solani* systematically induced by the plant-growth-promoting rhizobacteria Pseudomonas aeruginosa (1E-6S(+)) and *P. fluorescens* (CHAO*). J. Phytopathol.*, 150: 500-506.

Sikora, R. A. (1988). Interrelationship between plant health rhizobacteria, plant parasitic nematodes and soil microorganisms. Med. Fac. Landbouww, Rijksuniv, Gent.53/2b,867-878.

Silva, H. S. A., Romeiro, R. S. and Mounteer, A. (2003). Development of a root colonization bioassay for rapid screening of rhizobacteria for potential biocontrol agents. *J. Phytopathol.*, 151: 42-46.

Smalla, K., Wieland, G., Buchner, A., Zock, A., Parzy, J., Kaiser, S., Roskot, N., Heuer, H. and Berg, G. (2001). Bulk and rhizosphere soil bacterial communities studied by denaturing gradient gel electrophoresis: Plant-dependent enrichment and seasonal shifts revealed. *Appl. Environ. Microbiol.*, 67: 4742-4751.

Smith, K. P. and Goodman, R. M. (1999). Host variation for interactions with beneficial plant-associated microbes. *Ann. Rev. Phytopathol.*. 37: 473-491.

Smith, R. S. (1997). New inoculant technology to meet changing legume management. Pages 621-622 in: Biological Nitrogen Fixation for the 21st Century. C. Elmerich, A. Kondorosi, and W. E. Newton, eds. Kluwer, Dordrecht, The Netherlands.

Sorensen, J., Jensen, L. E. and Nybroe, O. (2001). Soil and rhizosphere as habitats for Pseudomonas inoculants: New knowledge on distribution, activity and physiological state derived from micro-scale and single-cell studies. *Plant Soil.* 232: 97-108.

Steddom, K., Menge, J. A., Crowley, D. and Borneman, J. (2002). Effect of repetitive applications of the biocontrol bacterium *Pseudomonas putida* 06909-rif/nal on citrus soil microbial communities. *Phytopathol.* 92: 857-862.

Subba Rao, N. S. (1993). In "Biofertilizer in Agriculture and Forestry",3rd edn Oxford and Ibh Pub, New Delhi, India.

Suslow, T. V. (1982). In "Role of root colonizing bacteria in plant growth". "*Phytopathogenic Procaeryotes*" (M. S. Mount and G. H. Lacy, Eds.), vol. 1, pp. 187-223. Academic press, New York, USA.

Thomas-Bauzon, D., Weinhard, P., Villecourt, P. and Balandreau, J. (1982). The spermosphere model. I. Its use in growing, counting and isolating N2-fixing bacteria from the rhizosphere of rice. *Can. J. Microbiol.*, 28: 922-928.

Thomashow, L. S. and Weller, D. M. (1995). Concept in the use of introduce bacteria for biological disease control. In "plant Microbes Interaction "(G. Stacey and N. Keen, Eds.),vol.1, pp.187-235.Chapman and Hall, New York, USA.

Toro, M., Azcon, R. and Barea, J. M. (1997).Improvement of arbuscular mycorhyza development by inoculation of with phosphate-solubilizing rhizobacteria to improve rock phosphate bioacailability (32-P) and nutrient cycling. *Appl. Environ. Microbiol.* 63: 4408-4412.

Vidhayasekaran, P. and Muthamilan, M. (1999). Evaluation of powder formulation of *pseudomonas fluorescens* Pf1 for control of rice sheath blight. *Biocontrol. Sci. Technol.*, 9: 67-74.

Viswanathan, R. and Samiyappan, R. (1999). Induction of systemic resistance by plant growth promoting rhizobacteria against red rot disease caused by *Colletotricum falcutum* went in sugarcane. Proc. *Sugar Technol Assoc.of India.* 61: 24-39.

Wei, G., Kloepper, J. W. and Tuzun, S. (1991). Induction of systemic resistance of cucumber to *Colletotricum orbiculare* by select strains of plant growth promoting rhizobacteria. *Phytopathology*. 80: 1508-1512.

Weller, D. M., Raaijmakers, J. M., McSpadden Gardener, B. B. and Thomashow, L. S. (2002). Microbial populations responsible for specific soil suppressiveness to plant pathogens. *Ann. Rev. Phytopathol.* 40: 309-348.

Yardin, M. R., Kennedy, I. R. and Thies, J. E. (2000). Development of high quality carrier materials for field delivery of key microorganisms used as bio-fertilisers and bio-pesticides. *Radiation Phys. Chem.*, 57: 565-568.

Young, S., Pharis, R. P., Reid, D., Reddy, M. S., Lifshitz, R. and Brown, G. (1991). Pgpr: Is their a relationship between plant growth regulators and the stimulation of plant growth or biological control activity?. *In* "second international workshop on plant growth promoting rhizobacteria" (C. Keel, B. Koller, and G. De' faso, Eds.), pp. 102-103. Bull. Srop. Interlaken, Switzerland, Oct. 14-19, 1990.

6

Sustainable Water Management

Pabitra Adhikary, Partha Sarathi Patra, Shyamal Kheroar Amrit Tamang **and** ***Subhra Shil***

Introduction

Water is a vital component of agricultural production. It is essential to maximise both yield and quality. Water has to be applied in the right amounts at the right time in order to achieve the right crop result. At the same time, the application of water should avoid waste of a valuable resource and be in sympathy with the environment as a whole. Understanding, measuring and assessing how water flows around the farm, and recognising how farming practices affect flows, will help farmers to manage water efficiently and reduce pollution risks. Land management and advanced technologies can help maximise the "crop per drop" that farmers produce worldwide. Water used in irrigation accounts for 90% of agricultural water demand. Improving irrigation systems is therefore crucial to making contributions to water savings in agriculture. In any agricultural production system the promotion of good agricultural practices and product stewardship is fundamental for sustainable agriculture. Industry is part of various multi stakeholder partnership initiatives that support efforts to protect water quality. These include reducing soil erosion, avoiding pesticide run-off and maintaining wildlife habitats as part of a holistic farm or land management approach.

Irrigation

Irrigation is the dominant form of water use, measures that improve the efficiency of water application and minimize water loss are most effective in conserving water and energy in regions facing limited supply. Water-use efficiency is a complex subject, however, with many different definitions (Molden *et al.*, 1998). Overall efficiency affecting performance is conveyance from the source to the farm, uniformity of application to the crop, and drainage losses following application. Globally, many systems perform poorly, resulting in only 30 to 50 percent of the water withdrawn actually being taken up by the crop (Faures *et al.*, 2007). However, water lost in that way from one farm might have a beneficial use elsewhere, so it is important to look at basin wide efficiency when estimating true water savings. In addition to managing water consumption, precipitation or water can be reused to improve the long-term sustainability of water use in agriculture.

Irrigation scheduling

Quantitative irrigation scheduling methods rely on one of two approaches: soil or crop monitoring or a combination of both; or soil water balance computations. For the monitoring methods, the soil water content or matric potential is measured at several places in the field to decide when to irrigate. Methods based on plant measurements generally involve monitoring leaf water potential or canopy temperature. Soil water balance calculations require estimates of soil storage capacity, rooting depth, allowable depletion, and crop evapotranspiration to develop an irrigation schedule (Martin *et al.*, 1990).

Impact and adaption of irrigation scheduling

Irrigation scheduling has been shown to significantly reduce water usage compared to traditional methods in a number of studies. For example, Mohammad and Al-Amoud (1993) achieved both higher wheat yields and a 25 percent decrease in water use when central pivot sprinkler irrigations were scheduled based on a calculation of evapotranspiration demand in Saudi Arabia. In New Mexico's Mesilla Valley, pecan farmers have been slow to adopt new soil-based or climate-based irrigation scheduling technologies because these technologies require high in-season labor input. Kallestad *et al.* (2008) developed a simple, practical irrigation scheduling tool specifically for flood-irrigated pecan production using 14 years of archived climate data and model-simulated consumptive water use. Eventually, the hope is that these farmers will convert to using Internet databases as their main resource for climate and irrigation information (Kallestad *et al.*, 2008).

Gravity Systems

Improving the uniformity of irrigation water application is the most effective means by which water can save. Non uniform irrigations are wasteful, because water has to be added at rates greatly exceeding those needed by the parts of the field receiving the most water to avoid yield decreases on the parts that are receiving less water. Gravity flow systems distribute water laterally across the entire field or into furrows. Various water management measures have been developed to reduce water losses under gravity-flow systems. Improved gravity systems generally involve on farm water conveyance upgrades that increase uniformity of applied water and reduce percolation losses and field runoff.

Improved water management practices for gravity irrigation remains an area of significant growth potential, with many available technology or management improvements such as alternate row irrigation, furrow modification, tailwater reuse, or soil amendments not in widespread use (Schaible and Aillery, 2006).

Sprinkler Irrigation

Sprinkler irrigation is a planned system in which water is applied by means of perforated pipes or nozzles operated under pressure to form a spray pattern (USGS, 2009b). Like gravity systems, pressurized systems improve irrigation uniformity. Pressurized systems include a variety of sprinkler and low-flow irrigation techniques to distribute water across a field. The applicators are generally attached to moving center pivot or linear advance lines to allow continuous advance over large areas.

Impact of sprinkler irrigation

Field application efficiencies for properly designed and operated sprinkler systems range from 50 to 95 percent, with most systems achieving 75 to 85 percent (USDA-NRCS,1997). The low-pressure center-pivot and linear-move systems combine high application efficiencies with reduced energy and labor requirements.

Trickle or Drip Irrigation

Trickle or drip irrigation applies water directly to the root zone of plants using applicators operated under low pressure. The applicators could be placed on or below the surface of the ground (USGS, 2009a). Shifting to trickle or drip irrigation has been the greatest strategic improvement in water-use efficiency and energy savings. Most orchards and vineyards are converting to these systems, and nearly all newly planted ones are using precision water application, as are a broad range of annual horticultural crops. The application tubes are placed in close proximity to the tree or vine of crop plants, and water is applied as needed, monitored by a host of newly engineered moisture and plant stress-sensing devices (Locascio, 2005).

Impact of trickle or drip irrigation

Energy savings for systems changes are region-and crop-specific. In the central coast region of California, the better vegetable growers are saving upwards of 25% in water pumping, fertilizer, and herbicide costs by using subsurface drip irrigation technologies (California Energy Commission, 2008).

Water Reuse

A variety of sources of water of marginal quality can potentially be used to augment the supply of water for agriculture. Domestic waste water, if properly reclaimed, can serve a variety of uses beneficial to agriculture, either as a source of irrigation water. At the present time, waste water provides only a small portion of the national water resource for agriculture. Saline water has some potential for augmenting water use in agriculture, primarily through reuse of drainage water on more salt-tolerant species or using cyclic rotations of good quality and saline water to grow a range of sensitive and more tolerant crops (Shennan *et al.*, 1995). The greatest concern with reuse is the biological and chemical quality of the reclaimed water. Removal of pathogens and anthropogenic chemicals is an important requirement for any wastewater treatment system. The reclaimed water would have to be monitored carefully to reduce chances of any harmful pollutants entering into an agricultural production system (Falconer *et al.*, 2006).

Protect and enhance water quality

Surface and ground water must be protected from direct and indirect pollution. In particular, the following must be considered:

(a) Siltation through sediments
(b) Pollution from nutrients
(c) Pollution from agrochemicals and other chemicals
(d) Pollution from fuels, lubricants or solvents (for example from workshops)
(e) Pollution from contaminated run-off.
(f) Pollution from livestock
(g) Pollution from human sewage
(h) Pollution from waste water

Protecting from contaminated run-off waters

Irrigation and rainfall can both cause soil erosion, and transfer agrochemicals and heavy metals to nearby water courses. Risk is highest where there are high intensity rainfall events, or where irrigation is poorly managed or inappropriate techniques used. Management practices include ensuring good soil structure to aid infiltration; the use of buffer strips, contour tillage, dividing long slopes, siltation pits/micro catchments to avoid or catch run-off; mulch and crop canopy cover and irrigation management. The risks of agrochemicals in run-off water are also addressed in the Agrochemicals and Fuels.

Protecting from livestock

When livestock have unrestricted access to watercourses, water can be contaminated by soil, nutrients, bacteria and other microorganisms. Various measures can reduce the risk of contamination. The most stringent way of protecting water courses is to fence off all watercourses on farm, and provide water for animals from a 'non-stream' source. Complete exclusion of animals may not be necessary in some cases, and other measures may be put in place to encourage animals not to spend too much time by the edge of the water. These include placing salt, minerals or supplemental feed away from the riparian area to attract livestock away, and providing shade away from the watercourse to stop them congregating in that area.

Protecting from human sewage

Pollution of waters from human sewage arises from seepage or drainage from toilet facilities and from areas where people do not have access to toilet facilities. Toilets in workplaces (processing factories, pack houses, farmyards, clinics etc.) and in farmhouses and farm worker accommodation should discharge into holding tanks/cesspits, septic tanks, composting facilities, geological structures or sewers. These structures must be located and managed in such a way that nutrient-and pathogen-rich sewage does not pollute ground or surface waters. Discharge from toilets into surface waters (*e.g.*, by placing them above, draining into, or very close to streams) is clearly unacceptable. Cesspits and holding tanks must not leak and must be emptied by a contractor licensed and competent to manage the wastes safely. Such facilities are relatively costly to maintain well and are usually inappropriate for farm worker accommodation or processing facilities. Septic tanks are very common in many parts of the world. The location of septic tanks in relation to local soil characteristics and drainage systems is therefore crucial for avoiding water pollution. For large-scale farm accommodation or processing facilities, professional advice on the location and management of septic tanks should be sought.

Protecting from Waste Water

Waste water can arise from irrigation and rainfall, from yard and workshop washing and from basic processing operations that may take place on-farm. Risks of pollution arising from discharge of waste water must be assessed and managed accordingly on a case-by-case basis. Waste from milk parlours, for example, has a high pollution potential and generally requires full treatment in local sewage treatment plants; Water used for vegetable washing may be recycled and/or returned to the land.

Dairy Guidance

There are a number of sources from which water pollution might arise, and a number of ways in which the risk of such pollution from livestock can be managed:

- Before any new sheds are constructed, ensure that they are located at a safe distance from ground water (wells) or surface water sources – this is often dictated by legislation.
- Manage animal holding areas and pastures to avoid effluent run-off, *e.g.,* by storing and spreading farm manures in accordance to local requirements. See manure management guidance on page 4 of this guide, and the advice below.
- Direct wastewater to manure storage areas.

Livestock yard management

Livestock yards (barnyards, holding areas, woodchip corrals, stand-off pads and feedlots) are concentrated areas of livestock, and hence their wastes, and are therefore vital to protection of water quality. These yards, especially when on permeable soils or near on-farm water sources, can cause nitrate and bacterial contamination in ground or surface water. To minimise the possibility of contaminants leaching to groundwater or running off to surface water, such yards should be located on concrete or fine-to-medium textured soils over 100 feet/30 metres from water sources such as wells, surface water, adjacent property, drainage ditches or other areas that could result in the run-off reaching water sources. Having a roof over the yard or otherwise diverting water from the yard is the best way to prevent run-off. Other practices, such as keeping the yard clean, diverting run-off to manure storage areas or collecting and reusing run-off (*e.g.,* as nutrients on fields), can minimise potential pollution to water sources.

Milk-house washings

Water used to clean the milking parlour and milking equipment contains high levels of organic matter, nutrients, chemicals and micro-organisms, which can contaminate water with ammonia, nitrate, phosphorus, detergents and disease-causing organisms if not disposed of properly. Milking parlour wastewater is made nutrient-rich by virtue of having high amounts of milk residues or being washed down the drain with manure and feed. This nutrient-rich water can lead to pollution if it is untreated before it reaches water supplies.

Irrigation water quality

Irrigation water quality must be monitored and managed where necessary to avoid crop damage, crop or soil contamination or soil damage through contamination or erosion. Knowledge of irrigation water quality is critical to understanding management for long-term productivity. In some regions, water quality can influence productivity more than soil fertility, crop variety, weed control and other factors. The main water quality related problems are:

Salinity

Salinity is the presence of soluble salts in or on soils, or in water applied to soils. Yield reductions occur when the salts accumulate in the root zone to such an extent that the crop is no longer able to extract sufficient water from the salty soil solution, resulting in a water stress for a significant period of time. In irrigated agriculture, salinity problems are exacerbated by shallow water tables. The measure to monitor with respect to salinity is the EC (Electrical conductivity, deciSiemens per metre (dS/m)) or TDS (total dissolved solids, reported in milligrams per litre (mg/l)).

Sodicity

Sodicity, a high proportion of sodium in the soil or water relative to other cations, degrades soil properties by making the soil more dispersible and erodible, restricting water entry and reducing the ability of the soil to conduct water. This reduces water availability and hence yield, but these factors also limit leaching so that salt accumulates over long periods, giving rise to saline subsoils. Furthermore, a soil with increased dispersibility becomes more susceptible to erosion by water and wind. The measure to monitor with respect to sodicity is the SAR (sodium adsorption ratio).

Toxic ions

Toxicity problems can occur if certain constituents (ions) in the soil or water are taken up by the plant and accumulate to concentrations high enough to cause crop damage or reduced yields. The degree of damage depends on the uptake and the crop sensitivity, and often occurs at relatively low ion concentrations for sensitive crops. It is usually first evidenced by marginal leaf burn and chlorosis, but if accumulation is great enough, reduced yields result. The ions of importance are usually chloride, sodium and boron. Boron sometimes results from perborate, a bleaching agent, used in some household products. Units are commonly milli Equivalents/litre (me/l).

Other contaminants

Other irrigation water quality problems may arise from anthropogenic pollutants deriving from agriculture, industry or domestic use of chelating agents, heavy metals, biocides or CPPs, or from human and animal pathogens (including enteric bacteria).

The integrated water resources management approach

Agriculture is the single largest user of freshwater. The challenge of producing enough food becomes especially delicate, as it must be met mainly on currently cultivated land since expansion and conventional intensification of agriculture comes at major environmental costs (local to global scale factors: erosion, biodiversity loss, salinization, water pollution and eutrophication, water scarcity, greenhouse gas emissions). Agricultural water management has played a key role in increasing global food supply and contributed substantially to poverty alleviation (Namara *et al.*, 2010).Furthermore, significant yield gaps exist across various farming systems, indicating a substantial scope for yield gains through mitigation of nutrient and water deficiencies.

Increasing production on existing agricultural land by managing available resources more efficiently, placing less pressure on the environment and sustaining future capacities, *i.e.,* sustainable intensification, is thus seen as an important part of a solution and high on the global policy agenda. The renewed sustainable development goals now stipulate sustainable agriculture as an agreed goal among all nations, but there is little quantitative evidence of how to achieve it. While most global strategies focus on improving soil fertility, urge an international high level consideration of integrated water management.

Water resource development

Water is a finite resource, the availability of which is declining with each passing day. The total amount of water on earth is about 1400 million cubic kilometres. The amount of fresh water is very small proportion (3%) of this enormous quantity which about 68.7% lies frozen in polar region and another 30.1% is present as ground water. The rest is available in lakes, rivers, atmosphere, soil and vegetations. Globally, current withdrawals of about 4,500 km^3 exceeded the availability of about 4200 km^3. By 2030 the demand is expected to increase to about 6900 km^3. Thus by 2030 the global of 40% is forecast. But in reality we may reach this threshold much earlier considering the wasteful ways of water usage. Low water rates, under assessment of irrigated area and water rate demand, and poor collection rate continue to deprive the irrigation sector from realizing potential revenue, critical for system's financial sustainability. Sustained efforts are needed for rehabilitating the irrigation infrastructure and initiating institutional reforms in water sector. One of the major strategies for achieving such massive increase irrigation potential could be inter basin transfer of water. Only then irrigation management transfer will become effective with system wide impacts to provide water security needed for sustainable food security.

Strengthening of water resources

Rainwater harvesting

Rainwater harvesting is the process of collecting, concentrating and improving the productive use of rainwater, and reducing unproductive depletion such as runoff, evaporation and seepage. Harvested rainwater may be used for providing pre-sowing irrigation, supplemental/life saving irrigation, recharging ground water and/or for domestic purposes. There is generally enough rainfall in most of the moist semi-arid and dry subhumid zones to meet the crop water needs (Agarwal, 2000).

In-situ water harvesting

Utilization of rainwater as and when/where it falls on the land, *i.e.*, the places of availability and utilization are the same. Options for *in-situ* rainwater harvesting include:

(a) Land leveling and field bunding with thick, strong and raised bunds impound rainwater *in-situ* and helps rainwater distribution in the entire field evenly and uniformly.

(b) Land shaping should be followed to concentrate runoff at the desired location, such as tree basins in orchards land shaping for flowing water etc.

(c) Development of Micro-catchment water harvesting by converting flat plots into ridge-furrow system, where crops are planted in furrows so that rainwater concentrates in planting area and root zone. It improves moisture conservation and enhances crop yields.

(d) Raised-sunken bed technology for inter-plot rainwater harvesting. High water requiring crops (*e.g.,* rice) are cultivated in sunken beds and low water requiring crops (*e.g.,* maize, soybean, vegetables etc.) on raised beds. This technology has been found to be useful in high rainfall areas also (Sharma 2003).

(e) *Terracing:* On sloping landscapes, and cultivating crops requiring well drained conditions (maize, soybeans, vegetables etc.) on upper terraces and high water demanding crops (rice, vegetables) on lower terraces where run-off concentrates. Divert water from upper terraces to lower terraces.

Ex-situ water harvesting

Place of availability of water resource is away from the place of its utilization, for which a suitable storage structure and water conveyance system are necessary. Construction of a suitable reservoir to collect and store excess rainfall/runoff, stream flow, spring water, roof-top water etc. Design of a water storage structure at micro watershed level depends on the following factors: amount of rainfall, characteristics of the catchment area that generates runoff, amount of runoff generated, command area to be irrigated with stored water, crop water requirement etc.

Watershed management

Watershed management is an effective technique of water harvesting. It aims to establish a workable and efficient framework for the integrated use, regulation and development of land and water resources in a watershed for socio-economic growth. It is an approach that aims at improving ecology of the region as well for resilience of climate change impacts, more so on rainfall (Sharma *et al.*, 2005). Local communities play a central role in the planning, implementation and funding of activities within participatory watershed development programmes.

Soil-moisture conservation

It is a technique to increase soil moisture storage, decrease evaporative losses, and enhance duration of water availability to crops. Non-productive soil moisture depletion can be alleviated through organic or plastic film mulching, and different conservation tillage systems. These techniques can improve grain yield remarkably through conserving soil moisture for additional plant transpiration, suppressing weeds, and improving cold tolerance (Liu *et al.*, 2014). Organic crop residues covering 50% of the soil surface can reduce soil evaporation by approx 25%, plastic mulching can reduce soil evaporation by 50%–90% (Bu *et al.*, 2013). Several materials such as organic or inorganic materials of natural (field crop residues, waste plant biomass, wood chips, saw dust, coco-coir etc.) or of synthetic origin (plastic sheets, soil conditioners), soil (dust mulch) or pebbles (stone mulch) can be used as mulch. The choice of a mulch material, however, depends on its cost, availability, ease of handling and efficacy etc. Mulching of chillies with paddy straw increased yield (17.2 t ha^{-1}) by 15% and saved 25% irrigation water, as compared with no mulching (14.8%) (Sekhon *et al.*, 2008).

The *alley cropping* is another system of raising mulch material. It is essentially an agro-forestry system in which food crops are grown in alleys formed by hedge-rows of trees and shrubs. Water holding capacity (WHC) and available water capacity (AWC) of soils are directly linked to soil organic carbon (SOC). Studies have shown that 1% increase in soil organic manure (OM) increases field capacity by about 2.2%, permanent wilting point by about

1%, and available water capacity by about 1.5% (Sharma and De Datta, 1994). Tillage practices have direct effect on soil water conservation. Conventional tillage (repeated primary and secondary tillage operations) invariably enhances SOM decomposition by breaking soil aggregates and exposing the soil to weathering agents, thus, weakening soil structure, and making soil loose to be washed away with runoff water. Conservation tillage (a non-inversion tillage that retains at least 30% of the soil surface covered by residues after planting, and is less intensive than conventional tillage) has been found to be a potential alternative to conventional tillage to reduce soil erosion by water or wind; improve and sustain soil health; and lower the cost of cultivation. Conservation tillage is environment-friendly. The term "conservation tillage" is an umbrella classification for several crop production systems, including no-till, reduced-till, minimum-till, direct seeding and other names. If tillage operations are inevitable, they should be performed across the slope so as to retain water at soil surface and cut down the flow velocity of run-off water.

Groundwater management

Surface water is inadequate to meet agricultural demand of the country and we have to depend on ground water. Increased dependence on groundwater for irrigation and other uses have over-exploited this natural water resource. In order to sustain groundwater resource it is essential to utilize groundwater judiciously and efforts are directed for recharging the ground aquifers. The exploitation of groundwater resources should be regulated so as not to exceed the recharging possibilities, as well as to ensure social equity. The detrimental environmental consequences of over-exploitation of groundwater need to be effectively prevented.

Structures generally used for recharging ground water are:

- ***Pits:*** Recharge pits are constructed for recharging the shallow aquifers. These are constructed 1 to 2 m wide and to 3 m deep which are back-filled with boulders, gravels and coarse sand.
- ***Trenches:*** These are constructed when the permeable stream is available at shallow depth. Trench may be 0.5 to 1 m wide, 1 to 1.5 m deep and 10 to 20 m long depending up availability of water. These are back-filled with filter materials.
- ***Dug wells:*** Existing dug wells may be utilized as recharge structure and water should pass through filter media before putting into dug well.
- ***Hand pumps:*** The existing hand pumps may be used for recharging the shallow/deep aquifers, if the availability of water is limited. Water should pass through filter media before diverting it into hand pumps.

- **Recharge wells:** Recharge wells of 100 to 300 mm diameter are generally constructed for recharging the deeper aquifers and water is passed through filter media to avoid choking of recharge wells.
- ***Recharge shafts:*** For recharging the shallow aquifers which are located below clayey surface, recharge shafts of 0.5 to 3 m diameter and 10 to 15 m depths are constructed and back-filled with boulders, gravels and coarse sand.
- ***Lateral shafts with bore wells:*** For recharging the upper as well as deeper aquifers lateral shafts of 1.5 to 2 m width and 10 to 30 m length depending upon availability of water with one or two bore wells are constructed. The lateral shafts are back-filled with boulders, gravels and coarse sand.
- ***Spreading techniques:*** When permeable strata start from top then this technique is used. Water in the streams/*Nalas* should be spread by making check dams, *nala* bunds, cement plugs, gabion structures or constructing a percolation pond.

Efficient utilization of conserved water

Farmers must follow improved agronomic practices (selection of crops and crop cultivars, mixed cropping, crop rotation, use of manures and fertilizers, irrigation methodology, crop protection, weed control etc.) to cultivate crops in order to sustain soil fertility, achieve maximum crop productivity and enhance water use efficiency. High yielding and less water-requiring crops (short duration, low transpiration, deep and well branched root system) are needed.

Several studies suggest that better management of rain water; soil moisture and supplemental irrigation hold the key to high water productivity. It is estimated that a 35% increase in water productivity could reduce additional crop water consumption from 80 to 20% (Molden 2007). Sharma *et al.* (2010) estimated that with one time supplemental irrigation of about 100 mm (using harvested run-off), along with adoption of improved agronomic practices, the overall productivity level could be enhanced from the existing aggregated level of 1.20 t ha^{-1} to a level of 2.65 t ha^{-1}. The efficiency of irrigation water can be increased by using water-saving technologies, such as optimum irrigation scheduling and proper sowing method. Scheduling irrigation to rice on the basis of soil matric potential can save 25% irrigation water (Kukal *et al.*, 2005) without any reduction in yield of rice. In cotton, broad-bed with spacing of 135 cm and planting cotton in furrows in paired rows increased cotton yield by 44% and saved 40% irrigation water, as compared with row spacing of 76.5 cm in flat/bed system. Planting of potato on both sides of narrow beds increased tuber yield (24 t ha^{-1}) by 25% and saved 20% irrigation water, as compared with the ridge-planting method (20 t ha^{-1}) (Thind *et al.*, 2007).

Multiple use of water provides challenging opportunities for increasing water productivity at various scales ranging from farm to basin scales, both under irrigated and rainfed conditions. Multiple use of stored water integrates crops, livestock, fisheries and horticulture for enhancing water productivity to improve food security, health and environment and ultimately the improved livelihood of the resource-poor farmers. The protected cultivation coupled with rainwater harvesting also is a viable and profitable technique of high-value crop production in small holdings.

Drainage water management systems

Drainage water management (DWM), often referred to as controlled drainage or water table management, in which the outlet from a conventional drainage system is intercepted by a water control structure. The drainage outlet's elevation is then artificially set at levels ranging from the soil surface to the bottom of the drains. Drainage water management systems are installed primarily to regulate drainage, thereby improve productivity, but they can be designed and managed to achieve additional environmental goals simultaneously (Evans *et al.*, 1996). At the field scale, the drainage outlet can be set at or close to the soil surface between growing seasons to recharge the water table. The recharge temporarily retains soil water containing nitrate nitrogen in the soil profile where it might be subjected to attenuating and nitrate transformation processes, depending on soil temperature and microbiological activity. In addition, it is possible to raise the outlet elevation after planting to help increase water availability to then-shallow plant roots, and to raise or lower it throughout the growing season in response to precipitation conditions. In some soils, water may even be added during very dry periods to reduce crop loss from drought, a related practice called subirrigation. Although there have been reported instances where subsurface DWM has resulted in reduced nitrate concentrations in the drainage outflow, the general consensus is that the dominant process leading to reductions in nitrate loads is a reduction in drain outflow. With less water leaving the field through the tile drain, significantly less nitrate flows out of the drain, even with no change in nitrate concentration (Cooke *et al.*, 2008).

Reference

Agarwal, A. (2000) Try capturing the rain: *Briefing Paper for Members of Parliament and State Legislatures*. Center for Science and Environment, New Delhi.

Bu L D, Liu J L, Zhu L, Luo S S, Chen X P, Li S Q, Lee Hill R and Zhao Y. 2013 The effects of mulching on maize growth, yield and water use in a semi-arid region. *Agricultural Water Management* (123): 71–8.

California Energy Commission. 2008. Energy in agriculture program on-farm irrigation. Available at http://www. energy.ca.gov/process/agriculture/farm_irrigation.html. Accessed on May 8, 2009.

Cooke, R.A., G.R. Sands, and L.C. Brown. 2008. Drainage water management: a practice for reducing nitrate loads from subsurface drainage systems. In Final Report: Gulf Hypoxia and Local Water Quality Concerns Workshop. St. Joseph, Mich.: American Society of Agricultural and Biological Engineers.

Evans, R., J.W. Gilliam, and W. Skaggs. 1996. Controlled Drainage Management Guidelines for Improving Drainage Water Quality. Raleigh: North Carolina Cooperative Extension Service.

Falconer, I.R., H.F. Chapman, M.R. Moore, and G. Ranmuthugala. 2006. Endocrine-disrupting compounds: a review of their challenge to sustainable and safe water supply and water reuse. *Environmental Toxicology* 21(2):181–191.

Faures, J.M., M. Svendsen, and H. Turral. 2007. Reinventing irrigation. In Water for Food, Water for Life: A Comprehensive Assessment of Water Management in Agriculture, D. Molden, ed. London and Colombo: Earthscan and IWMI.

Kallestad, J.C., J.G. Mexal, T.W. Sammis, and R. Heerema. 2008. Development of a simple irrigation scheduling calendar for Mesilla Valley pecan growers. *HortTechnology* 18(4):714–725.

Kukal, S.S., Hira, G.S. and Sidhu, A.S. (2005) Soil matric potential-based irrigation scheduling to rice (*Oryza sativa*). *Irrigation Science* 23, 153-159.

Liu E K,HeW Q and Yan C R. 2014 'White revolution' to 'white pollution' - agricultural plastic film mulch in China *Environ. Res. Lett.* 9 091001.

Locascio, S.J. 2005. Management of irrigation for vegetables: past, present, and future. *HortTechnology* 15(3): 482–485.

Martin, D.L., E.C. Stegman, and E. Fereres. 1990. Irrigation scheduling principles. In Management of Farm Irrigation Systems. St. Joseph, Mich.: American Society of Agricultural Engineers.

Mohammad, F.S., and A.I. Al-Amoud. 1993. Water conservation through irrigation scheduling under arid climatic conditions. *Agricultural Water Management* 24:251–264.

Molden D (Editor) (2007) *Water for Food, Water for Life: A Comprehensive Assessment of Water Management in Agriculture*. Earthscan, London and International Water Management Institute (IWMI), Colombo, Sri Lanka.

Molden, D.J., R. Sakthivadivel, C.J. Perry, C.J. de Fraiture, and W.H. Kloezen. 1998. Indicators for Comparing Performance of Irrigated Agricultural Systems. Colombo, Sri Lanka: International Irrigation Management Institute.

Namara, R., Hanjra, A., Castillo, G., Ravnborg, H., Smith, L., Van Koppen, B., 2010. Agricultural water management and poverty linkages. *Agricultural Water Management* 97(4): 520–527.

Schaible, G., and M. Aillery. 2006. Irrigation water management. In Agricultural Resources and Environmental Indicators, 2006 Edition/EIB-16. Washington, D.C.: U.S. Department of Agriculture Economic Research Service.

Sekhon, N.K., Singh, C.B., Sidhu, A.S., Thind, S.S., Hira, G.S. and Khurana, D.S. (2008) Effect of straw mulching, irrigation and fertilizer N levels on soil hydrothermal regime, water use and yield of hybrid chilli. *Archives of Agronomy and Soil Science* 54, 163-174.

Sharm, P.K. and De Datta, S.K. (1994) Rainwater utilization efficiency in rainfed lowland rice. *Advances in Agronomy* 52, 85-120.

Sharma, B.R., Rao, K.V., Vittal, K.P.R., Amarasinghe, U.A. and Singh, A. (2010) Estimating the potential of rainfed agriculture in India. Prospects of water productivity improvements. *Agricultural Water Management* 97, 23-30.

Sharma, B.R., Samra, J.S., Scott, C.A. and Wani, S.P. (Editors) (2005) *Watershed Management Challenges: Improved Productivity, Resources and Livelihoods.* International Water Management Institute, Colombo, Sri Lanka, 334 pp.

Sharma, P.K. (2003) Raised-sunken bed system for increasing productivity of rice-based cropping system in high rainfall areas of Himachal Pradesh. *Journal of the Indian Society of Soil Science* 51, 10-16.

Shennan, C., S.R. Grattan, D.M. May, C.J. Hillhouse, D.P. Schachtman, M. Wander, B. Roberts, R.G. Burau, C. McNeish, L. Zelinski, and S. Tafoya. 1995. Feasibility of cyclic reuse of saline drainage in a tomato-cotton rotation. *Journal Environmental Quality* 24:476–486.

Thind, S.S., Sidhu, A.S., Sekhon, N.K., Singh, C.B., Tarinder-Kaur and Amanpreet-Kaur. (2007) Effect of different irrigation geometry on water and nutrient use efficiency of potato-sunflower sequence. *Proceedings of Environmental and Livelihood Security through Resource Management in Northern India* (ELSTRM 2007).

USDA-NRCS (U.S. Department of Agriculture National Resource Conservation Service). 1997. National Irrigation Guide. East Lansing, Mich.: Author.

USGS (U.S. Geological Survey). 2009a. National Water-Quality Assessment Program. Available at http://water. usgs.gov/nawqa/. Accessed on August 6, 2009.

USGS (U.S. Geological Survey). 2009b. Some irrigation methods. Available at http://ga.water.usgs.gov/edu/irquicklook.html. Accessed on July 17, 2009.

7

Soil Management Vis-a-Vis Production Sustainability Blooming Animal and Human Health

Amrit Tamang, Partha Sarathi Patra, Sagar Tamang, Hriday Komal Tarafdar, Tarik Mitran, Nirmalendu Basak* and *Ashok Choudhury

Is it possible to feed a world population projected to reach 8.5 billion by 2030, 9.7 billion by 2050 and exceed 11 billion in 2100? India expected to surpass China as the most populous country in the world within seven years (UN report July 2015). Besides that there is an immense decline in yield and quality over the past three decades may be at the point of diminishing returns (Bouis 1993; Cassman *et al.* 1995; Flinn and De Datta 1984). Furthermore, the soil is partially under tremendous pressure to sustain yield and production. The excessive and exhaustive agriculture practices have not only declined the soil quality but it's also reported for different human and animal health issues. Thus, keeping pace with population growth and health safety appropriate soil management practices must be addressed. Now burgeoning population load, declining yield, soil health diminution, and effect on human and animal health arises the opportunity for sustainable soil management to cater the problem at the right time. Figure 1 clearly indicates the tremendous load on soil and subsequent results. Burgeoning increase in population and a load of food production intensification, sustainable soil management will become increasingly important in future. Soil being the powerhouse of the mineral nutrients also retains most living organisms within it. The appropriate soil management practices ensure the critical level of mineral elements which finally come into the human body.

Management of soils is important, both directly and indirectly, for crop productivity, environmental sustainability and human health. For maintaining the self-sufficiency in food grain, the sustainable soil management will be the more challenge, through proper nutrient management practices and appropriate soil conservation techniques. The most essential is awareness at farmers level to check further degradation through erosion or chemical contamination or by other means to produce safe and nutritious food. The aim of the author is to sensitize the current scenario to assure food security whilst preserving natural resources.

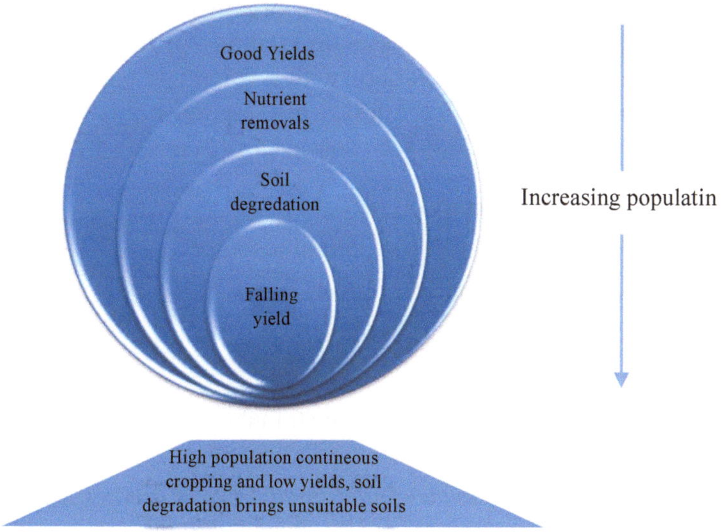

Figure 1: Low population, long fellows, short cropping periods

How soil behaves

Soils are a fundamental base for farming. It supplies essential nutrient and water to the crops, provides physical supports, helps control pests and protects the quality of drinking water, air, and wildlife lives. A prime target of the soil management is to protect soil and restore its fertility, so you can have profitable agriculture with safe environment for future. To give shelf life to the soil with profitable and environmental safe farming is actual sustainable soil management. The special issues are to know the fact file of human and animal health because of poor soil health and in management techniques.

Countries with limited land and water resources, particularly developing nation like India cannot easily finance increased food imports, will face serious problems. Land degradations is adversely affecting different soil resources and declining yield and which is alarming for food security in coming years.

Some researcher has reported that the adoption of agroforestry systems can check soil losses, and may also increase soil organic matter (SOM), improve soil physical properties, and increase water holding capacity in light textured soil. Moreover, a technique like zero or minimum tillage, intercropping, mulching and crop rotation can be used to increase SOM in the soil. The single largest threat to productivity and productive land is losses of top soil either through mineral imbalance or erosion.

Soil management is central part of Sustainable Agriculture (SA)

Sustainable agriculture is farming in sustainable manners maintaining agricultural ecosystem, an ecology of the organism, restoring soil health with optimum production. It has been defined as "an integrated system of plant and animal production practices with a scientific method which will sustain for the coming future", the following relationship is very important in terms of SA.

- Food and Nutritional Security
- Sustain environmental quality and natural resource
- Soil Health, soil quality, and soil fertility
- Sustain economic viability farming practices
- Human and animal health

Though there are various definitions of sustainable agriculture, the following are integrated ideas which define about SA:

"An agriculture that can evolve indefinitely toward greater human utility, greater efficiency of resource use and a balance with the environment that is favourable both to humans and to most other species" (Harwood, 1990)

"Sustainability lies in the dynamic nature of its fundamental components: ecological (spatial and temporal relations, diversity, stability, and resilience; economic (resource distribution and allocation); and social (equity, access, stewardship and institutions)." (Miller and Wali, 1995)

Soil quality and sòil health in SA

Soil quality and soil health is one important impartial part of sustainable agroecosystem management, similar to water and air quality. Assessing and managing soil quality and soil health may help to become more sustainable in terms of recourses and agriculture.

Soil quality or soil health is one important aspect of sustainable agriculture. The Figure 2. clearly shows that agricultural sustainability can be achieved through maintaining soil quality at optimum level with safe air and water in the environment.

Sustainable agriculture with reference to soil health and quality, environmental quality and human and animal health at the same time is challenging. Soil management becomes primary to machination to restructure or maintain sustainability in agricultural production and productivity keeping all above factor constant. The know-how that will aware the farming community is the most important issue than new technology to attain sustainable agriculture. Soil management directly correlates with agriculture sustainability but the fact establishment and awareness are lacking.

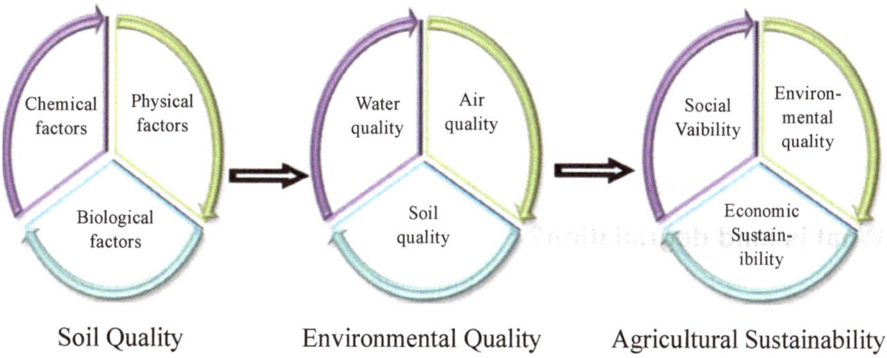

Figure2: Soil quality influences environment quality and sustainable agriculture

Challenges of SA in relation to the soil (Land degradation)

Nearly, 13,069 million ha (Glasod, 1994) land area in the world are affected by various forms of degradation. In which Asian continent account almost 28.1% of total land degradation in the world that is also the highest by any other continent. Poor and inappropriate soil management systems are the main causes of physical and chemical degradation of cultivated lands. Decreasing yields and loss of vegetation cover and erosion improvement are the typical demonstrations of mismanagement practices.

In, India the total geographical area is 328.73 million hectare (mha), 304.89 m ha comprise the reporting area and 264.5 mha only is under use for agriculture, forestry, pasture and other biomass production. According to the latest estimates (Sehgal and Abrol 1994), about 187.8 mha (57% approximately) out of 328.73 mha of land area has been degraded in one way or the other. Indeed, the recent emphasis by FAO on food production in support of food security, issues related to land degradation and its negative impacts on food production are now receiving a special attention. Resuming soil degradation and sustaining crop production through appropriate soil management practices are, therefore, important components of food security. The scenario of land degradation in India by a different organization in different years is illustrated in Table 1.

Tablel: Land degradation in India estimated by different organizations.

Organizations	Assessment Year	Degraded Area (Mha)
National Commission on Agriculture	1976	148.1
National Remote Sensing Agency	1985	53.3
Ministry of Agriculture	1994	107.4
NBSS & LUP	1994	187.7
NBSS & LUP (revised)	2004	146.8

Ranjan Bhattacharya *et al.,* (Sustainability 2015)

What is land degradation?

Land is defined as the terrestrial bio-productive system that comprises soil, vegetation, and water resources, other biota and the ecological and hydrological processes that operate within the system. It refers to all natural resources that contribute to agricultural production, including livestock and forestry (FAO, 1976; UNEP, 1992).

Land degradation is the temporary or permanent lowering of the productive capacity of land the (UNEP, 1992). It refers to the reduction in productivity of croplands, range and pastures, forest and woodlands, resulting from land uses or from processes arising from human activities such as: erosion; deterioration of physical, chemical and biological properties of soils; long-term loss of natural vegetation. A new definition was recently given to land degradation as "Land which due to natural processes or human activity is no longer able to properly sustain an economic function and/or the original natural ecological function".

Causes of land degradation

Causes of land degradation are numerous and multifaceted. Improper crop rotations, overuse of agrochemicals, and mismanagement of the irrigation system have created some serious soil degradation. Moreover, "shifting cultivation" practiced in many parts of the country is responsible for deforestation and the expansion of agriculture to less productive lands. India supports 18% of the world human population, 15% of the global livestock population, but endowed with only 2.4% of world land area. Land degradation is a big challenge to policy makers who need to balance the multiple goals of poverty eradication, food security and sustainable land management. The following reasons are accounting for serious land degradation;

(a) Geo-accelerated degradation

It is also known as natural degradation or loss of a land. The different topographical features such as stream channels, water gullies, valleys, etc. are the results of geologic degradation or erosion. Acceleration is a further stage of geological erosion. It is activated by nature or due to human activities which changes the natural cover and soil conditions. Accelerated erosion takes place by the action of water, wind, gravity and glaciers. So, the geo-accelerated degradation is combining action or process of geologic and accelerated degradation or erosion on land.

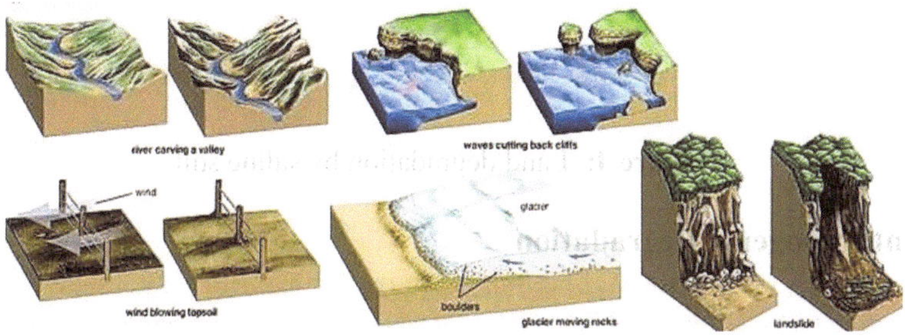

Figure 3: Various types of water and wind erosion creating land degradation

The different geo-accelerated degradation is; water erosion, wind erosion, glacial erosion (due to the mass of ice moving very slowly), droughts, earthquakes, volcanic eruptions, floods and snow erosion (due to slow and creeping movement of snow towards slope).

(b) Physical degradation

This degradation reflects the adverse changes in land or soil due to some physical activities which counter the productive potential of the land. The important and clear physical degradation are the reduction in organic matter, soil compaction, soil crusting, water logging conditions and soil sealing. Physical deterioration almost accounts for 11.6 million hectares out of 328.7 million hectares of a geographical area in India.

(c) Chemical degradation

Chemical soil degradation after geo-accelerated degradation (mainly water and wind erosion) is the second most abundant form of soil degradation. This poses a threat to the finite soil resource, as it tends to render less productive. Therefore it's necessary to understand the means by which soils are degraded chemically.

Chemical degradation is one way by which soils degrade chemically is through soil contamination; either by diffuse contamination or from localized sources. Such as salinization, alkalinization, acidification of soils, losses of organic matter and accumulation of pollutants and toxic elements.

Figure 4: Land degradation by saline soil

(d) Anthropogenic Degradation

There are various mismanagement practices in agriculture due to lack of proper knowledge which account for different degradation of land. Sustainability agriculture stands alone to cope with the problem for better management practices. Indeed, the sustainable agriculture ensures the food chain system which ultimately improves the human health. Some Human-induced land degradations are: construction of roads, parking lots construction, buildings construction, mismanagement with forests, over-grazing, deforestation, industrial wastes, nutrient mining, proper crop rotation, urbanization and industrialization.

Figure 5: Industrial waste and urbanization increasing land degradation

Land degradation in relation to human and animal health

The world food production is 2.62 billion metric tonnes during 2016-17. Undeniably, the world population requires ever increasing food supplies. The fact 99.7% of human food (calories) comes from the land (FAO, 2004) and less than 0.3% comes from oceans and other aquatic ecosystems. To continue the balance food supply which depends on the productivity and quality of the soil. Again, the food production of India is 275.11 million tonnes during 2016-17 which almost suffice the need of the country. Challenges are, burgeoning population

@ 2.3% per annum which may reach 1.5 billion by 2030 and also subsequent decreasing in agricultural land due to multiple factors. The degradation not only accounts for food production but it has countered the human and animal health. To strive the stress of combined effect of increasing population and to protect soil degradation for sufficient food production the agriculture practices are running in faulty manner at farmers level. The over stress or faulty agricultural practices have so many after effects on human and animal health. A few reported examples are arsenic, cadmium and nitrate toxicity in ground water, heavy metal contamination in soil and some soil pollutants. The unscientific agriculture practices will lead us to different soil, human and animal health degradation. The below mentioned are few health issues of human and animal which is which is directly or indirectly created by soil degradation.

The relationship between human health and land degradation is a result of the interaction between biological, physical environment, social structure and behavioral phenomena. Land degradation has a crucial role in influencing the human health by activating disease ecology and by human vulnerability to different diseases. Some direct association due to land degradation is malaria due to water stagnation and cholera because of brackish wetlands. So, soil and its degradation can be a human health factor in many ways;

What does land degradation mean for health?

Numbers of social and environmental activities are stressing the arable and pastures lands for good quality food, water and air. Land degradation can affect human health through complex pathways. As land becomes degraded for example in some places deserts, drought, flood, wind and water erosion, toxicants accumulation after reaches maximum limits, consequently, food production is reduced, water sources dry up and populations are pressured to move to another potential area. The impacts of degradation on health include;

- higher threats of malnutrition from reduced food and water supplies;
- occurrence of water-and food-borne diseases will increased;
- atmospheric pollution by wind erosion and other air pollutants may cause respiratory diseases;
- the spread of infectious diseases as populations migrate from one to another places.

There are basically two types of effect (a) Direct effect and (b) Indirect effect.

(a) Direct effects of soil degradation

The farmers, construction workers or miners, who work closely with soil, are at a greater risk of health problems that involve direct contact with soil, but the health of all peoples are affected by soil to some extent. Even, dust generated from soil can travel thousands of kilometers and affect people health. Wind erosion can reduce visibility due to the dust that pollutes the air, causes automobile accidents, fouls machinery, and imperils animal and human health are few examples of direct effect. Now water erosion can wash away part of your lawn, or part of an uphill lawn may wash down. It cost the government money to rebuild roadways that are damaged by erosion, especially dirt roads, but even highways can be damaged by large scale erosion that ultimately counters health due to disease spread from dirt and damage roads. Soil erosion is a natural event, but human caused erosion to fulfill their needs, along with the results of contaminant runoff has serious consequences on people and the environment.

There are three ways that humans will become contact with soil materials: (i) ingestion, (ii) respiration and (iii) skin absorption or penetration (Brevik, 2013).

Ingestion can occur deliberately, known as geophagy such as during hand to mouth contact (particularly children) or animals when raw fruits or vegetables are consumed without adequate washing. Ingestion of soil is very common in children, some animals and tribal peoples. This can potentially supply essential nutrients, but it can also lead to exposure to heavy metals, organic chemicals or pathogens and in large amounts can cause an intestinal obstruction (Henry & Cring, 2013).

The process of areal lifting of soil materials is called deflation, and these are referred to as dust. Airborne dust contains different mineral particles, trace elements; biogenic particles, and harmful viable biota. Respiration involves inhaling soil materials. There are various serious problems due to inhalation of dust, such as acute inflammation of the bronchial passages, marked an increase in asthma and other atopic diseases, pulmonary disorders, chronic bronchitis, valley fever and cardiovascular diseases.

Absorption or penetration of the skin can expose an individual to pathogens and soil chemicals which may cause many bacterial or fungal diseases like podoconiosis, which is a non-infectious disease found in subsistence farmers who frequently go barefoot.

(b) Indirect effects of soil degradation

Soil erosion is second only to population growth as the biggest environmental problem the world faces. Erosion is one of those problems that may lead us to poor agriculture. One rainstorm can wash away 1 mm of soil. It doesn't sound like much, but when you consider a hectare, it would take 13 tons of topsoil and if you calculate it takes 20 years with natural processes to replace that loss, such kind of loss occurs year after year by wind and rain around the world (Pimentel, 2006). It causes two problems for farmers, one being that the soil will not have enough structure for plants to have root support and the other being a loss of nutrients. That will reduce crop yields when valuable cropland is lost, so it can raise prices or reduce food availability. Soil erosion also destroys watershed, rivers and reservoirs. In addition, if erosion water carries toxicant to fish pond that secondarily causes human health after consumption. The prime reason is soil provides many of the essential elements we require and can pass on harmful substances through the food that we eat. Wind erosion removes a most fertile portion of the soil. Wind erosion reduces the soil, air, and water quality and threatens the sustainability of agriculture and health hazards to a living being.

The water logging is soil degradation, at farmer's field, on roads and lanes of city localities which cause open invitation of water-borne diseases like gastroenteritis, diarrhea, fever and stomach ailments, allergic problems, malaria and dengue on human and animal health. The following are some negative effect of soil degradation in a different form which causes health issues to humans and animals.

Element toxicity or heavy metals

There are many ways that heavy metals can adversely affect human health. The soil may contaminate either naturally or through anthropogenic activities. The chemical elements and substances that are in toxic amounts when ingested or inhaled may cause a serious problem in health. The heavy metals of which may affect human health include: As, Pb, Cd, Cr, Cu, Hg, Ni, NO_3^- and Zn. Heavy metals enter into the soils by weathering of rocks, but they are also established within soils through human activity. They are basically the by-products of mining ores, and present in mine spoils and in the immediate surroundings of metal processing plants. Urban soils are particularly subjected to accumulations of heavy metals from automobile exhaust, dumped sludge-slurry and industrial waste.

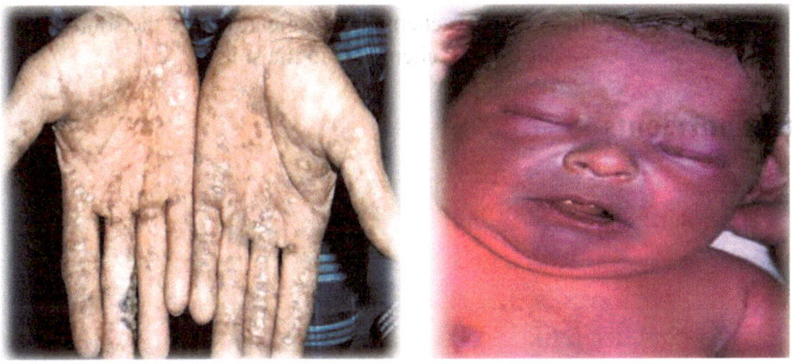

Arsenic toxicity (Arsenicosis) Nitrate toxicity (Blue baby syndrome)

Figure 6: Element toxicity of heavy metals due to excess use of chemicals

In agricultural land, the unscientific application of fertilizers, manures, and pesticides has contributed heavy metals accumulation. Particularly heavy metals toxicity in humans, including Cd, Pb, Hg, and As, are very alarming with no biological advantages that interrupt enzymatic activities mainly affecting the brain and kidneys.

Organic Chemicals

Organic chemicals are carbon-based compounds that are very resistant to biological decomposition and are usually toxic to organisms even at a small amount. The common sources for these organic chemicals are due to an application of pesticides in both rural and urban areas, and imbalanced applied chemicals into the soil. The municipal and domestic system waste, farm animal waste, and other biowastes are sources at urban areas. The recent health concern is pharmaceutical waste derived from medicines, hormones, and some drugs used to treat humans and animals. The common types of organic chemicals found in soil are polyhalogenated biphenyls, aromatic hydrocarbons, insecticides, herbicides, fossil fuels, and the by-products of fossil fuel combustion. Because of very long half-lives of these organic chemicals, they are referred as "persistent organic pollutants."

Figure 7: Different agro-chemicals applications are the source of organic chemicals in the soil.

Essential nutrients

There are 17 essential elements important for plant growth subjected to the soil, and almost all of these elements are essential for human health. Hydrogen, oxygen, carbon, major nutrient and chlorine occupy 99.9% of the atoms in the human body. However, the remaining 0.1% consists of micronutrients or trace elements that are equally essential in small amounts to maintain human health. Hence, soils that provide plants with the proper nutrients for growth also contain many of the elements that are necessary for human health as well. Degradation in soil at any condition or stage may short fall the essential element in plants we eat, which brings deficiency of particular nutrient in the human body.

Dust

Airborne dust can impact human health, especially when the particles sizes are less than 10 microns. The different particle size and type will determine its level of toxicity. Dust particles which are small enough to be inhaled may causes:

- irritation of the eyes
- coughing
- sneezing
- hayfever
- asthma attacks.

The persons with respiratory problems like asthma and chronic obstructive airways disease (COAD) a small increase in the dust concentration can make their health condition worse. Infact, various medical findings say, breathing in high concentrations of dust over a long period may reduce lung function and gives the disorders like chronic bronchitis and heart and lung function disorders in human and animals. Moreover, airborne dust may carry harmful pathogens, harmful gases, heavy metals, organic chemicals, insects and radioactive materials that can cause health problems in human and animals.

Soil Pathogens

The soil is habitat for millions of the organisms but a majority of them are not harmful to humans, except few species of bacteria, fungi, protozoa, viruses and prions can cause disease. Bacteria found most abundantly in the soil. Saprophytes fungi that absorb nutrients from the decomposition of dead organisms, out of the approximately 100,000 total fungi species present in soil only 300 soil fungi species are known to be harmful to humans which cause diseases. Protozoa found in soil can cause human parasitic diseases such as diarrhea and amoebic dysentery. Viruses are also known to survive in soil. Some pathogenic viruses are bought into the soil by human sewage waste. Viruses present in the soil may cause polio, conjunctivitis, hepatitis, gastroenteritis, smallpox etc in human and animals.

Challenges of sustainable soil management (SSM)

Soils are a natural dynamic body having different chemical, physical and biological properties. The management practices are also different with various degree of response to management. There are some key threats that hamper the goal of SSM like, soil erosion due to water and wind, depletion in SOM, soil fertility, salinization, toxicity of the chemicals, acidification, soil crusting, waterlogging and soil compaction. These different threats may have different intensity and dominance depending upon climatic condition which must be taken care in right time and the right place to attain sustainable soil management.

Strategies for SSM

The main aim of "sustainable soil management" should be improvements in the soil quality which is basic media for plant growth. Soil degradation can be checked by SSM, especially through soil conservation measures. The principles and practices of SSM must be promoted by agricultural extension worker/scientist to achieve SSM. An agricultural soil system is poor if the crops grown are attacked by weeds, insects, and pathogens. Particularly, sustainable soil management is not enough for sustainable production as an output, but it must consider crop, soil, nutrient, water, pest, and farm power management in space and time to maintain ecology and economic viability.

The conservation of soil vis-a-vis degraded soil management broadly signifies the safeguard against nutrient depletion. The productivity enhancement with proper natural resources uses is the key factor is SSM. To reorganize the soil within a time in terms of physical characteristics and to restore the nutrient for soil fertility improvement will support proper management. Basically, SSM is achieved by many systems of practice but broadly following measures are more effective to rectify the soil degradation ultimately attains sustainable agriculture;

(a) Mechanical or physical measures

The main mechanical measures are different types of bund like contour bunds, graded bunds, across the slope to check the soil losses through erosion. Some technics of terraces like bench and half-moon terraces at the foothills are also effective to check soil losses. Water harvesting ponds is also another management practice to check the soil erosion but it also supplies water for different crop cultivation when there are no sources of irrigation. Some barriers such as embankments to avoid extra water flow or to check flood and windbreaks may be trees or of other structures and soil husbandry is major techniques to control soil erosion. A guard against sea floods can be managed by constructing sea walls, sea shore barriers, beach defenses with structural boulders and planting some forests. All these physical measures will check the fertile soil loss or will protect soil fertility partially managing soil in good health for sustainable production.

(b) Agronomical or agro-technical measures

The various agro-technical measures not only look after single dimension of soil conservation practices but it also enhances soil fertility through improving different physio-chemical and biological characteristics of the soil. The most popular agronomic methods that help in soil erosion are choice of different crop, cropping system to maintain soil fertility also, crop geometry of different layout according to slope and gradient. Intercropping and the cover crop are also effective for soil erosion check. Tillage is also important for weed control, again its low intensity favours better soil structure, good infiltration capacity of soil and better pore size distribution. Again a zero-or minimum-tillage technique is successful to avoid different hard pan formation at soil depth, which inhibits root development, water and air movement in the soil. Profile inversion and pedoturbation are some useful techniques to manage crusting, hardpans and compactions at different soil depth which also improves soil physical properties. Subsoiling is another technique to improve physical properties of soil and proper moisture balance at soil depth. Agroforestry is advance methods to manage soil losses or soil fertility losses by growing multipurpose tree species with crops.

(c) Chemical measures or amendments

Application or addition of organic matter not only maintaining soil functions but also prevent soil degradation. Soils are largest carbon sink regulating climate and mitigating climate change. A loss of soil organic carbon (SOC) due to inappropriate land use can cause a decline in soil quality or health. Which ultimately poor the soil structure and increase soil erosion but ultimately emit carbon into the atmosphere. Now to prevent, minimize and mitigate the soil salinization and alkalinization is another important challenge of soil management. The acid soil is another problem soil which reduces the crop production and accounts 49 million ha of Indian soil. Use of gypsum and sulfur is advised to ameliorate the salinity and alkalinity of different extent for better crop production. Agricultural liming materials are advised to manage the soil acidity for proper crop production. So, a problem soil or chemical degraded soil can be technically managed for better agriculture or yield. Indeed, integrated nutrient management (INM) is a judicious use of organic fertilizer with an inorganic chemical is worldwide popular soil management practice. The concept of INM not only ensure the nutrient supply but also prevent soil tends to its degradation. Different types of organic manures, organic mulches, FYM, oil cakes, vermicompost along with chemical fertilizer can improve soil quality and soil health which lead us to sustainable production is well documented. Moreover, the organic manures improve the soil physical properties and also enhance microbial population at rhizosphere zone.

(d) Other measures

Soils system is the largest reservoirs of biodiversity on earth. Little is known to scientific society regarding the soil biodiversity. Soil health and environment should be maintained through maximum vegetative cover, using various organic materials, optimal nutrient use, less soil disturbance and restoring vegetation. Integrated or organic pest management should be encouraged to avoid unauthorized pesticide use in the agricultural system. So, this is the management practices lies within the arm of scientific society for production sustainability. Modern disaster detection techniques should be should be installed or developed to check losses from earth quick, sea cyclone or sea floods, floods and drought. Reforestation approaches by policy make or prevention law will also reduce the land degradation. A minimum disturbance to land use system for different purposes like highway construction, industrialization, urbanization, dams construction, hydro-power project etc. must be scientific and modern to avoid soil degradation. A better soil management practices not only maintain soil health or quality but it also minimizes the deposition of eroded soil material off-site, water pollution due to inefficient use of fertilizers and pesticides, and increases in greenhouse gases effect. These are the few techniques that can be implemented by law makes or adopted by farmers to achieve sustainability in production and sustainability in soil management.

Conclusion

It may be concluded that a great wealth of knowledge has been acquired about soils degradation, its management and ill effect on human and animal health. There is now a primary understanding of the prevention and management principles for sustainable soil management. In spite of the fact and knowledge, it is not practiced mainly in rural agriculture which serves us various degradation problems resulting in different infected diseases to human and animal health. Indeed the judicious use of inorganic fertilizer, pesticide, herbicide and other chemicals with proper scientific way will ensure safe residual toxicity in soil and plant. Now, at the same time of either static or declining agricultural productivity scenario, the use of organic fertilizer, pesticide, herbicide and amendments is challenging. A sustainable system promotes the use of fertilizers and other inputs, including the use of organic based fertilizer, pesticide, manures and plant residues. Thus, the use of organic amendment for nutrient supply will also reduce the cost and ensure better soil health. Moreover, an organic amendment is highly benefited to ameliorate some chemical soil degradation. So, to sustain soil health or quality for better crop production organic amendment is indispensable. That will restructure or restore the soil fertility and also minimize the degradation problem in the land, which may further establish a sustainable agricultural system in long run. Such scientific practices of using organic as a fertilizer or pesticide or herbicide not only rejuvenate the soil quality and environmental safety but it will make certain for sustainable agriculture system with healthy food production for the living being.

Reference

Bhattacharyya R., Ghosh B. N., Mishra P. K., Mandal B., Rao C. S., Sarkar D., Das K., Anil K. S., Lalitha M., Hati K. M. and Franzluebbers A. J.(2015). Soil Degradation in India: Challenges and Potential Solutions. *Sustainability* 2015, 7, 3528-3570.

Bouis, Howarth E. (1993). Measuring the sources of growth in rice yields: Are growth rates declining in Asia? *Food Research Institute Studies* 22(3).

Brevik, E.C. (2013). Soils and human health: an overview. In: Soils and Human Health (eds E.C. Brevik & L.C.Burgess), pp.29–56. CRC Press, Boca Raton, FL.

Cassman, K. G., S. K. De Datta, D. C. Olk, J. Alcantara, M. Samson, J. Descalsota, and M. Dizon. (1995). Yield decline and the nitrogen economy of long term experiments on continuous, irrigated rice systems in the tropics. In Soil management: Experimental basis for sustainability and environmental quality, ed. R. Lal and B. Stewart. Boca Raton, Fla., USA: CRC Press.

D. Pimentel 2006. Soil erosion: A food and environmental threat. *Environment, Development and Sustainability.* 8: 119-137.

FAO. (19760. A framework for land evaluation. FAO Soils Bulletin 32. FAO, Rome.

FAO. (2004) FAO Food Balance Sheets. FAOSTAT, Food and Agriculture Organization of the United Nations, Available online: http://faostat.fao.org/site/368/default.aspx#ancor (accessed on 1 August 2013).

Flinn, J. C., and S. K. De Datta. (1984). "Trends in irrigated rice yields under intensive cropping at Philippine research stations". *Field Crops Research* 9.

Harwood, R.R. (1990). "A history of sustainable agriculture" Book in. *Soil and Water Conservation Society*, Ankeny, IA.USA. pp.3-19., ISBN : 093573421X.

Henry, J.M. & Cring, F.D. (2013).Geophagy: an anthropological perspective. In: Soils and Human Health (eds E.C. Brevik & L.C. Burgess), pp. 179–198. CRC Press, Boca Raton, FL.

Miller, F.P. and M.K. Wali. (1995). "Soils, land use and sustainable agriculture: a review". *Canadian Journal of Soil Science.* 75 (4): 413-422.

Sehgal, J. and Abrol, I.P. (1994) *Soil Degradation in India: Status and Impact.* Oxford and IBH Publishing Co., New Delhi, 80 pp.

UNEP (1992b). World Atlas of Desertifcation, Desertification Control Bulletin 21, Arnold, London.

8

Sustainable Biodiversity Management

Lakshmi Hijam, Moumita Chakraborty, Rocky Thokchom and *Suvendu Kumar Roy*

"Come forth into the Light of Things, Let Nature be your teacher."
William Wordsworth

Biodiversity is a magnificent contribution of nature on our planet. The great variety of life forms on earth has catered for man's needs over thousands of years. This diversity of living creatures which forms a support system has been used by each civilization for its growth and development. Those that used this "bounty of nature" carefully and sustainably survived. Those that overused or misused it disintegrated. The attempt to classify and categorize the variability existing in nature for over a century has led to an understanding of its organization into communities of plants and animals. This information has helped in utilizing the earth's biological wealth for the benefit of humanity and has been integral to the process of development. This includes better health care, better crops and the use of these life forms a raw material for industrial growth which has led to a higher standard of living for the developed world. However, this has also produced the modern consumerist society, which has had a negative effect on the diversity of biological resources upon which it is based. The diversity of life on earth is so great that if we use it sustainably we can go on developing new products from biodiversity for many generations. This can only happen if we manage biodiversity as a precious resource and prevent the extinction of species.

The sustainable management of biodiversity is one of the most important challenges facing humanity. Biological diversity, abbreviated to biodiversity, refers to the variety of life forms at all levels of organization, from the molecular to the landscape level. These are the legacy of billions of years of evolution, shaped by natural processes and increasingly, by the activities of humans. Losing biodiversity has eroded the basis for social and ecological resilience, which has resulted in reducing the capacity for adaptive responses in a rapidly changing world. However, human activities still continue to have a large impact on the environment and on biodiversity. Agricultural, urban and infrastructural developments and climate change has exacerbated biodiversity loss. As a result, ecosystem functions and services are affected with serious consequences. The Sustainable Development Goal should be devoted to "protect, restore and promote sustainable use of terrestrial ecosystems, sustainably manage forests, combat desertification, and halt and reverse land degradation and halt biodiversity loss". It is high time to realize the linkages between the severity of global biodiversity loss and degradation of ecosystems and the negative impact that this situation have resulted on food security, nutrition, access to water, health of the rural poor and people worldwide.

Introduction

Biological diversity or Biodiversity, a portmanteau of "bio" (life) and "diversity", generally refers to the variety and variability of life on Earth. It can be described as the totality of genes, species, and ecosystems within a region. Biological diversity deals with the degree of nature's variety in the biosphere. This variety can be observed at three levels; the genetic variability within a species, the variety of species within a community, and the organisation of species in an area into distinctive plant and animal communities who only constitutes ecosystem diversity. The wealth of life on earth today is the product of hundreds of millions of years of evolutionary history. Over the course of time, human cultures have emerged and adapted to local environments, discovering, using, and altering their biotic resources. Many areas that now seem "natural" bear the marks of millennia of human habitation, crop cultivation, resource harvesting, and waste production. The domestication and breeding of local varieties of crops and livestock have further affected biodiversity.

Raymond F. Dasmann, a wildlife scientist and conservationist first used the term biological diversity in the year 1968 in the lay book, " A Different Kind of Country" advocating conservation. The term however, was introduced to the scientific community by Thomas Lovejoy, in the foreword to the book Conservation Biology. Its abbreviated form, "biodiversity" was apparently coined by W.G. Rosen in 1985 during the 1986 National Forum on Biological Diversity organized by the National Research Council (NRC). It first appeared in a publication in 1988 when sociobiologist E. O. Wilson used it as the title of the proceedings (Wilson, E.O. 1 January 1988) of that forum (Global Biodiversity Assessment: Summary for Policy-makers (https://books.google.com/books?id=VKnimg,EACAAJ). Cambridge University Press. 1995). Since this period the term has achieved widespread use among environmentalists, biologists, political leaders and concerned citizens. The United Nations Conference on Environment and Development (UNCED) held in 1992 at Rio de Janeiro (Rio Summit or Earth Summit) has further elevated the importance of Biodiversity.

Biodiversity: Concept and Definition

'Biological diversity' or biodiversity is considered an umbrella term which refers to organisms found within the living world, *i.e.*,, the number, variety and variability of living organism. Thus, it may be assumed as a synonym for ' Life on Earth', 'variety of life and its processes' (Keystone Centre, 1991),' condition of being different' (Gove *et al.* 1996) or what Darwin (1859) exclaimed as ' Life's endless forms'. In other words, biodiversity refers to the variability's among species of plants, animals and microorganisms; ecosystems; ecosystem including terrestrial, aerial, marine and other aquatic system and ecological complexes of which they are part. In simpler terms, biodiversity is the assemblage of different life forms. If reflects the number of different organisms and their relative frequencies in an ecological system. It includes the organisation of organisms at many levels ranging from complete ecosystems to the chemical components that form the molecular basis of heredity. Biodiversity, food and nutrition intersect on a number of key issues. Biodiversity contributes directly to food security, nutrition and well-being by providing a variety of plant and animal foods from domesticated and wild sources. Biodiversity can also serve as a safety-net to vulnerable households during times of crisis, present income opportunities to the rural poor, and sustain productive agricultural ecosystems. Thus, biodiversity is sum of all the genes, varieties, species, populations in different ecosystems and their relative abundance. Biological resources not only provide us nourishment, clothing, housing, fuel and medicine but also meet our several other requirements. Therefore, the knowledge of biodiversity is of immense utility in planning sustainable livelihood and conserving the natural resources.

Biological diversity deals with the degree of nature's variety in the biosphere. This variety can be observed at three levels; the genetic variability within a species, the variety of species within a community, and the organisation of species in an area into distinctive plant and animal communities constitutes ecosystem diversity.

Genetic diversity

Genetic diversity also referred to as *within species diversity* or *intra or inter specific diversity* describes the variation in the number and types of genes as well as chromosomes present in different species. 'Genetic diversity pertains to the range of diversity in the genetic resources of the organisms'. Every individual member of a plant or animal species differs from other individuals in its genetic constitution. Each individual has specific characters, which is due to the genetic makeup or code. The genes present in the organisms can form infinite number of combinations that causes genetic variability. For example, we find that each human, who is representative of the same species, *i.e., Homo sapiens*, is distinct from another. Similarly, there are many varieties within the same species such as rice, wheat, apples, mangoes, roses, tomatoes etc. that differ from one another in shape, size, colour of flowers and taste of fruits and seeds due to the variations at the genetic level. The term 'gene pool' has been used to indicate the genetic diversity in the different species. This also includes the diversity in the wild species, which through intermixing in nature over millions of years have given rise to newer varieties. The domesticated varieties of agricultural crops and animals have also evolved from the wild gene pool. Genetic diversity allows species to adapt to changing environments. This diversity aims to ensure that some species survive drastic changes and thus carry on desirable genes. The genetic variation arises by gene and chromosome mutation in individuals and in sexually reproducing organisms and it is spread in the population by recombination of genetic materials during cell division after sexual reproduction.

Genetic diversity has the following importance

(a) It represents variation between individuals of the same species.
(b) It helps in speciation or evolution of new species;
(c) It is useful in adaptation to changes in environmental conditions;
(d) It is important for agricultural productivity and development.
(e) It represents the raw material for evolution and adaptation.

Species diversity

Species diversity refers to biodiversity at the most basic level and is the 'variety and abundance of different types of individuals of a species in a given area'. It is a measure of the diversity within an ecological community that incorporates both **species richness** (the number of species in a community) and the evenness of species. It includes all the species on Earth, ranging from plants such as bacteria, viruses, fungi, algae, bryophytes, pteridophytes, gymnosperms, angiosperms and all the species of animals including unicellular protozoans to mammals. Taxonomic diversity is more precise as it considers the relationship of species to each other. For example an area with four species of wheat and two species of barley is considered more diverse than an area with six species of wheat even though they both contain the same number of species. The richness of a species gives an idea about the extent of biodiversity of a site and provides a means for comparing different sites.

The species richness is mainly dependent on climatic conditions. The number of individuals of different species with in a region represents species evenness or species equitability. The product species richness and species evenness give species diversity of a region.

In nature, species differ markedly from one another in their genetic makeup and do not inter-breed. However, closely-related species have in common much of their hereditary characteristics. The total number of plant and animal species described so far according to the IUCN (2004), is slightly more than 1.5 million and a large proportion of the species waiting to be discovered are in the tropics. More than 70 per cent of all the species recorded are animals, while plants (including algae, fungi, bryophytes, gymnosperms and angiosperms) comprise no more than 22 per cent of the total. Among animals, insects are the most species-rich taxonomic group, making up more than 70 per cent of the total. That means, out of every 10 animals on this planet, 7 are insects. The number of fungi species in the world is more than the combined total of the species of fishes, amphibians, reptiles and mammals. In general, there is a gradual reduction in species diversity as we move away from the equator towards the poles. With very few exceptions, tropics (latitudinal range of 23.5° N to 23.5° S) harbour more species than temperate or polar areas. India, with much of its land area in the tropical latitudes, has more than 1,200 species of birds.

Certain regions on earth appear to support more diverse populations than others. Regions endowed with nutrients and with well balanced climatic factors, such as moderate temperature, proper light and adequate rainfall, show high degree of diversity in their life forms. The tropical areas thus support more diverse plant and animal communities than the desert and polar areas. As for instance, tropical forest has higher species diversity as compared to a timber plantation. Conservation scientists at present are able to identify and categorized about 1.8 million species on earth. However, many new species are being identified, especially in the flowering plants and insects. Areas that are rich in species diversity are called 'hotspots' of diversity. India is among the world's 15 nations that are exceptionally rich in species diversity. When a species is confined entirely to a particular area, it is termed as endemic species. The ecological state of a species being unique to a defined geographic location, such as an island, nation, country or other defined zone, or habitat type is termed Endemism.

Species diversity has the following importance

(a) It represents variety of species in a given region or area.

(b) Species represent distinct units of diversity.

(c) It is relatively easy to identify by eye in the field. It represents the raw material for evolution and adaptation.
(d) It can be determined simply by counting the number of different species.

Ecosystem diversity

Ecological diversity reflects the variability among the species of plants and animals living together and connected by flow of energy and cycling of nutrients in different ecosystems or ecological complexes. In other words, it refers to communities of plants and animals, along with the physical components of their environment (geology, soil and climate). In addition it also includes variability within the same species and variability among the different species of plants, animals and microorganisms of an ecosystem. Thus, it pertains to the richness of flora, fauna and microorganisms within an ecosystem or biotic community. The ecosystem diversity is due to diversity of niches, trophic levels and ecological processes like nutrient cycling, food webs, energy flow, role of dominant species and various related biotic interactions. Such type of diversity can generate more productive and stable ecosystems or communities capable of tolerating various types of stresses *e.g.,* drought, flood etc. There are a number of ecosystems like grasslands, forests, semi arid deserts, marine, freshwater, wetland swamp, marshlands etc. and each one is having its distinct floral, faunal and microbial assemblages. It is also referred to as Land escape diversity because it includes placement and size of various ecosystems.

Ecological diversity represents an intricate network of different species present in local ecosystems and the dynamic interaction among them. The ecological diversity is of great significance that has developed and evolved over millions of years through interactions among the various species within an ecosystem. At the ecosystem level, India, for instance, endowed with deserts, rain forests, mangroves, coral reefs, wetlands, estuaries, and alpine meadows reflects greater ecosystem diversity than a Scandinavian country like Norway. India being one of the 12 centres of diversity and also the origin of several cultivated plants in the world makes it exceptionally rich in its ecosystem diversity. It is estimated that around 15,000 species of plants occur in India with the flowering plants comprising 15,000 species and out of which several hundred (5000-7500) species are endemic to India. The region is also rich in fauna, containing about 65,000 species of animals. An ecosystem when relatively undisturbed by human activities is referred to as "natural." When it is changed to other types of uses, such as farmland or urban areas it is known as "modified." Ecosystems are most natural in wilderness areas. When natural ecosystems are overused or misused their productivity decreases eventually and they are then said to be "degraded."

Ecosystem level diversity also deals with species distributions and community patterns. In addition it also associated with the role and function of key species and combines species function and interactions. Among the three types of biodiversity, this is the least understood level due to the complexity of the interactions.

Ecosystem diversity has the following importance

(a) It represents communities of plants and animals with physical structures of their environment.
(b) It is the least understood and the measurement is more difficult than genetic diversity and species diversity.
(c) Many complex measurements are required to be taken over a long period of time for the measurement of ecosystem diversity.
(d) It is a tedious and time consuming task.

Measurement of Biodiversity

According to Whittaker three common metrics are used to measure species-level biodiversity, encompassing attention to species richness or species evenness:

- Species richness-the simplest of the indices available.
- Simpson index
- Shannon-Wiener index

Another new index called the Mean Species Abundance Index (MSA) is a new addition and this index calculates the trend in population size of a cross section of the species which is in accordance to the CBD 2010 indicator for species abundance [MSA Index (page 4)].

(http://wayback.archiveit.org/all/20080510063151/http://www.unepwcmc.org/GLOBIO/PDF/Flyer_%20IB_Project_LR.pdf)

Species richness which is the simplest of the indices available measures the diversity of species in the following ways:

Alpha diversity: This is the diversity in species, *i.e.,* the number of species within a community. This depends on the interaction between the biotic and abiotic factors and also takes into account immigration from other locations.

Beta diversity: This is the change in the composition of the species with reference to the changes in the environment.

Gamma diversity: This refers to the overall diversity and is applied to larger areas in which both alpha and beta diversity is measured.

Biogeographic classification of India

India is one of the twelve mega diversity nations of the world, rich in biodiversity and associated traditional knowledge. Based on the geography, climate and pattern of vegetation seen and the communities of mammals, birds, insects, reptiles, amphibian and other invertebrates found, our country can be conveniently divided into ten major regions, with each regions representing a variety of ecosystems such as forests, lakes, rivers, grasslands, wetlands, mountains and hills having specific plant and animal species.

India's Biogeographic Zones

1. The cold mountainous snow covered Trans Himalayan region of Ladakh.
2. The Himalayan ranges and valleys of Kash-mir, Himachal Pradesh, Uttarakhand, Assam and other North Eastern States.
3. The Terai, the lowland where the Himalayan Rivers flow into the plains.
4. The Gangetic and Bhramaputra plains.
5. The Thar Desert of Rajasthan.
6. The semi-arid grassland region of the Deccan plateau Gujarat, Maharashtra, Andhra Pradesh, Karnataka and Tamil Nadu.
7. The Northeast States of India.
8. The Western Ghats in Maharashtra, Karnataka and Kerala.
9. The Andaman and Nicobar Islands.
10. The long western and eastern coastal belt with sandy beaches, forest and mangroves.

Values and uses of Biodiversity

Mankind has been blessed with the most precious gift of nature *i.e.,*, Biodiversity. The presence of life manifested through the diversity in flora and fauna contributes to the uniqueness of our planet Earth. The value of biodiversity in the life of all the organisms including humans is enormous as all the organisms in an ecosystem are interlinked and interdependent. In addition to its ecological and environmental value, biodiversity has significant socio-economic values. Today the world acknowledges that the loss of biodiversity has contributed to global climatic changes. The loss of forest cover which is the main mechanism for the conversion of carbon dioxide into carbon and oxygen, coupled with the increasing release of carbon dioxide and other gases through industrialization has contributed to the greenhouse effect. Resources like food, clothing, housing, energy, medicines, are all resources that are directly or indirectly linked to the biological variety present in the biosphere. This fact is most obvious in the tribal communities who gather resources from the forest or fisher folk who catch fish in marine or freshwater ecosystems. For others, such as agricultural communities, biodiversity is used to grow their crops to suit the environment. Urban communities generally use the greatest amount of goods and services, which are all indirectly drawn from natural ecosystems. It has become imperative to preserve the biological resources which are essential for the well-being and the long-term survival of mankind. This diversity of living organisms which is present in the wilderness, as well as in our crops and livestock, plays a major role in overall human 'development'. The wilderness is an outcome of a long evolutionary process that has created an unimaginably large diversity of living species, their genetic differences and the various eco-systems on earth in which all living creatures live including mankind as well. Think about this and we cannot but want to protect out earth's unique biodiversity. We are highly dependent on these living resources. The preservation of 'biodiversity' is therefore integral to any strategy that aims at improving the quality of human life.

The value of biodiversity can be grouped under the following heads:

Productive values

Biodiversity represents the original stock from which new varieties are being developed and produces a number of products harvested from nature and marketed commercially. Almost all the present day agricultural crops have originated from wild varieties. Similarly, all our domesticated animals came from wild living ancestral species. Through modern scientific breeding techniques animals are giving better yield of milk, meat etc. Indeed the commonly used animal products used by the modern society come from the advances made in the fields of poultry farming, pisciculture, silviculture, dairy farming, etc. Even the fossil fuels like coal and petroleum are the products of biodiversity from the geological past. Indirectly our people are getting economic benefits which include water quality soil protection, equalisation of climate, environmental monitoring, scientific research, recreation etc.

Consumptive value

The consumptive value are related to the natural products that are used directly for food, fodder, timber, fuel wood, etc. and which may be consumed locally but do not figure in national and international market. Approximately around 40,000 species of plants and animals are used by humans on a daily basis. The biodiversity held in the ecosystem provides forest dwellers with all their daily needs, food, building material, fodder, medicines and a variety of other products. Similarly, fishermen in the coastal areas are dependent on the marine resources and know where and how to catch fish and other edible aquatic animals and plants.

Social value

The environment in totality *i.e.,* flora, fauna, etc. were held in high esteem in ancient times, especially in India. Trees like Banyan Peepal, and Tulsi are still worshipped and offering water to Tulsi daily is considered good till today. Also there are festivals when sacred threads are tied around Peepal and Banyan trees and prayers are offered for the welfare of families. Similarly, certain animals and birds were represented as vehicles of Gods and were duly respected. Thus, the different facets of biodiversity were closely linked to the social values in many regions. The life of the indigenous people in many parts of the world still revolves around the forests and environment, even in the modern times. It is ironic that the societies, whose whole life is intricately associated with the forests, are at present are unable to use the natural resources for their sustenance. The biodiversity in different parts of the world has been largely preserved by the traditional societies. Since the indigenous people always protect the forests for their own benefit, the Government should formulate plans to involve such people for environmental protection.

Aesthetic value

Biological diversity adds to the quality of life and provides some of the most beautiful aspects of our existence. The earth's natural beauty with its colour and hues, thick forest, and graceful beasts has inspired the human beings from their date of birth to take necessary steps for its maintenance. Similarly botanical and zoological gardens are the means of biodiversity conservation and are of aesthetic values. In many societies, the diversity of flora and fauna has become a part of the traditions and culture of the region and has added to the aesthetic values of the place. In India, for example, the richness of flora and fauna are depicted in many paintings, animals are represented as vehicles of Gods, the dances and festivals are intricately associated with nature. Biodiversity which beautifies the natural landscape has led to the promotion of a type of tourism called ecotourism, which has now become a major source of income in many countries. Eco-tourism includes visiting wildlife sanctuaries, national parks, coral reefs, exotic islands, safaris and trekking in the mountainous and forested areas. In some countries like Nepal, Bhutan, Kenya, Rwanda, eco-tourism has now become the major source of foreign currency income.

Ethical and moral value

Indian civilization has been preserving nature over several generations through local traditions. We have in our country a large number of sacred groves or 'deorais' preserved by tribal people in several States. These sacred groves around ancient sacred sites and temples serve as gene banks of wild plants. As such Biodiversity must be seen in the light of holding ethical value which is based on the principle of 'live and let others live'. Morality and ethics teaches us to preserve all forms of life and not to harm any organism unnecessarily. It should be the prime responsibility and moral obligation of man to preserve and conserve other organisms which will favour the existence of the man directly or indirectly. Sometimes people degrade and pollute the environment by their unethical actions. Therefore, the people's conscience against such practices must be raised through proper education and awareness.

Optional value

This refers to the value of biodiversity that is yet unknown, but needs to be explored for future possibilities and use. Keeping future possibilities open for their use is called option value. About 1.75 million species have been discovered and named by scientists, which is of utmost importance. We should preserve the entire world's biodiversity that can be used by the future generations. It is impossible to predict which of our species or traditional varieties of crops and domestic animals will be of great use in the future. As such for continuous improvement of cultivars and domestic livestock, we need to return to wild relatives of crop plants and animals.

Biodiversity and its Conservation

Biodiversity is being depleted by the loss of habitat, fragmentation of habitat, over exploitation of resources, climatic changes, human, pollution, invasive exotic spices, diseases, shifting cultivation, poaching of wild life etc. Conservation of biodiversity should aim at protection, upliftment and scientific management of biodiversity so as to maintain it at its threshold level and derive sustainable benefits for the present and future generation. Since the human beings are enjoying all the benefits from biodiversity, they should take proper care for the preservation of biodiversity in all its form and good health for the future generation *i.e.,*, the human being should prevent the degradation and destruction of the habitats thereby maintaining the biodiversity at its optimum level.

Basic objectives of biodiversity conservation includes

(a) To maintain essential ecological processes and life supporting systems.

(b) To preserve the diversity of species.

(c) To make sustainable utilisation of species and ecosystems.

Why Is Biodiversity Conservation Important?

"At least 40 per cent of the world's economy and 80 per cent of the needs of the poor are derived from biological resources. In addition, the richer the diversity of life, the greater the opportunity for medical discoveries, economic development, and adaptive responses to such new challenges as climate change."

The Convention about Life on Earth, Convention on Biodiversity web

One way of explaining biodiversity is the phrase-'when one tugs at a single thing in nature, he finds it attached to the rest of the world.' All living creatures need other creatures and plants in one way or another-even if the connection is not so clear. Biodiversity is all about the diversity-the range of different living things and systems in an area. More the number of plant, insect and animal species in one area the greater the biodiversity and the healthier the ecosystem. The conservation of Biodiversity is a basic apprehension of mankind to facilitate a sustainable outlook and that Biodiversity conservation is unlike any other sustainable development issue, because loss of biodiversity is irreversible. The biodiversity we have today is a result of concerning 3.5 billion years of evolution. But the over-exploitation of natural resources and unsustainable expansion by human are leading to the sixth extinction crisis. Forests are a living ecosystem. India is the seventh largest country in the world. Forest loss and adaptation is as major issue in decline along with predicted extinction of forest birds in South-East Asia. According to expert the approximation of current species extinction rate is between 1,000 and 10,000 times higher than it would as expected be alive. Thus, there is a threat to the life of each and every one species together with humans and the need to conserve biodiversity is vital to continue the planet earth. Quantifying loss of genetic diversity is rather easier said than done, but it is understandable that the extinction of species along with declines of population lead to a loss of genetic diversity. The Convention on Biological Diversity, 1992, in its Article 2 stresses biological diversity as the going on variety of living organism moreover it includes terrestrial, marine and all further resources as well as assortment between species and ecosystems. An ecosystem in this regards means dynamic complex of plants, animal in addition to microorganism communities as well as their nonliving surroundings interacting as a purposeful component. The economic and social growth of humanity is depended on the Earth's biological diversity and consequently the biological diversity is a global advantage for the present and future generations.

Our reliance on biodiversity as a resource for human medicine is equally important. Of the top 150 prescription drugs used in the United States, 118 are based on natural resources: 74% from plants, 18% from fungi, 5% from bacteria, and 3% from one vertebrate (snake) species. Nine of the top ten drugs in this list are based on natural plant products. Globally, about 80% of the human population relies on traditional medical systems, and about 85% of traditional medicine involves the use of plant extracts. In addition, natural products extracted from many hundreds of species provide a wide variety of diverse inputs to human industries: gums and exudates, essential oils and flavourings, resins and oleoresins, dyes, tannins, vegetable fats and waxes, insecticides, and multitudes of other compounds. It is clear that the survival of our own species is vitally dependent on the services of other species and ecosystems. In addition to the production of ecosystem goods, such as seafood, wild game, forage, timber, biomass fuels and many pharmaceuticals, ecosystems also provide all human beings with the critical functions for life as follows:

- Pollination of crops and natural vegetation;
- Dispersal of seeds;
- Generation and preservation of soils and renewal of their fertility;
- Purification of air and water;
- Mitigation of droughts and floods;
- Detoxification and decomposition of wastes;
- Cycling and movement of nutrients;
- Control of the vast majority of potential agricultural pests;
- Protection of coastal shores from erosion by waves;
- Protection from the sun's harmful ultraviolet rays;
- Partial stabilization of climate;
- Moderation of weather extremes and their impacts; and
- Provision of aesthetic beauty and intellectual stimulation

It is clear from indebt discussion that human health and survival are dependent upon other species and key ecosystem services in the long-term. The conservation of specific information (genes), their libraries (species), and support systems (habitats) should be given urgent concern taking into consideration the current changes in extinction rates caused by human impacts. With more and more keystone species disappearing, these often being the ones most vulnerable to habitat destruction, one may well ask as a result of changing roles and growing numbers, whether one have positioned themselves as the ultimate keystone species, both in a position to determine the survival of so many other species and dependent on them for our own survival. We now find ourselves with the unenviable responsibility for affecting not only the rate of extinction, but also for the maintenance of the many essential ecological processes on which we depend for our survival.

Unfortunately, the biodiversity knowledge and much needed information are at the same place, politically and socially, just as environmental issues were in the early 1970s. The public is still very uninformed, scientific consensus is largely lacking as specialists continue to argue at what level to place the emphasis, that is, species, populations, or ecosystems which leaves the political decision-makers with no meaningful information on what to base their decisions, even after knowing the criticality of the issue. Another major barrier to effective decision-making for the conservation of biodiversity is the lack of ecological literacy, particularly among political decision-makers worldwide. How to rapidly increase their literacy and communicate the urgency of biodiversity loss is a major issue for biodiversity conservation.

Defining sustainable development

There are many ways to define sustainable development.

"Improving the quality of human life while living within the carrying capacity of supporting ecosystems."

The World Conservation Union (IUCN), United Nations Environment Programme (UNEP) and World Wide Fund for Nature/World Wildlife Fund (WWF)

"Sustainable development is development that meets the needs of the present without compromising the ability of future generations to meet their own needs."

United Nations in "Our Common Future, the Brundtland Report"

"Sustainable development involves the simultaneous pursuit of economic prosperity, environmental quality and social equity. Companies aiming for sustainability need to perform not against a single, financial bottom line but against the triple bottom line."

World Business Council on Sustainable Development

Connecting biodiversity conservation to sustainable development

Biodiversity benefits people through more than just its contribution to material welfare and livelihoods. Biodiversity contributes to security, resiliency, social relations, health, and freedom of choices and actions. Biodiversity is the foundation of ecosystem services to which human well-being is intimately linked. No feature of Earth is more complex, dynamic, and varied than the layer of living organisms that occupy its surfaces and its seas, and no feature is experiencing more dramatic change at the hands of humans than this extraordinary, singularly unique feature of Earth. Biodiversity, as described in previous discussion, is a key component of the "environmental health" pillar of sustainable development.

Biodiversity forms the foundation of the vast array of ecosystem services that critically contribute to human well-being. Biodiversity provides people with basic ecosystem goods and services. It provides goods such as food, fibre and medicine, and services such as air and water purification, climate regulation, erosion control and nutrient cycling. Biodiversity also plays an important role in economic sectors that drive development, including agriculture, forestry, fisheries and tourism. More than three billion people rely on marine and coastal biodiversity, and 1.6 billion people rely on forests and non-timber forest products (*e.g.,* the fruits from trees) for their livelihoods. Many people depend directly on the availability of usable land, water, plants and animals to support their families. In fact, ecosystems are the base of all economies.

Strategies for Conservation of Biodiversity

The following strategies should be undertaken in order to conserve biodiversity

1. Conservation of all the possible varieties (old or new) of food, forage and timber plants, live stock, agriculture animals and microbes.
2. Identification and Conservation of all the economically important organisms in protected areas.
3. Identification and safeguarding the critical habitats for each species.
4. Priority should be given to preserve unique ecosystems.
5. There should be sustainable utilisation of resources.
6. International trade in wild life should be highly rgulated.
7. The poaching and hunting of wildlife should be prevented as far as practicable.
8. Care should be taken for the development of reserves and protected areas.
9. Efforts should be made to reduce the level of pollutants in the environment.
10. Public awareness should be created regarding biodiversity and its importance for the living organisms.
11. Priority should be given in wildlife conservation programme to endangered species over vulnerable species and to vulnerable species over rare species.

12. The habitats of migratory birds should be protected by bilateral and multilateral agreement.
13. The over exploitation of useful products of wild life should be prevented.
14. The useful animals, plants and their wild relatives should be protected both in their natural habitat (*in-situ*) and in zoological botanical gardens (*ex-situ*)
15. Efforts should be made for setting up of National parks and wild life sanctuaries to safeguard the genetic diversity and their continuing evolution and Environmental laws should be strictly followed.

Conservation Methods

There are two types of conservation methods namely *in-situ* and *ex-situ* conservations. Let us discuss the different conservation methods along with their importance. *In situ* and *ex situ* are the Latin words for "on-site" and "off-site". They are two different, but complementary approaches to biodiversity conservation; each plays a distinct and important role.

In-situ conservation

The conservation of species in their natural habitat or natural ecosystem is known as *in-situ* conservation. In the process, the natural surrounding or ecosystem is protected and maintained so that all the constituent species (known or unknown) are conserved and benefited. The factors which are detrimental to the existence of species concerned are eliminated by suitable mechanism. *In-situ* conservation occurs in nature, for instance in a protected area, traditional farm, nature reserve or national park. *In-situ* conservation helps to guarantee the survival of a species in its natural habitat. It is important for observing the behaviour of a species, understanding how individuals interact with other members of their species and with other species, and classifying a species as endemic (*e.g.,* exists exclusively in a particular region), rare or under threat of extinction. *In-situ* conservation also enables researchers to determine the distribution of a species throughout the world, to assess traditional communities' contribution to conservation, and to inform local conservation initiatives.

Biodiversity at all its levels, genetic species and as intact ecosystems can be best preserved *in-situ* by setting aside an adequate representation of wilderness as 'Protected Areas'. The protection and management of biodiversity through *in situ* conservation involve certain specific areas known as protected areas which include National parks, Sanctuaries and Biosphere reserves.

Protected areas

The 1994 IUCN (International Union for the Conservation of Nature and Natural Resources) Guidelines for Protected Area Management Categories (IUCN 1994a,b) define a protected area as follows: An area of land and/or sea specially dedicated to the protection and maintenance of biological diversity and of natural and associated cultural resources and managed through legal or other effective means. In simpler terms, the protected areas are bio geographical areas where biological diversity are protected, maintained and managed through legal and administrative measures along with natural and cultural resources. In each area the demarcation of biodiversity is determined on the basis of climatic and physiological conditions. In these areas, firewood collection, hunting, timber harvesting etc. are strictly prohibited so that the wild plants and animals can grow and multiply freely without any hindrance. Protection of areas is a recent phenomenon dating back only to about a century in most countries, but centuries in India, Thailand and China. Some protected areas include: Cold desert (Ladakh and Spiti), Hot desert (Thar), Saline Swampy area (Sunderbans and Rann of Kutch), Tropical moist deciduous forest (Western Ghats and North East) etc. Protected areas include national parks, sanctuaries and biosphere reserves. There are 37,000 protected areas throughout the world. As per World Conservation Monitoring Centre, India has 581 protected areas, national parks and sanctuaries.

Wildlife Sanctuaries and National Parks of India

There are 589 Protected Areas in India of which 89 are National Parks and 500 are Wildlife Sanctuaries. They include a variety of ecosystems and habitats. Some have been created in order to protect highly endangered species of wild plants and animals found nowhere else in the world National parks. **National parks** are the small reserves meant for the protection of wild life and their natural habitats. These are maintained by government. The area of national parks ranges between 0.04 to 3162 km. The boundaries are well demarcated and circumscribed. The activities like grazing forestry, cultivation and habitat manipulation are not permitted in these areas. There are about 89 national parks in India. Sanctuaries are the areas where only wild animals (fauna) are present. **Sanctuaries** aim to protect animals and eventually release them into the wild. Most of the animals are confiscated from poachers (illegal hunters), pet traders, etc. The activities like harvesting of timbers, collection of forest products, cultivation of lands etc. are permitted as long as these do not interfere with the project. That is, controlled biotic interference is permitted in sanctuaries, which allows visiting of tourists for recreation. The area under a sanctuary remains in between 0.61 to 7818 km.

The Great Himalayan National Park is the largest sanctuary in this ecosystem and is one of the last homes of the beautiful snow leopard. Dachigam Sanctuary is the only place where the rare Hangul or Kashmir stag is found. There are several Sanctuaries in the Terai region, Kaziranga National Park, Assam is the most famous which has the endangered one horn rhino, elephant, wild buffalo, gaur, wild boar, swamp deer, and hog deer, in large numbers, as well as tiger and leopard. Its bird life is extremely rich and includes ducks, geese, pelicans and storks. The Manas Sanctuary, in addition to the above Terai species, also includes the rare golden langur and the very rare pygmy hog, the smallest wild boar in the world. The florican is found only in a few undisturbed grasslands in the Terai sanctuaries. The Keibul Lamjao National Park, Manipur is the last natural habitat of the marsh-friendly brow –antlered deer (Sangai) of Manipur. It is the only floating national park in the world. In the sal forests of Madhya Pradesh, there are several Protected Areas. Kanha, Madhya Pradesh offers a wonderful opportunity to observe wild tigers from elephant back. It is the only Protected Area in which a sub species of the Barasingha is found. Bharatpur is one of the most famous water bird sanctuaries in the world. Thousands of ducks, geese, herons, and other wading birds can be seen here. This is the only home of the very rare Siberian crane which migrates to India every winter. During the last 20 years, the 30 or 40 Siberian cranes have dwindled to only 2 or 3. During 2002-2003 no cranes were seen and it is possible that this beautiful bird will never again come to India. In the Thar desert, the wild life is protected in the Desert National Park. Here large numbers of black buck, neelgai and chinkara can be seen. The Great Indian Bustard lives in these arid lands. The Great and the Little Rann of Kutch have been made into sanctuaries to protect the very rare wild ass, the flamingo, the star tortoise and the desert fox. In Gujarat, the Gir Sanctuary protects the last population of the majestic Asiatic lion. This thorn and deciduous forest is also the home of large herds of chital, sambhar, and nilgai. The Sanctuaries of the Western Ghats and associated hill ranges protect some of the most diverse forest types in the country. The few examples of highly threatened species include the Malabar giant squirrel, the flying squirrel and a variety of hill birds, several species of amphibians, reptiles and insects. These regions are also rich in highly endemic plant life. Sanctuaries such as Bhimashankar, Koyana, Chandoli and Radhanagari preserve this rich flora in Maharashtra, Bandipur, Bhadra, Dandeli, Nagarhole, etc. in Karnataka, and Eraviculum, Perambiculum, Periyar, Silent Valley, in Kerala. In the Nilgiri Hills the rich forest Sanctuaries protect some of the last pockets of the Indian elephant in South India. Examples include Bandipur, Madhumalai, Wynad and Bhadra. During the last 10 years, a large number of the great tusker elephants of this region have been ruthlessly killed for their ivory. Now very few of these magnificent animals are left in these jungles. Two important sanctuaries meant for preservation of coastal ecosystems are the Chilka Lake and Point Calimere. The Sunderbans protect the largest mangrove delta in India. The Marine National Park in Gujarat protects shallow areas in the sea, islands, coral reefs and extensive mudflats. Over a hundred Protected Areas have been created in the Andaman and Nicobar Islands to preserve their very special island ecosystems.

Biosphere reserves

Biosphere reserves or natural reserves are multipurpose protected areas with boundaries circumscribed by legislation. The main aim of biosphere reserve is to preserve genetic diversity in representative ecosystems by protecting wild animals, traditional life style of inhabitant and domesticated plant/animal genetic resources. These are scientifically managed allowing only the tourists to visit. The Biosphere Reserve programme is a worldwide one of international scientific co-operation. It combines biodiversity conservation with scientific research, monitoring of environment, training and demonstration in bio resources management, environmental education and local co-operation and co-ordination. The real purpose of this programme is to promote a harmonious 'marriage' between conservation and sustainable development. The Biosphere reserve programme was essentially the outcome of the efforts of UNESCO; this organisation started the Man and Biosphere (MAB) programme (Gregg, 1988).

Biosphere reserves are important from the following point of views:

(a) These help in the restoration of degraded ecosystem.
(b) The main role of these reserves is to preserve genetic resources, species, ecosystems, and habitats without disturbing the habitants.
(c) These maintain cultural, social and ecologically sustainable economic developments.
(d) These support education and research in various ecological aspects,

Ex-situ conservation

Ex-situ conservation deals with the conservation of biological diversity outside of their natural habitats. This involves conservation of genetic resources, as well as wild and cultivated or species, and draws on a diverse body of techniques and facilities. It involves maintenance and breeding of endangered plants and animals under partially or wholly controlled conditions in specific areas like zoo, gardens, nurseries etc. That is, the conservation of selected plants and animals in selected areas outside their natural habitat is known as *ex-situ* conservation. The stresses on living organisms due to competition for food, water, space etc. can be avoided by *ex-situ* conservation there by providing conditions necessary for a secure life and breeding. *Ex situ* conservation should be used as a "last resort" or as a supplement to *in situ* conservation. *Ex situ* conservation is rarely enough to save a species from extinction. However, it is a key element for environmental and species education programmes because it provides the public with an opportunity to observe rare species from around the world in one location.

***Ex-situ* Biodiversity conservation can be done as following**

- By forming Gene banks: In this banks seeds, sperm & ova are stored at extremely low temperature and humidity.
- It is very helpful to save large variety of species of plants & animals in a very small space. *e.g.,* sperm and ova banks, seed banks.
- Forming Zoo and botanical garden: for research purpose and to increase public awareness collecting living organisms for aquaria, zoos and botanic gardens.
- Collections of *in vitro* plant tissue and microbial culture.
- Captive breeding of animals and artificial propagation of plants, with possible reintroduction into the wild.

Botanical gardens are meant for plant research, display of specimens and training. Seed banks are more like museums; they house plant material that can be used as a source for planting if seed reserves – in cultivation or in nature – are destroyed or extinct. Seed banks also provide researchers and breeders with crop seeds important for agriculture.

The strategies for *ex-situ* conservations are: Identification of species to be conserved, adoption of different *ex-situ* methods of conservation, long-term captive breeding and propagation for the species which have lost their habitats permanently, short-term propagation and release of the animals in their natural habitat, animal translocation, animal reintroduction, advanced technology in the service of endangered species. *Ex-situ* biodiversity conservation strategy also plays a key role in recovery programmes for endangered species. The Kew Seed Bank in England has 1.5 per cent of the world's flora-about 4,000 species-on deposit. In agriculture, *ex-situ* conservation measures maintain domesticated plants which cannot survive in nature unaided. It provides good platform for research opportunities on the components of biological diversity. Some of the institutions also play a major role in public education and in increasing awareness among public by bringing members of the public into contact with plants and animals they may not normally come in contact with. It is estimated over 600 million people visit zoos every year worldwide.

Conservation on productive landscape

So far protected areas and parks remain a cornerstone for biodiversity conservation worldwide, conservation efforts should not be limited to these places. Lots of biodiversity can also be found in large productive landscapes having no specific conservation objective. Agricultural lands, timber forests, grasslands, rivers and marine areas are productive landscapes that are also important for biodiversity. These landscapes are managed with the aim of producing and harvesting food, timber, energy and marine resources. Although biodiversity conservation is not the main objective, the management of these landscapes should also be given due importance. If it's not, resource exploitation can harm the long-term health of the ecosystem and its ability to supply food, timber, energy and other resources. This recognition has led to the promotion of sustainable agriculture, sustainable forestry, sustainable grassland management and sustainable fisheries.

Monitoring

Another important conservation practice is Close monitoring of biodiversity is; it takes into account regular checking of the overall health of ecosystems and the species living within it. The data collected from ongoing monitoring programmes can help inform management plans and improve the sustainability of activities in productive landscapes. Monitoring is especially important when the activities are carried out on an industrial scale because their impact is greater than the impact of similar activities when carried out on a smaller subsistence scale by local communities.

Law and community enforcement

Law or community enforcement also comes under conservation mechanisms. A Biodiversity conservation officer ensures the communities relying on the site's natural resources to totally involve in conservation initiatives. Officers enforce the laws and record the details of community participation. When the laws are not respected, illegal logging, mining and bushmeat hunting erode the benefits of conservation efforts.

Traditional knowledge and practices

Traditional knowledge is vital for sustainability of natural resources including forests, water, and agro ecosystems across landscape continuum spanning from households through farms, village, commons and wilderness. In many cases, traditional knowledge has contributed to protecting wildlife and ecosystems and to ensuring a "natural balance". Traditional knowledge comprises of "knowledge, innovations and practices of indigenous and local communities around the world, developed from experience gained over centuries and adapted to the local culture and environment, which is transmitted orally from generation to generation", according to the Convention on Biological Diversity (CBD). Traditional knowledge is collectively owned and takes the form of stories, songs, folklore, proverbs, cultural values, beliefs, rituals, community laws, local languages and agricultural practices, including the development of plant species and animal breeds. Article 8(j) of the CBD calls for countries to respect, preserve and maintain knowledge, innovations and practices of indigenous and local communities embodying traditional lifestyles relevant to the conservation and sustainable use of biodiversity.

In order to be effective, efforts on biodiversity conservation can learn from the context-specific local knowledge and institutional mechanisms such as cooperation and collective action; intergenerational transmission of knowledge, skills and strategies; concern for well-being of future generations; reliance on local resources; restraint in resource exploitation; an attitude of gratitude and respect for nature; management, conservation and sustainable use of biodiversity outside formal protected areas; and, transfer of useful species among the households, villages and larger landscape. These are some of the useful attribute of local knowledge systems (Pandey, 2002a). Traditional knowledge on biodiversity conservation in India is as diverse as 2753 communities (Joshi *et al.* 1993) and their geographical distribution, farming strategies, food habits, subsistence strategies, and cultural traditions. Biodiversity conservation practitioners must therefore ensure that the communities relying directly on natural resources are involved in conservation initiatives, and guarantee their active participation during the whole conservation process. Some form of community engagement is essential for the success of any biodiversity conservation project. There are also many good examples of community conserved areas around the world. These sites have been managed by communities for generations for the sustainable use of natural resources such as medicinal plants and water springs, or even for religious purposes. These sites may or may not have government protection or written management regulations.

Role of research & technology for biodiversity conservation

As we've seen, Biodiversity conservation doesn't happen in a vacuum, it requires the participation of many different groups of people, working with various conservation mechanisms both *in situ* and *ex situ*. Biodiversity policy and conservation activities are informed, enhanced and driven by research and technology. Researchers like biologists, ecologists and social scientists play various roles in conservation, they identify species and their habitats, they locate areas of high ecological value, they pinpoint threats, and they propose innovative strategies and solutions. Various methods are used by researchers such as field surveys, observations and experiments, and technologies including remote sensing devices, data analyses, software and laboratory tests. Research results are very important for biodiversity conservation. They can be woven into community development programmes. Local communities in fact can be important contributors to biodiversity conservation research, and should be made to involve in all steps of the research and conservation processes. The creation and application of technologies are also benefits of biodiversity conservation research. When technologies are invented, selected, evaluated, tested and applied to solve specific problems they can also be transferred from rich countries to poor countries and vice versa. However, before using any technology, it's critical to have a clear understanding of its characteristics, so that the intervention does not harm local livelihoods, traditions, cultures or the environment.

Conclusion

Biodiversity conservation and sustainable development are two important inter-related branches which focuses on social progress, economic growth and environmental protection on one side, and ecosystem conservation on the other. Conservation includes the efforts carried out in protected areas such as national parks and community reserves, and in other areas with rich and important biodiversity where conservation is not the main focus. It is in these latter productive landscapes where sustainability is needed most. Sustainable agriculture, sustainable fisheries and sustainable management of natural resources are the main approaches for preserving these landscapes for long-term social, economic and ecological benefits. To achieve greater progress toward biodiversity conservation to improve human well-being and reduce poverty, it will be necessary to strengthen response options that are designed with the conservation and sustainable use of biodiversity and ecosystem services as the primary goal. These responses will not be sufficient, however, unless the indirect and direct drivers of change are addressed and the enabling conditions for implementation of the full suite of responses are established. Short-term goals and targets are not sufficient for the conservation and sustainable use of biodiversity and ecosystems. Given the characteristic response times for political, socio economic, and ecological systems, longer-term goals and targets (such as for 2050) are needed to guide policy and actions. Improved capability to predict the consequences of changes in drivers for biodiversity, ecosystem functioning, and ecosystem services, together with improved measures of biodiversity, would aid decision-making at all levels. Science can help ensure that decisions are made with the best available information, but ultimately the future of biodiversity will be determined by society.

Reference

Aditya, and A. Bhartiya (2013).Genetic variability, Correlation and Path Analysis for Quantitative Characters in Rainfed Upland Rice of Uttarakhand Hills. *Journal of Rice Research* 2013, Vol. 6 No. 2.

Global Biodiversity Assessment: Summary for Policy-makers (https://books.google.com/books?id=VKnimgEACAAJ). Cambridge University Press. 1995. ISBN 978-0-521-56481-6. Annex 6, Glossary. Used as source by "Biodiversity", Glossary of terms related to the CBD (http://www.biodiv.be/glossary_keywords/B), Belgian Clearing-House Mechanism. Retrieved 2006-04-26.

Darwin, C.1859. On the Origin of Species by means of Natural Selection. John Murry, London.

Gove, J.N., Patil, G.P., and Talle, C. 1996. Diversity measurement and comparison with examples. In: Szaro, R.C,. and Johnson, D.W. (Eds). Biodiversity in Managed Landscapes. Oxford Univ. Press, Oxford,pp.157-175.

MSA Index (page 4) (http://wayback.archive-it.org/all/20080510063151/http://www.unep-wcmc.org/GLOBIO/PDF/Flyer_%20IB_Project_LR.pdf)

Pandey, D.N. (2002a). Cultural resources for conservation science. *Conservation Biology*

Joshi, N. V., Gadgil, M., and Patil, S. 1993. Exploring cultural-diversity of the people of India. *Current Science* 64: 10-17.

Gregg, W.P.Jr.1988, On wilderness, national parks and biosphere reserves. In; Fourth World Wilderness Congress on Worldwide Conservation. Proceedings of the Symposium on Biosphere Reserves. U.S. Dept. Of Interior, Washington DC.

IUCN, 1994a. United Nations List of National Parks and Protected Areas. WCMC, Cambridge and IUCN Commission on National Parks and Protected Areas. Gland, Switzerland.

IUCN, 1994b. Guidelines for Protected Area Management Categories. WCMC, Cambridge and IUCN Commission on National Parks and Protected Areas. Gland, Switzerland.

Wilson, E.O. (1 January 1988). Biodiversity ((https://books.google.com/books?id=DSGF1xQBYi8C). National Academies Press. ISBN 978-0-309-03739-6. online edition (http://darwin.nap.edu/books/0309037395/html/R2.htm l) Archived (https://web.archive.org/web/20060913154719/http://darwin.nap.edu/books/0309037395/html/R2.html) 13 September 2006 at the Wayback Machine.

Websites

https://www.omicsonline.org/open-access/managing-biodiversity-with-emphasis-on-sustainable-development-2157-7625-S5-008.php?aid=72903

https://en.wikipedia.org/wiki/Biodiversity

https://www.infinityfoundation.com/mandala/t_es/t_es_pande_conserve.htm

http://www.biologydiscussion.com/biodiversity/biodiversity-concept-types-and-other-details-with-diagram/7132

https://www. ecosystems and human well-being/Biodiversity Synthesis/

https://www.liverpool.ac.uk> Sustainability(/sustainability/) > On campus (/sustainability/oncampus/)> Biodiversity(/sustainability/on-campus/biodiversity/) > Managing Biodiversity Loss

9

Agroforestry as a Tool for Biodiversity Conservation Carbon Sequestration and Sustainable Livelihood Development

Nazir A. Pala, Abha Manohar K, Biplov C. Sarkar, Megna Rashid Vineeta and *Sumit Chakravarty*

Agroforestry: At a Glance

Agroforestry as a landuse system is gaining wide importance and acceptance in the current global scenario to mitigate global climatic conditions. It is a highly efficient system of sustenance for the rural population with ample livelihood opportunities. The benefits from agroforestry is immense for socio-economic as well as in ecological perspective. From the vivid ecosystem services like climate regulation, carbon sequestration, soil productivity and biodiversity conservation to food security and sustainable livelihood enhancement, agroforestry envisages immense services. The implication of the agroforestry system and practice varies from area to area. Traditional agroforestry can be modified by implementing different species combinations suitable for vivid regions thus enhancing the long term productivity. An important specification of agroforestry is with a minimal labour input maximum production and utilization of resource could be attained. National Agroforestry policy, implemented is a big step forward for new agroforestry initiatives in the country. Conduction of region specific agroforestry studies and policy implementations with involvement of rural populations will be an effective breakthrough approach for a sustainable future.

The Land Use System

The practice of Agroforestry dates back to the time immemorial. Only thing new is the name Agroforestry itself. The term agroforestry was coined in the late 1970s, reflecting a significant change in agricultural development. The creation of the International Centre of Research on Agroforestry (ICRAF) was an indication of the international development community's recognition of agroforestry as an important land use practice deserving intensified research. With these research efforts, modern agroforestry can use science to improve already well-established local traditions. Agroforestry is a collective name for land-use systems and practices where woody perennials are deliberately integrated with crops and/or animals on the same land management unit (ICRAF, 1982). The integration can be either in spatial mixture or temporal sequence. Lundgren and Raintree (1983) defined Agroforestry as a collective name for land use systems and technologies where woody perennials (trees, shrubs, palm, bamboo etc.) are deliberately used in the same land management unit as agriculture crops and or animals either in same form of spatial arrangement or temporal sequence.

Agroforestry farming systems has got a unique way to contribute for the upliftment of rural livelihood. Depending on the component combinations of tree, animal and crop, agroforestry systems are classified into Agri-silviculture (crops, including shrubs/vines combined with trees); silvo-pastoral (pasture/animals and trees) and agro-silvo-pasture (crops, pasture/animal with trees); silvo-pastoral (pasture/animals and trees); and agro-silvo-pastoral (crops, pasture/animals and trees) (Nair, 1985) (Table 1).

Table 1: Major approached to classification of agroforestry systems and practices

Categorization of system Based on their structure and function			**Grouping of Systems (according to their spread and management)**	
Structure (nature and arrangement of components especially woody one)		**Function (role and/or input of component especially woody ones)**	**Agro-ecological environment adoptability**	**Socio-economic and management level.**
Name of component	**Arrangement of component**			
Agriculture (Crops and tree incl. Shrubs/trees)	In space (spatial) mixed dense (*e.g.*, Homegarden)	Productive function Food	Sustain in/for lowland humid tropics	Based on level of technology input
Silvopastoral (pasture/animal, and trees)	Mixed sparse (*e.g.*, Most systems of tree in pasture)	Fodder	Highland humid tropics (above 1,200 m.a.s.l. Malaysia)	Low input (marginal)
Agrisilvopastoral (Crop, pasture/animals and trees)	Strip (width of strip to be more than one tree)	Fuel wood Other woods	Lowland subhumid tropics (eg Savanna zone of Africa, Carrato of South America)	Medium input High input

Categorization of system Based on their structure and function			Grouping of Systems (according to their spread and management)	
Structure (nature and arrangement of components especially woody one)		Function (role and/or input of component especially woody ones)	Agro-ecological environment adoptability	Socio-economic and management level.
Name of component	Arrangement of component			
Other (multipurpose tree lots, apiculture with trees, aquaculture with trees etc.)	Boundary (trees on edges of plots/ field, in time (temporal)	Other products Protection function Wind break	Highland subhumid tropics (tropics Highlands) (eg in Kenya, Ethiopia)	Based on cost/ benefit relation
	*Coincident *Overlopping *Sequential (separate) *Interpolated	Shelter belt Soil conservation Moisture conservation Soil improvement Shade for(crop, animal and man)		Commercial Intermediate Subsistence

Source: Nair (1985)

Biodiversity and Agroforestry

Biodiversity is not simply a measure of the world's species; rather, it also encompasses genetic variability within and between populations, species evolutionary histories, and other measures of the diversity of life. An Agroforestry practice system varies from one region to another because of ecological fragmentation and conserves high diversity according to pattern of practices and knowledge. Agroforestry includes agrisilvicultural system, silvopastoral system and agro-silvicultural system which support and conserve large number of biodiversity. However, one multifunctional approach for our food system is agroforestry, the intentional combination of trees and shrubs with crops or livestock. Agroforestry has been recognized for nearly half a century as a sustainable agricultural practice (Nair and Garrity, 2012). Agroforestry system maintain rich biodiversity and provide essential goods and services which support ecosystem functioning, resilience and productivity. However, agroforestry system and the heterogeneous mosaic landscapes of which they are part have recently attracted the interest of conservation biologists and other investigators working on the interface between integrated natural resource management and biodiversity conservation (Gajaseni *et al.,* 1996; Perfecto *et al.,* 1996; Rice and Greenberg 2000).

In agroforestry system we can find diverse plant diversity, labour input and income generation derived in the regions where population densities are high (Soemarwoto, 1987; Padoch and de Jong, 1991). They are complex, multi-storied environments with very high species diversity and a wide range of very varied ecological micro-niches (Galluzzi *et al.*, 2010; Agbogidi, 2011) traditionally integrated within a larger surrounding ecosystem (Gliessman, 1990; Eyzaguirre and Linares, 2004). In the last three decades, agroforestry has been widely promoted in the tropics as a natural resource management strategy that attempts to balance the goals of agricultural development with the conservation of soils, water, local and regional climate, and, more recently, biodiversity (Izac and Sanchez, 2001). The occurrence of a good number of plant species in agroforestry system is due to result of farmer's attempt to have as much high crop plant diversity in their gardens (Feleke, 2000). The people living near to forest depend for food, fodder and fuel wood but increasing population, illicit collection and other hand decreasing of forest area leads to animal conflict is the major problem of buffer zone. In many regions agroforestry system conserve high in-situ biodiversity. The introduction of agroforestry practices into buffer zones around protected forest areas has been suggested as a technology option which may not only reduce pressures on forest resources but which also can improve the living standards of the rural population living around these protected areas (van Orsdol, 1987). The tree growing in India outside forests was estimated to have between 14,224 million (Ravindranath and Hall, 1995) and 24,602 million (Prasad *et al.,* 2000), spread over an equivalent area of 17 million ha (NFAP, 1999), supplying 49% of the 201 million tonnes of fuelwood and 48% of the 64 million m^3 of timber consumed annually by the country (Rai and Chakrabarti 2001). The total of 289 species of plants were observed in 106 suburban home gardens in the Thimbirigaskatuwa suburban area, western Sri Lanka, of which 51% were ornamental plants, 36% food plants, and 12% medicinal plants (Kumari *et al*, 2009). In India, there are studies which determine the importance and role of home gardens in the life of local people, especially in terms of livelihood and economic significance (Samati, 2004; Das and Das, 2005; Srivastava and Heinen, 2005; Tangjang and Arunachalam, 2009; Tynsong and Tiwari, 2010).

The enset-coffee home gardens of the Wolayta are home for rich diversity of overall 159 plant species as well as 112 useful species (Seta *et al.*, 2013). From the same area 60 crop plant species were also reported (Zemede and Zerihun, 1997). In Gimbi Wereda, Western Ethiopia 52 plant species in 12 home gardens (Etana, 2007). Tesfaye *et al.* (2010) identified a total of 78 cultivated crop species and an average of 16 crop species per home garden in Sidama, Southern Ethiopia. A total of 69 plant species (44 woody and 25 herbaceous) belonging to 40 families were recorded from the home gardens of Jabithenan district of North-Western Ethiopia (Mekonnen *et al.*, 2014).

MPT's in Agroforestry

Multipurpose tree species includes all the woody perennials which provide more than one usage/production/service function. The various functions of multipurpose tree species include food, fodder, shelter, energy, raw materials for processing, environmental protection, soil conservation etc. Some of the general characteristic features of MPTs (Chundawat and Gautam, 1993) for agroforestry systems adopted are (Table 2 and 3):

(a) Adaptability to local climatic conditions
(b) Light open crown that lets sunlight through
(c) Ability to sprout quickly after pruning, coppicing or pollarding
(d) Ability to assist in nitrogen fixation
(f) Resistance to drought, flooding, soil variability and other climatic hazards
(g) Deep thrusting taproot system
(h) Easy to manage
(i) Cheap to establish
(j) Higher demand and better value for the produce

Table 2: Multipurpose tree characteristics and agroforestry systems

Tree attributes	Relationship to performance in agroforestry systems
Height	Ease of harvesting leaf, fruit, seed and branch wood; shading or wind effects
Stem form	Suitability for timber, posts and poles; shading effects
Crown size, shape and density	Quantity of leaf, mulch and fruit production; shading or wind effects
Multistemmed habit	Fuel wood and pole production; shading or wind effects
Rooting pattern (deep or shallow, spreading or geotrophic)	Competitiveness with other components, particularly resource sharing with crops; suitability for soil conservation
Physical and chemical composition of leaves and pods	Fodder and mulch quality; soil nutritional aspects
Thorniness	Suitability for barriers or alley planting
Wood quality	Acceptability for fuel and various wood products
Phenology (leaf flush, flowering and fruiting) and cycle (seasonality)	Timing and labor demand for fruit, fodder and seed harvest; season of fodder availability; barrier function and windbreak effects

(Contd.)

Tree attributes	Relationship to performance in agroforestry systems
Di = or monoeciousness	Sexual composition of individual species in community (important for seed production and pollen flow)
Pest-and disease-resistance vigor	Important regardless of function; biomass productivity, early establishment
Site adaptability and ecological range	Suitability for extreme sites or reclamation uses
Phenotypic or ecomorphological variability	Potential for genetic improvement, need for culling unwanted phenotypes
Response to pruning and cutting management practices	Use in alley farming, or for lopping or coppicing
Possibility of nitrogen fixation	Use in alley farming, planted fallows, or rotational systems

Source: Wood (1990)

Table 3: List of some agroforestry tree species and their major uses

Species	Family	Major Uses
Mangifera indica	Anacardiaceae	Fodder, food, fuel, timber
Nypa fruticans	Arecaceae	Industrial uses
Phoenix dactylifera	Arecaceae	Food, fuel, soil amelioration
Tecomella undulata	Bignoniaceae	Fuel, soil amelioration, timber
Terminalia species	Combretaceae	Fodder, fuel, soil amelioration, timber
Shorea robusta	Depterocarpaceae	Industrial uses, soil amelioration, timber
Mallotus phillipinensis	Euphorbiaceae	Fodder, food, fuel, nitrogen fixing.
Ricinus communis	Euphorbiaceae	Food, fuel,
Hardwickia binata	Fabaceae	Fodder, Fuel, nitrogen fixing, soil amelioration
Leucaena leucocephala	Fabaceae	Fodder, fuel
Parkinsonia aculeate	Fabaceae	Fodder, Fuel, soil amelioration
Pithecellobium dulce	Fabaceae	Fodder, food, fuel, industrial uses, nitrogen fixing, soil amelioration
Pongamia pinnata	Fabaceae	Fodder, fuel, nitrogen fixing, soil amelioration, timber
Prosopis cineraria	Fabaceae	Fodder, food, fuel, nitrogen fixing, soil amelioration,
Prosopis juliflora	Fabaceae	Fodder, fuel, nitrogen fixing, soil amelioration
Pterocarpus dalbergiodes	Fabaceae	Fodder, fuel, nitrogen fixing, soil amelioration, timber

Species	Family	Major Uses
Pterocarpus marsupium	Fabaceae	Fodder, fuel, nitrogen fixing, soil amelioration, timber
Robonia pseudoacacia	Fabaceae	Fodder, fuel, nitrogen fixing, soil amelioration, timber,
Samanea saman	Fabaceae	Fodder, fuel, nitrogen fixing, soil amelioration, timber
Sesbania aculeate	Fabaceae	Fodder, fuel, nitrogen fixing, soil amelioration,
Sesbania framosa, grandiflora, javanica	Fabaceae	Fodder, food, fuel, nitrogen fixing, soil amelioration
Sesbania robusta, sesban	Fabaceae	Fodder, fuel, nitrogen fixing, soil amelioration
Tamarindus indica	Fabaceae	Fodder, food, fuel, industrial use, nitrogen fixing, soil amelioration, timber
Quercus spp.	Fagaceae	Fodder, fuel, soil amelioration, timber
Lagerstromia hypolenea	Lythraceae	Food, nitrogen fixing, soil amelioration
Lawsonia alba	Lythraceae	Food
Punica granatum	Lythraceae	Food, fuel,
Theobroma cocoa	Malvaceae	Fodder, food, fuel
Thespesia populanea	Malvaceae	Fodder, fuel, soil amelioration, timber
Melia azedarach	Meliaceae	Fodder, fuel, Industrial use, soil amelioration, timber
Morus alba	Moraceae	Fodder, food, fuel, industrial uses
Moringa oleifera	Moringaceae	Fodder, food, fuel,
Myristica fragrans	Myristicaceae	Food, fuel
Psidium guajava	Myrtaceae	Food, fuel
Syzygium aromaticum	Myrtaceae	Food, industrial uses, timber
Syzygium cumini	Myrtaceae	Fodder, food, fuel, timber
Zizyphus mauritiana	Rhamnaceae	Fodder, food, fuel
Zizyphus nummularis	Rhamnaceae	Fodder, food, fuel
Rhizophora apiculata, mucronata, stylesa	Rhizophoraceae	Fodder, fuel, industrial uses, soil amelioration, timber
Prunus armeniaca, Prunus cerasoides, domestica, persica	Rosaceae	Fodder, food, timber
Pyrus communis	Rosaceae	Food, fuel
Populus alba, deltoides	Salicaceae	Food, Industrial uses, timber

(Contd.)

Species	Family	Major Uses
Salvadora oleoides, persica	Salvadoraceae	Fodder, food, fuel, soil amelioration
Santalum album	Santalaceae	Fodder, fuel, soil amelioration, timber
Litchi chinensis	Sapindaceae	Food, fuel,
Madhuca indica	Sapotaceae	Fodder and fuel
Manilkara littoralis	Sapotaceae	Fodder, Fuel, soil amelioration, timber
Tamarix aphylla, articulata	Tamaricaceae	Fodder, fuel, soil amelioration
Tectona grandis	Verbanaceae	Fuel, soil amelioration, timber

Soil Productivity and its Characteristics

Soil is the most important basic factor for the plants to survive on this earth. Agroforestry has potential for conserving the soil and maintaining its productivity. Agroforestry carries much potential for conserving soil, soil fertility enhancement and productivity aspects in the landholdings. The ability of certain plants and trees to restore soil fertility is understood by several studies. The primary reason for agroforestry systems to improve soil properties is the protection; a tree cover gives the soil against surface compaction, runoff and erosion (Young, 1986). ICRAF reviews pointed several data showing that the presence of trees decreases soil erosion (Nair, 1984). Protecting the soil cover by mulching is one of the widely adapted techniques and it helps in moisture storage also. Certain plants or trees selectively store some nutrients even if the soil containing very low quantity of the same. For example Palm litter is rich in potassium (Folster *et al*., 1976), Tree ferns stores nitrogen (Mueller-Dombois *et al.*1984), *Gmelina arborea* stores calcium (Sanchez *et al.*, 1985), *Cecropia* species accumulates calcium and phosphorous on acidic sites (Odum and Pigeon, 1970), Leguminous trees can be used in the fields to enhance the fertility level. They can bring both organic matter and nitrogen nutrients. Along with translocation of nutrients from soil layers beyond the reach of annual crops or pasture species, enhancement of nutrient status beneath the tree crown has also been attributed to canopy capture of precipitation inputs (Kellman, 1979). Agroforestry is the best methodology which can help in conserve soil on land by reducing soil erosion, increasing soil organic matter and nutrient recycling. A reported research work of various scientists on tree species and their role in enhancing soil productivity has been given in Table 4.

Table 4: Plants used for Enhancing Soil Productivity

Species	Features	Reference
Gliricidia sepium	A fast growing tree. Leaves are rich in protein and highly digestible, and low in fibre and tannin. Helps in nitrogen fixing.	Sierra *et al.*, 2002
Fedherbia albida	Use for adding soil fertility. The fall of leaves at the beginning of the rainy season naturally enriches the soil and improves the micro-environment.	Mokgolodi, *et al.*, 2011
Tephrosia vogelli	A shrub that can improve soil fertility thanks to its important nutrients-rich leaf-fall. It can also be used as an insecticide. It provides good improvements of soil fertility from the second year	Rutunga *et al.*, 2008
Leucaena leucocephala	Leaves of *Leucaena* are highly nutritious for ruminants. Capable of producing a large volume of a medium-light hardwood for fuel wood. Leucaena hedges are useful as windbreaks and firebreaks, the latter due to the suppression of understorey grass growth.	Shelton and Brewbaker, 1994
Glycine max	Important sources of oil and protein. Soy meal is very rich protein feeding stuff for livestock for which there is an increasing demand. Used as a cover crop. Nitrogen fixing crop.	Kanchana *et al.*, 2012

Agroforestry for climate change and carbon sequestration

Agroforestry is making an important contribution to climate change mitigation. The agroforestry systems vary under varied geographic, edaphic and climate regimes of the world and it enables agricultural lands to withstand severe weather conditions and natural calamities. The practice of agroforestry is renowned to have the potential for mitigating climate change effects through the moderation of microclimate. The climate inside and around the agroforestry system is different (Verchot *et al.*, 2007). In a study by (Lal, *et al.*, 2001) from the South Asia reported that summer monsoon will be delayed and will become less certain, and that temperature increases will be most intense during the winter season. Average temperature and soil moisture in agroforestry systems vary with variation due to creation of microclimatic environment created even by the application of mulch (Table 6). Several studies have shown that combining spatial analysis with an analysis of the physiological effects of changes in Carbon dioxide, rainfall and temperature have been done in South Asia to assess the impact of climate change on crop production (Aggarwal and Sinha 1993; Rao and Sinha 1994; Kropff, *et, al.*, 1996; Berge *et al.*, 1997; Saseendran *et al.*, 2000; Aggarwal and Mall, 2002). These studies have shown a decrease in the growing season and yield of most crops as temperature increases. Such reductions were only partially offset by a positive response to increased CO_2 concentrations. Enhancing microclimates like lowering soil surface temperature, reducing evaporation loss from soil surface through mulching and planting shade plants. Carbon sequestration by trees helps in climate change mitigation. Because of greater efficiency of integrated systems in resource (nutrients, light, and water) capture and utilization than single-species systems agroforestry is a good gas–mitigation strategy through carbon sequestration (Nair *et al.* 2009). Several positive and negative implications of agroforestry adoption to climate change have been reported with an example cited in table 5.

Table 5: Examples of positive or negative implications of agroforestry practice for adaptation to or mitigation of climate change

		Mitigation	
		Positive	**Negative**
Adaptation	**Positive**	Soil carbon sequestration, improved water holding capacities, use of manure instead, mixed agroforestry for commercial products, income diversification with trees, reduced nitrogen fertilizer, fire management.	Dependence on biomass energy, overuse of ecosystem services, increased use of mineral fertilizers, poor management of nitrogen and manure, over extraction of non-timber products, timber extraction.
	Negative	Integral protection of forest reserves, limited rights to agroforestry trees, forest plantation excluding harvest.	Use of forest fires for pastoral and land management, tree exclusion in farming lands.

Source: Mbow *et al.*, 2014

Table 6: Average temperature and soil moisture content over a 60-day period after adding three different mulches at a rate of 5000 kg dry matter ha

Treatment/ mulch material	**No. of observations at 15.00 h**	**Average temperature at 5 cm (°C)**	**Average% soil moisture over 0-5 cm**
Unmulched soil	40	37.1	4.8
Leucaena leucocephala	40	34.2 (-2.9)	7.1 (+ 2.3)
Gliricidia sepium	40	32.5 (-4.6)	8.7 (+ 3.9)
Flemingia macrophylla	40	30.5 (-6.6)	9.4 (+ 4.6)
LSD		1.20	1.84

Source: Budelman (1989)

Note: Values in parentheses: the difference relative to an unmulched soil

Carbon sequestration involves the removal and storage of carbon from the atmosphere in carbon sinks (such as oceans, vegetation, or soils) through physical or biological processes. The incorporation of trees or shrubs in agroforestry systems can increase the amount of carbon sequestered compared to a monoculture field of crop plants or pasture (Sharrow and Ismail 2004; Kirby and Potvin 2007). The potential of agroforestry systems to sequester carbon varies depending upon the type of the system, species composition, age of component species, geographic location, environmental factors, and management practices (Table 7 and 8). Agroforestry may involve practices that favour the emission of GHGs including shifting cultivation, pasture maintenance by burning, paddy cultivation, nitrogen fertilization and animal production (Dixon, 1995; Le Mer and Roger, 2001). However, several studies have shown that the inclusion of trees in the agricultural landscapes often improves the productivity of systems while providing opportunities to create C sinks (Winjum *et al.*, 1992; Dixon *et al.*, 1993; Krankina and Dixon, 1994; Dixon, 1995; Montagnini and Nair, 2004). A large number of studies have appeared in recent years that report carbon sequestration potential of agroforestry systems globally.

Global anthropogenic emissions of carbon dioxide (CO_2) to the atmosphere increased to about 9.7 Pg carbon (C) (1 Pg=1015 g) in 2012 mainly due to an increase in fossil fuel combustion (Peters *et al.* 2013). To reduce the emissions, agroforestry was included in global programs such as reducing emissions from deforestation and forest degradation including the role of conservation, sustainable management of forests, and enhancement of forest C stocks (REDD+) related to climate change adaptation and mitigation (Nair and Garrity 2012). Several studies on global carbon sequestration potential of agroforestry systems has done. For example, Dixon (1995) estimated a total of 585–1,215 million ha of land in Africa, Asia and the Americas under agroforestry and a global potential to sequester 1.1–2.2 Pg of carbon (vegetation and soil) over 50 years. Nair *et al.* (2009) estimated a land area of 1,023 million ha under agroforestry worldwide. Using the median carbon sequestration potential used by Dixon (94 Mg ha^{-1}), the land area of 1,023 million ha represents a carbon sequestration potential of 1.9 Pg of carbon over 50 years. Many observers believe that the Clean Development Mechanism (CDM) offered by the Kyoto Protocol could reduce rural poverty by extending payments to low-income farmers who provide carbon storage through land-use systems such as agroforestry (Smith and Scherr, 2002). The SOC sequestration depends primarily on the soil C input and soil stabilization processes. Plant root and rhizosphere inputs, in particular, make a large contribution to SOC (Schmidt *et al.* 2011).

Table 7: Total biomass and carbon stock estimated for agroforestry systems in villages of Karnataka and Tamil Nadu, excluding the forest and plantation.

Villages	Area under cropland (ha	Total biomass (t/ha)	Total carbon stocks (tC/ha)	Total carbon stocks in villages (tC/ha)
Karnataka villages Sirsimakki	72.00	9.47	4.73	340.56
Hallusarige	198.80	1.62	0.81	161.03
Hegle	240.69	3.96	1.98	476.57
Lukkeri	58.00	8.51	4.25	246.50
Tamil Nadu villages Kempanaickenapalayam	501.00	13.17	6.59	3301.59
Sellipalayam	389.00	10.15	5.07	1972.23
Thalakudi	77.00	3.51	1.76	135.52
Valadi	183.00	1.33	0.66	120.78
Total	1719.49			6754.77

Source: Murthy *et al.*, 2013

Table 8: Carbon storage potential of agroforestry systems in different eco-regions of the world

Continent	Eco region	System	Potential (Mg Cha-1)
Africa	Humid tropical high	Agrosilvicultural	29-53
South America	Humid tropical low dry lowlands		39-102 39-195
Southeast Asia	Humid tropical dry lowlands		12-228 68-81
Australia	Humid tropical low	Silvipastoral	28-51
North America	Humid tropical high humid tropical low dry lowlands		133-154 104-198 90-175
Northern Asia	Humid tropical low		15-18

Source: Murthy *et al.*, 2013

Livelihood Security and Agroforestry

From the very beginning agroforestry has been a livelihood survival strategy for the needy especially in the tropics. Rural food security could be met through diversification of suitable species in the farms thus generating income and livelihood opportunity. The diverse products are obtained from these systems like homegardens to ensure food diversity too (Tiwari *et al.*,

2017). Homegardens are land use system involving deliberate management of multipurpose trees and shrubs in intimate association with annual and perennial agricultural crops and invariably livestock within the compounds of individual houses, the whole tree-crop, and animal unit is being intensively managed by family labour (Kumar and Nair, 2006). Home gardens are logical production systems for crop plants that are consumed fresh, used on a daily basis, consumed in small quantities and requiring specific attention (Tesfaye *et al.*, 2010). The study of Tiwari *et al.*, (2017) pointed that farmers' physical assets are important indicator of wealth and is a source of coping shocks in the rural livelihoods. In Asia, about 450 million people rely on forest and trees resources for their livelihoods and in Africa more than 60 percent of the population depends on forests for their food and health (Tiwari *et al.*, 2017). Home garden is generally maintained by small holding farmers. It is believed that traditional home garden practices depend on the type of human community, tradition, needs and beliefs (Tangjang and Arunachalam, 2009). Homegardens are the sites that have long been considered as signs of prestige and pride (Zemede, 1997) by the community on top of their key economic roles. Farmers adopting agroforestry practices have achieved good contribution of their income from agroforestry landuse as is evident from the reported study of Tiwari, *et, al.*, (2017) and details given in Table 9.

Table 9: Agroforestry farmer's income

Revenue	Total Income	Average Revenue per year(Rs/year)	Percentage (%)
Agroforestry			91.43
Agriculture	652,204,118	15,528,670	74.98
Wood	66,060,05	1,572,858	7.59
Livestock	77,067,500	5,137,833	8.86
Non Agroforestry			8.57
Trade	44,580,000	7,430,000	5.13
Labour	21,400,000	3,057,143	2.46
Service	8,520,000	4,260,000	0.98

Source: Tiwari *et al.*, 2017

Conclusion

In present scenario agroforestry systems has been developing scientifically to improve the quality and quantitative of same piece of land to provide better yield and enhance the sustainability. Now, agroforestry system is the major sink after forest to mitigate the carbon, agroforestry systems can at least partially meet the energy needs of billions of people all over the world, by a prudent use of agricultural residues and biomass generated in agroforestry systems. The scientific agroforestry systems with traditional knowledge of rural people can improve the wealth and employment necessary for livelihood improvement and sequester large amount of carbon in a decentralized manner. Such a strategy would also ensure ecological, economic and social well-being.

Reference

Aggarwal P.K, Mall R.K. (2002). Climate change and rice yields in diverse agro-environments of India. II. Effect of uncertainties in scenarios and crop models on impact assessment. *Climate Change*. 52:331–343.

Aggarwal P.K, Sinha S.K. (1993). Effect of probable increase in carbon dioxide and temperature on productivity of wheat in India. *J Agric Meteorol* 48:811–814

Berge H.F.M, Aggarwal P.K, Kropff M.J. (1997). Applications of rice modelling. Elsevier Publishers, Netherlands, p 161

Budelman, A. (1988). The decomposition of the leaf mulches of *Leucaena leucocephala,Gliricidia sepium* and *Flemingia macrophylla* under humid tropical conditions. *Agroforestry Systems*. 7:33-45.

Chakraborty, M., Haider, M.Z. and Rahaman, M.M. (2015). Socio-Economic Impact of Cropland Agroforestry: Evidence from Jessore District of Bangladesh. *Inter. J.Res. in Agric. and For.*, 2(1):11-20.

Chundawat, B.S. and S.K. Gautam, (1993). Textbook of Agroforestry, Oxford & IBH Publishing CO. PVT. LTD. New Delhi, 188p.

Dixon, R. K. (1995). Agroforestry systems: sources or sinks of greenhouse gases? *Agroforestry Systems* 31: 99-116.

Dixon, R. K.; Andrasko, K. J.; Sussman, F. A.; Lavinson, M. A.; Trexler, M. C. and Vinson, T. S. (1993). Tropical forests: their past, present and potential future role in the terrestrial carbon budget. *Water Air Soil Pollution* 70: 71-94.

Dixon, R.K. (1995). Agroforestry systems: sources or sinks of greenhouse gases? *Agroforestry System* 31: 99–116.

Etana, T. (2007). *Use, Threat and Conservation of Traditional Medicinal Plants by Indigenous People in Gimbi Woreda, Western Wollega, Western Ethiopia.* Master's Thesis, University of Addis Ababa, Ethiopia.

Exceutive summary of the synthesis report http://www.agassessment.org/docs/ SR_ Exec_Sum_280508_English.htm. Accessed on March 27, 2009.

Feleke, W. (2000). *A study on biodiversity Management in Daaddegoyo (Traditional Homegardens) by Kaficho people of Bonga area (Southwestern Ethiopia).* Master's Thesis, University of Addis Ababa, Ethiopia.

Folster, H., G. de las Salas and P. Khanna. (1976). A tropical evergreen forest site with perched water table. Magdalena Valley, Colombia. Biomass and bioelement inventory of primary and secondary vegetation. Oecol. *Plant.* 11: 297-320.

Gajaseni, J., Matta-Machado, R. and Jordan. C. F. (1996). Diversified agroforestry systems: buffers for biodiversity reserves, and landbridges for fragmented habitats in the tropics. Pages 506–513 in R. C. Szaro and D. W. Johnston (eds.), Biodiversity in managed landscapes: theory and practice. Oxford, UK: Oxford University Press.

Galluzzi, G.; Eyzaguirre, P. and Negri, V. (2010). Home gardens: neglected hotspots of agro-biodiversity and cultural diversity. *Biodiversity Conservation* 19: 3635-3654.

Garrett, H.E., McGraw, R.L. (2000). Alley cropping practices. In: Garrett HE, Rietveld WJ, Fisher RF (eds) North American agroforestry: an integrated science and practice. ASA, Madison, pp 149–188.

Garrity D.P. (2004). Agroforestry and the achievement of the millennium development goals. *Agroforestry System.*61:5–17.

Gliessman, S. R. (1990). Understanding the basis for sustainability for agriculture in the tropics: experiences in Latin America. In: Sustainable agricultural systems, eds. Edwards, C. A.; Lal, R.; Madden, P.; Miller, R. H. and House, G. St. Lucie Press, Delray Beach, Florida. Pp. 378-390.

GOI, (1999). National Forestry Action Programme, Government of India, Ministry of Environment and Forests, New Delhi, vol. 1 & 2.

Izac, A.M. N. and. Sanchez, P. A. (2001). Towards a natural resource management paradigm for international agriculture: the example of agroforestry research. *Agricultural Systems* 69: 5–25.

Kanchana P, Lakshmi Santha M, Dilip Raja K. (2015). A review on *Glycine max. World Journal of Agroforestry and Pharmaceutical Science.* 5(1): 356-371.

Kellman, M. (1979). Soil enrichment by neotropical savanna trees. *J. Ecol.* 67: 565-577.

Kirby, K.R. and Potvin, C. (2007). Variation in carbon storage among tree species: implications for the management of a smallscale carbon sink project. *Forest Ecology Management.*246:208–221.

Krankina, O. N. and Dixon, R. K. (1994). Forest management options to conserve and sequester terrestrial carbon in the Russian Federation. *World Resource Review* 6: 88-101.

Kropff M.J, Teng P.S, Aggarwal P.K, Bouman B, Bouma J, van Laar H.H.(1996). Applications of systems approaches at the field level, vol. 2 Kluwer Acad. Pub., Netherlands, p 465

Kumar BM and Nair PKR. (2006). Tropical Home Gardens: A Time-Tested Example of Sustainable Agroforestry.India: Environmental Experts.

Kumari, M. A. S.; Kansuntisukmongkol, K. and Brockelman, W. Y. (2009). Plant Diversity in Home Gardens and Its Contribution to Household Economy in Suburban Areas in Sri Lanka. *Environment and Natural Resources Journal* 7: 12-30.

Lal M, Nozawa T, Emori S, Harasawa H, Takahashi K, Kimoto M, Abe-Ouchi A, Nakajima T, Takemura T, Numaguti A.(2001). Future climate change: implications for Indian summer monsoon and its variability. *Curr Sci* 81:1196–1207

Le Mer, J. and Roger, P. (2001). Production, oxidation, emission and consumption of methane by soils: a review. *European Journal of Soil Biology,* 37: 25-50.

Mbow, C., Smith, P., Skole, D., Duguma, L., and Bustamante, M. (2014). Achieving mitigation and adaptation to climate change through sustainable agroforestry practices in africa. *Curr. Opin.Environ. Sustainability*. 6: 8–14. doi: 10.1016/j.cosust.2013.09.002

Mekonnen, E. L.; Asfaw, Z. and Zewudie, S. (2014). Plant species diversity of home garden agroforestry in Jabithenan District, North-Western Ethiopia. *International Journal of Biodiversity and Conservation* 6: 301-307.

Mokgolodi N, Setshogo M, Shi L, Liu Y, Ma C.(2011). Achieving food and nutritional security through agroforestry: a case of *Faidherbia albida* in sub-Saharan Africa. Forestry Studies China 13(2):123-131.

Mueller-Dombois, D., P.M. Vitousek and K.W. Bridger. (1984). Canopy dieback and ecosystem processes in Pacific forests. Hawaii Botanical Science Paper 44, University of Hawaii, Manoa.

Murthy IK, Gupta M, Tomar S, Munsi M, Tiwari R. (2013). Carbon Sequestration Potential of Agroforestry Systems in India. *J Earth Sci Climate Change*. 4: 131.

Nair, P.K.R. (1984). Soil productivity aspects of agroforestry. Nairobi: ICRAF.

Nair, P.K.R. (2009). Agroforestry as a strategy for carbon sequestration. *J. Plant Nutr. Soil Sci.* 172: 10–23.

Nair, P.K.R. and Garrity, D. (2005). Agroforestry—The Future of Global Land Use, Advances in Agroforestry; Springer: Dordrecht, The Netherlands, 2012; Volume 9. Millennium Ecosystem Assessment. Ecosystems and human well-being: biodiversity synthesis. World Resources Institute, Washington, DC.

Nair, P.K.R. and Garrity, D. (2012). Agroforestry research and development: the way forward. In: Nair PKR, Garrity D (eds) Agroforestry—the future of global land use. Springer, Dordrecht, pp 515–531.

Nair, P.K.R.; Kumar, B.M.; Nair, V.D. (2009). Agroforestry as a strategy for carbon sequestration. *Journal of Plant Nutrient and Soil Science* 172:10–23

Odum, H.T. and R.F. Pigeon (eds.). (1970). A tropical rainforest. Vol. III. Washington, D.C.: Office of Information Services. U.S. Atomic Energy Commission.

Perfecto, I., R. A. Rice, R. Greenberg, and M. E. van der Voort. (1996). Shade coffee: a disappearing refuge for biodiversity. *BioScience* 46:598–608.

Rice, R. A., and R. Greenberg. (2000). Cacao cultivation and the conservation of biological diversity. *Ambio* 29:167–173.

Peters, G.P.; Andrew, R.M.; Boden, T.; Canadell. J.G.; Ciais, P; Le Quéré, C.; Marland, G.; Raupach, M.R.; Wilson, C. (2013). The challenge to keep global warming below 2 °C. *Natural Climate Change* 3: 4–6.

Rai, S.N., Chakrabarti, S.K. (2001). Demand and supply of fuelwood and timber in India, Indian

Rao G.D, Sinha S.K.(1994). Impact of climatic change on simulated wheat production in India. In: Rosenzweig C, Iglesias I (eds) Implications of climate change for international agriculture: Crop Modelling Study. EPA, USA, pp 1–10

Ravindranath, N.H. and Hall, D.O., (1995). Biomass, Energy and Environment: A developing country perspective from India, Oxford University Press, New York, 376 pp.

Samati, H. (2004). Kitchen garden plants of Pnar tribe of Jantia hill district. *Ethnobotany* 16: 125-130.

Saseendran S.A, Singh K.K, Rathore L.S, Singh S.V, Sinha S.K. (2000). Effects of climate change on rice production in the tropical humid climate of Kerala, India. *Clim Change*. 44:495–514

Schmidt, M.W.I.; Torn, M.S.; Abiven, S.; Dittmar, T.; Guggenberger, G.; Janssens, I.A.; Kleber, M.; Kögel-Knabner, I.; Lehmann, J.; Manning, D.A.C.; Nannipieri, P.; Rasse, D.P.; Weiner, S.; Trumbore, S.E. (2011). Persistence of soil organic matter as an ecosystem property. *Nature* 478:49–56.

Seta, T.; Demissew, S. and Asfaw, Z. (2013). Home gardens of Wolayta, Southern Ethiopia: an ethnobotanical profile. *Academia Journal of Medicinal Plants* 1: 14-30.

Sharrow, S.H. and Ismail, S. (2004). Carbon and nitrogen storage in agroforests, tree plantations, and pastures in western Oregon, USA. *Agroforestry System* 60:123–130.

Shelton H M, Brewbaker J L. (1994). *Leucaena leucocephala*-the most widely used forage tree legume. *In*Forage Tree Legumes in Tropical Agriculture Eds. R C Gutteridge and H M Shelton. pp 15-29. CAB International, Wallingford. 389 p

Sierra J, Dulormne M, Desfontaines L. (2002). Soil nitrogen as affected by *Gliricidia sepium* in a silvopastoral system in Guadeloupe, French Antilles. *Agrofor Syst* 54:87–97.

Smith, J. and Scherr, S. J. (2002). *Forest Carbon and Local Livelihoods: Assessment of Opportunities and Policy Recommendations*. CIFOR Occasional Paper 37, Centre for International Forestry Research, Jakarta, Indonesia.

Soemarwoto, O. (1987). Homegardens: a traditional agroforestry system with a promising future. In: *Agroforestry: A Decade of Development*, eds. Steppler, H. A. and Nair, P. K. R. ICRAF, Nairobi Kenya. Pp. 157-172.

Srivastava, R. J. and Heinen, J. T. (2005). Migration and Home Gardens in the Brahmaputra Valley, Assam, India. *Journal of Ecological Anthropology* 15: 20-34.

Tangjang, S. and Arunachalam, A. (2009). Role of traditional home garden systems in Northeast India. *Indian Journal of Traditional Knowledge* 8: 47-50.

Tesfaye, A.; Wiersum, K. F. and Bongers, F. (2010). Spatial and temporal variation in crop diversity in agroforestry homegardens of southern Ethiopia. *Agroforestry Systems* 78: 309-322.

Tiwari, P., Kumar, R. Thakur, L. and Salve, A., (2017). Agroforestry for Sustainable Rural Livelihood: A Review, *Int. J. Pure App. Biosci.* 5(1): 299-309.

Tynsong, H. and Tiwari, B. K. (2010). Plant diversity in the homegardens and their significance in the livelihoods of War Khasi community of Meghalaya, north east India. *Journal of Biodiversity* 1: 1-11.

Van Orsdol, K. G. (1987). Buffer zone agroforestry in tropical forest regions. Washington, DC:Forestry Support Program, U.S. Agency for International Development.

Venant Rutunga, Nancy K. Karanja, Charles K.K. and Gachene.(2008). Six month-duration Tephrosia vogelii Hook.f. and Tithonia diversifolia (Hemsl.) A.Gray planted-fallows for improving maize production in Kenya. Biotechnol. *Agron. Soc. Environ.* 12(3), 267-278

Verchot, L.V., M. Van Noordwijk, S. Kandji, T. Tomich, C. Ong, A. Albrecht, J. Mackensen, C. Bantilan, K.V. Anumpama, and C. Palm. (2007). Climate change: linking adaptation and mitigation through agroforestry. *Mitigation and Adaption Strategies for Global Change* 12:901-918.

Williams-Guille-n K, Perfecto I, Vandermeer J. (2008). Bats limit insects in a tropical agroforestry system. *Science* 320:70.

Winjum, J. K.; Dixon, R. K. and Schroeder, P. E. (1992). Estimating the global potential of forest and agroforest management practices to sequester carbon. *Water Air Soil Pollution* 64: 213-228.

Wood, P.J. (1990). Principles of species selection for agroforestry. In: MacDicken, K.G. and Vergara, N.J. (eds.), Agroforestry: Classification and Management, pp. 290-309. John Wiley, New York, USA.

Young, A. (1986). Effects of trees on soils. In R.T. Prinsley and M.J. Swift (eds.), Amelioration of soils by trees: A review of current concepts and practices. London: Commonwealth Science Council.

Zemede, A. and Zerihun, W. (1997). Crop Associations of Homegardens in Wolaytaand Gurage in Southern Ethiopia. *SINET: Ethiopian Journal of Science* 20: 73-90.

10

Organic Waste Management

Pabitra Adhikary, Partha Sarathi Patra, Arju Sayed Ahmed
Subhra Shil and *Rajesh Saha*

Introduction

During the past 60 years, conventional agriculture has become increasingly dependent upon petroleum-based. Chcmically-synthesized fertilizers and pesticides for crop protection and to supply plant nutrients. Certainly, these energy intensive technologies have contributed greatly to this Nation's agricultural productivity. However, sharply escalating production costs associated with the increasing cost and uncertain availability of energy *i.e.*, fuel and fertilizers, have generated considerable interest in less expensive and more environmentally compatible production alternatives such as organic farming (USDA, 1980). The apparent decline in soil productivity throughout the world from excessive soil erosion, nutrient runoff and loss of soil organic matter; the impairment of environmental quality from sedimentation and pollution of natural waters by agricultural chemicals and the potential hazards to human and animal health and food safety from heavy use of pesticides, have also stimulated interest in organic farming systems of food production.

Use of organic wastes and residues in agriculture

Most countries have traditionally utilized various kinds of organic materials to maintain or improve the fertility, and productivity of their agricultural soils. However, several decades ago organic recycling practices were largely replaced with chemical fertilizers which were applied to high yielding cereal grains that responded best to a high level of fertility and adequate moisture, including irrigation. Soil cultivation was also intensified to improve weed control and seedbed conditions. Consequently, the importance of organic matter to crop production received less emphasis, and its proper use in soil management was neglected, or even forgotten. As a result of this and failure to implement effective soil conservation practices, the agricultural soils in a number of developed and developing countries have undergone serious degradation and decline in productivity.

The most realistic approach for many of the developing countries to achieve sufficiency in food production and maximum crop yield potential is to accelerate efforts to halt that decline in soil productivity and to restore the productivity of degraded soils in the shortest possible time. A recent FAO assessment of organic recycling states that the improvement of soil productivity as a whole is expected to contribute about 60 percent of the increased food production that is currently needed worldwide (Hauck, 1981). Much of this goal can be achieved through proper management of agricultural and municipal organic wastes on land to protect agricultural soils from wind and water erosion and to prevent nutrient losses through runoff and leaching. Efficient and effective use of these materials as soil conditioners also provides one of the best means we have for maintaining and restoring soil productivity. The beneficial effects of organic wastes soil physical properties on as evidenced by increased water infiltration, water-holding capacity, water content, aeration and permeability, soil aggregation and rooting depth, and by decreased soil crusting, bulk density and runoff and erosion are widely known (USDA, 1957).

Important manurial resources

Cattle wastes

Cattle play an important role in the Indian economy as they provide an abundant source of manure and valuable items in the diet, as well as being a main source of motive power for agriculture. A cow or buffalo produces from 6 to 9 t of manure per year. The dung *i.e.,* collected from the stalls and pathways and is competed with crop wastes, weeds, graze and silt. To a limited extent the dung is also used in bio-gas plants. Droppings and urine from goat and sheep also provide rich sources of plant nutrients. Poultry, duck and rabbit droppings constitute a minor source of plant nutrients. The nutrient contents of different animal manures, are different.

Human habitation wastes

Night soil has been collected and used for the fertilization of crops for centuries and is a rich source of plant nutrients. Since night soil is considered a valuable source of crop nutrients, its application to the fields could not be banned on the grounds of being a public health hazard lent ill health be replaced by famine. After liberation, nautical modifications were made to exiting practices. The principle of giving priority to prevention has been adopted and there have been health campaigns aimed at the extermination of pests and diseases. The storage chambers are now widely used in rural areas. It is subjected to anaerobic digestion which effectively kills parasitic micro-organism. Night soil is also used as a source of food for fish. Public latrines are built over the fish ponds. In cities, night soil is efficiently collected and dispatched to the rural areas in closed carte or by boat and in mixed with city or crop wastes for high temperature composting, or it is stored in covered pit for about 1-2 months before being distributed to production brigades/teamer. City garbage is effectively reclaimed. Disposal of city garbage in this manner not only improves urban environmental hygiene but also provides large amounts of high quality organic fertilizer for agriculture. The city garbage is partly used for feeding pigs.

Crop wastes

Crop wastes or residues like straw, stalks, stubble, trash, husks, etc. also serve as fertilizer when incorporated into the soil. A mineral fertilizer in appropriate quantities is added to the soil to speed decomposition and make the nutrients available. However, the quantity of these wastes available for manurial purpose may be small because a large proportion is used as animal feed or fuel. Such surplus wastes as are available are used for compost making together with animal/pig/human excreta. Long fibrous material° like sorghum and maize stalks are first chopped mechanically and then composted with nitrogenous rich materials (human/pig excreta) in certain proportions to ensure an appropriate C:N ratio (25:1) for speedy and efficient decomposition. Green vegetation from the mountains, weeds, leaves and turf (non-crop residues) are also applied to the fields after processing.

Aquatic plants

In areas served with-networks of waterways, water weeds are cultivated to provide fodder for livestock and manure for the fields. The weeds provide a major portion of fodder requirements and to that extent pressure on land for fodder rising is relieved. Fish culture is also stimulated under water weeds.

Other resources

The other manurial resources in use in India are (1) Fish wastes, (ii) Silk wastes, (iii) Animal bones, (iv) Oil cakes, (v) Ashes, (vi) Poultry droppings, (vii) Goat and sheep droppings and (viii) Industrial wastes. When appropriately tested and processed these relatively minor sources provide concentrated organic fertilizers.

Fish wastes

Aquaculture in India is a part of the overall agricultural system and considerable quantities of fish waste become available for use as manure, apart from the waste products derived from fish that accumulate in ponds and/or lakes.

Silk waste

Silk wastes are fed to fish which in turn supply waste/droppings for fertilization of cropland. It is basically a natural cyclical pattern and appears to be ecologically sound.

Animal bones

The bones from slaughtered and fallen animals are crushed for production of bone-meal which is a rich source of phosphatic fertilizer.

Oil cakes

As by-products from crushing for oil, cakes and meals from soybean, groundnuts, rape seeds, sesame seeds, make a substantial source of concentrated manure in India. These are used mostly for cash crops such as fruits and vegetables.

Ashes

Ashes from burnt straw, stalks, bushes, wood and coal are used as fertilizer, either alone or mixed with human excreta.

Poultry droppings

The rapid expansion of poultry farming during recent years has considerably increased the availability of poultry droppings for application to the soil as manure.

Sheep and goat droppings

The number of sheep and goats has been increasing in India. They produce substantial quantities of droppings rich in plant nutrients.

Industrial waste

Industrial wastes are usually recovered in substantial quantities and are being re-used mainly in the form of effluents. The effluent, a source of plant nutrients is applied to the land by irrigation. Factories, enterprises, government officials, schools and the general population are involved either as a part of their daily routine or through participation in drives for collecting wastes for re-use in industry or agriculture.

Constraints and potential for increased use of organic wastes

Organic wastes have not been used extensively on land because of certain competitive uses, high costs of collection, processing, transportation and application and because of constraints on usage related to certain chemical and physical properties. For example, (a) cotton gin trash and sugarcane bagasse are now increasingly sought as sources of fuel for burning, (b) some food processing wastes may have extremely high acidity or alkalinity that may adversely affect soil pH, (c) some sewage sludges contain excessive amounts of heavy metals and organic chemicals that may be toxic to plants or endanger the food chain alter absorption and accumulation, and (d) shredded municipal refuse may contain considerable amounts of solid fragments (glass, plastic, and metal) that do not readily biodegrade and might detract aesthetically when applied to land.

The USDA (1978) report pointed out that there is a growing shortage of good quality organic wastes for use in maintaining and improving the productivity of our agricultural soils. The report cited a number of ways in which our limited amounts of organic wastes might be used more effectively as soil amendments. These include:

(a) Improving methods of collection, storage, and processing (*e.g.*, composting) of animal manures to minimize the loss of nitrogen that often occurs in these operations.
(b) Applying manures to land that are presently being wasted.
(c) Applying crop residues to land that are not now being fully utilized.
(d) Increasing the use of sewage sludge on land.
(e) Increasing the use of the organic/compostable fraction of municipal refuse.

Composting to enhance the usefulness of organic wastes

Composting is an ancient practice whereby farmers have converted organic wastes into resources that provide nutrients to crops and enhance the tilth, fertility and productivity of soils. Through composting, organic wastes are decomposed; nutrients are made available to plants. Pathogens are destroyed and malodors are abated. Composts provide a more stabilized form of organic matter than raw wastes and can vastly improve the physical properties of soils. For example, addition of sludge compost to sandy soils will increase their ability to retain water and render them less droughty. In heavy-textured clay soils, the added organic matter will increase permeability to air and increase water infiltration thereby minimizing surface runoff and increasing water storage. Addition of sludge compost to clay soils has been shown to reduce soil compaction, lower the bulk density, and increase the rooting depth (Hornick,*et al.,* 1984).

Some organic wastes may have chemical, physical, and or microbiological properties that would greatly limit the extent to which they could be composted alone. For example some wastes may have an extremely acidic or alkaline pH, others may have an unusually high or low C:N ratio, and still others may vary widely in their solids content. In such cases, selective composting of these wastes with sewage sludges, and night soil, municipal solid waste (*i.e.,*. garbage or refuse), crop residue, animal manures, food processing wastes and certain industrial wastes, may alleviate these deficiencies and provide a readily compostable mixture and higher quality product.

Bio-gas towards recycling of organic wastes

The technology and utilization of bio-gas has recently attracted attention all over the world. The rationale behind it is seen in the ecological benefits of the utilization of animal and agricultural wastes, conservation of natural resources, pollution control, environmental sanitation and, above all, exploitation of new resources of energy. The local resources are largely unused; local fuel and organic manures in the form of animal and vegetable wastes are partly wasted and are partly burned in inefficient furnaces which capture little of the energy content and preserve almost none of the fertilizer value. Through bio-gas technology, these wastes alone can provide fuel for domestic use, energy for irrigation and other agricultural operations, organic manures for farms and light and electricity in remote villages.

Bio-gas plants have several benefits. Briefly these are:

1. The manure obtained from bio-gas plants is richer in nitrogen content and more in quantity than that obtained by conventional composting. It is reported that fermentation increase the ammonia content by 120 percent and the amount of quick acting phosphorus by 150 percent.
2. Bio-gas manure is free from the offensive odour normally associated with manure pits/heaps. The parasites normally present are also killed during the process of digestion and storing of slurry.
3. Bio-gas plants conserve local fuel, wood or imported coal and kerosene, and upgrade into an excellent organic fertilizer the vegetable refuse which would be otherwise burned inefficiently and animal and human wastes which might otherwise be a serious threat to health.
4. Bio-gas can be used to power internal combustion engines and to substitute for diesel oil in small electric generators.
5. The gas provides a clean and convenient fuel for household cooking and a good light for reading. The rice straw which is otherwise burnt as fuel is saved for use as fodder or for silage making.
6. Bio-gas saves time in cooking and thus reduces the housekeeping load for women.
7. Bio-gas plants are complementary to piggery development.

Bio-gas Technology

The process of bio-gas generation, which is essentially the digestion under anaerobic condition of organic wastes, produces a gas which in largely a mixture of methane and carbondioxide. The methane is a combustible gas. The digestion is a two-part process and each part is performed by a specific group of organisms. The first part is the breakdown of complex organic matter into simple organic compound by acid forming bacteria. Several species of acid forming bacteria grow and reproduce rapidly. Those are not very sensitive to their environment. They break down complex organic substances and produce primarily acetic acid and propionic acid, together with ammonia and carbon dioxide. The second group of microorganisms, the methane-formers, break down the acid into methane and carbon dioxide. This group is composed of a relatively few species of bacteria that grow and reproduce slowly and are quite sensitive to their environment. In a properly functioning digester, the two group of bacteria must be balanced so that the methane-former just use the acid produced by the acid-formers. If the acid formers get ahead of the methane-formers, acids begin to accumulate, the pH drops, the methane-formers are inhibited and ultimately digestion ceases.

Organo-mineral fertilizers

India has been producing organo-mineral fertilizers though perhaps on a limited scale. Superphosphate and ammonia are generally used together with organic materials in making these fertilizers. Agronomically, the benefits from balanced use of organic manures and mineral fertilizers are well established. An organo-mineral fertilizer has the following advantages:

1. It supplies all the nutrients in balanced quantities and therefore improves the percentage utilization of nutrients added in fertilizers as required by crops;
2. It prevents losses of nutrients because of the high ion exchangeable capacity of organic matter;
3. It supplies nutrients readily and steadily over the entire growing period of crops;
4. It helps maintain the soil organic matter at a steady level, thereby ensuring good soil physical properties and fertility status;
5. It has significant residual effects on succeeding crops and helps maintain productivity;
6. It is more economical to transport the material over greater distances since it is enriched in N, P and K and contains more plant nutrients per unit volume, as compared to bulky organic manures.

Thus, such an organo-mineral complex would help in removing the shortcomings of organic manures on the one hand and mineral fertilizers on the other. There is, therefore, a need to popularize production of organo-mineral fertilizers in developing countries. There could be several approaches to production of enriched organic manures. It has been found that the rate and extent of liberation of nitrogen from applied organic manures is adequate for normal plant growth when their C:N ratio is less than 10:1 or the nitrogen content is more than 2.5%. In practice, the C:N ratio of organic wastes can be reduced by adding the requisite amount of nitrogen in the form of nitrogenous fertilizer. Thus a low quality manure containing 0.5% N with a C:N ratio of 20:1 can be treated with ammonium sulphate solution to add 2% more nitrogen. This would reduce the C:N ratio to 8:1 and the enriched manure can then be utilized effectively. This method is well suited to the direct utilization of large quantities of organic wastes of wide C:N ratios such as paddy and groundnut husks, wheat chaff, tobacco wastes and any material in a suitable physical condition to be easily incorporated in soil. This method enables the process of composting to be eliminated, thug saving time and labour.

There is a chemical process also for disintegrating bones, leather wastes and obtaining a sterilized bone-meal powder which can be used as manure or as a cattle feed. The dry bones are treated with diluted caustic soda (5.0%) until they disintegrate completely to form a paste (12-15 days). The disintegrated material is then washed with water and treated with 10% hydrochloric acid solution to neutralize it. The dried material yields bone-meal in powder form.

Aquatic Plants

Aquatic plants are considered as weeds in most countries and all possible means are taken to eliminate them, while in India these plants are cultured in a systematic way to provide material both for compost making and fodder for live stocks. Feeding pigs with various aquatic plants such as water hyacinth, water lettuce or water chestnuts as a substitute for concentrate feed is a useful innovation. These plants could also serve as raw material for bio-gas plants.

Techniques for controlled cultivation of water plants adopted in India are worth emulating by other countries. A simple method is to drag out the plants from waterways and put them on the banks for semi-drying; they are then reaped in alternate layers with mud or silt in equal proportions. To hasten decomposition, the heap is turned two or three times and when turning, it must be ensured that the undecomposed plants are completely covered and not left exposed on the surface.

Tank, pond and river silt

The beneficial effect of silt or mud on crop growth and soil improvement is well known to Indian farmers. The extensive use of silt in agriculture that country also helps in de-silting waterways, tanks and ponds. The mechanical device for sucking mud from the bottom of waterways could be adopted with ease in other countries.

Reference

Hauck. F.W. (1981). The Importance of Organic Recycling In the Agricultural Programmes of Developing Countries. In Organic Recycling in Asia, pp: 17-28.

Hornick. S.B.. Murray, I., Hainey. R.J., Sikora. I.J., Parr. J.F., Burge. W.D., Willswn. G.B. and Tester. C.F. (1984). Use of sewage sludge cotmpost for soil improvement and plant growth. Science and Education, Administration and Agricultural Reviews and Manuals. Northeastern Region Series 6. U.S. Department of Agriculture: Beltsville. MD. pp.10.

U. S. Department of Agriculture (1957). Soil Year Book of Agriculture. Government Priting Olice; Washington. D.C.

U. S. Department of Agriculture (1978). Improving Soil with Organic Wastes. Report to the Congress and response to Section 1461 of the Food and Agriculture Act of' 1977 (PL 95-113).I.S. Government Priting Olice; Washington. D.C. 157 pp.

U.S. Department of Agriculture (1980). Report and Recommendation on Organic Farming. Specialreport prepared for the Secretary of Agriculture. U S. Government Printing Oficc; Washington, D.C. pp. 94.

11

IPM: An Eco-friendly Approach for Pest Management

Biswajit Patra **and** ***Biplab Tudu***

Introduction

Chemical pesticides play an important role in plant protection in India and most of the developing countries of the world. These synthetic organic insecticides provide spectacular control of insects. Chemical control is one of the effective and quicker methods in reducing pest population. However, over reliance and injudicious use of pesticides resulted in a series of problems in the agricultural ecosystem mainly, the development of resistance in insects to insecticides, pest resurgence, outbreak of secondary pests into primary nature, environmental contamination and residue hazards, destruction of natural enemies of insect pests etc. All these problems aggravate the pest problem in agricultural crops. Thus, the importance of eco-friendly pest management techniques is being realized more and more in the recent past. The modern concept of pest management is based on ecological principles and involves the integration of different control tactics into a pest management system. This is not altogether a new concept. It was practiced before the advent of modern chemicals. Integrated Pest Management is an ecological approach in which utilization of all available techniques of pest control to reduce and maintain the pest population at levels below economic injury level. The UN's Food and Agriculture Organisation defines IPM as the careful consideration of all available pest control techniques and subsequent integration of appropriate measures that discourage the development of pest populations and keep pesticides and other interventions to levels that are economically justified and reduce or minimize risks to human health and the environment. IPM emphasizes the growth of a healthy crop with the least possible disruption to agro-ecosystems and encourages natural pest control mechanisms. A well-defined Integrated Pest Management (IPM) is a program that should be based on prevention, monitoring, and control which offers the opportunity to eliminate or drastically reduce the use of pesticides, and to minimize the toxicity of and exposure to any products which are used. So, IPM is an eco-friendly approach which encompasses cultural, mechanical, biological and need based chemical control measures.

Objectives of IPM

The major objectives of IPM are as follows:

- Maximize crop production with minimum input costs;
- Minimize environmental pollution in soil, water and air due to application of pesticides;
- Minimize occupational health hazards due to chemical pesticides;
- Preserve ecosystem and maintain ecological equilibrium;
- Judicious use of chemical pesticides for reducing pesticide residues,
- To improve farming systems.

Utilization of economic threshold levels (ETL)

One of the fundamental concepts of integrated pest management is that each pest species has a definable relationship in terms of damage to the plant it attacks. For effective management of target pest, the level of pest population is very important consideration. The concepts of 'Economic Injury Level' (EIL) and 'Economic Threshold Level'(ETL) are considered keystones of the Integrated Pest Management (IPM) programme (Stern *et al.*, 1959; Pedigo, 1991; Stejskal, 2001;). Originally Stern *et al.* (1959) defined the 'Economic Injury Level' (EIL) as the lowest population density that will cause enough economic damage to justify the cost of artificial control measures. The economic threshold is the pest density at which control measures should be determined to prevent an increasing pest population from reaching economic injury level. The determination of these thresholds is a pre-requisite to the development of any pest management strategy.

IPM decision making

There is a 6 step process to decision-making in IPM

1. Identify the problem or pest
2. Determine the severity of the problem (scouting, traps, past history)
3. Assess the management options (do nothing, cultural, biological, chemical control)
4. Select and apply one or more options
5. Measure the success of options employed
6. Record the results

IPM components/strategies

The available techniques for management of insect pests are as follows:-

1. **Selection of variety:** Choosing beneficial crop varieties, such as those with disease and pest resistance, has always been an important consideration of IPM. These varieties can be derived from traditional cross-breeding or modern biotechnology: pest-resistant and herbicide-tolerant varieties, for example, may reduce the need for other crop protection measures.
2. **Cultural Practices:** Cultural methods of pest control are simple modification or adaptations of regular farm operations in such a way which either destroy the pests or prevent them from causing economic loss. An area-wide implementation of cultural practices is essential for successful pest management. Some of the examples of cultural practices are as follows-
   - The manipulation of planting time helps to minimize pest damage by producing asynchrony between host plant and the pest or synchronizing insect pests with their natural enemies. Adjustment of time of sowing and harvesting is necessary to escape peak season of pest attack.
   - Community approach is required to sow the crops simultaneously (synchronized sowing) in vast area so that pest may not get different staged crops suitable for its population build up. If pest appears in damaging proportion, control operation could also be applied effectively in whole area.
   - Use of appropriate seed rate ensures proper stand, spacing, crop canopy. Spacing may influence the population build up of many insect pests by modifying the micro-environment. Closer spacing has been reported to increase the incidence of brown plant hopper in rice.
   - Selection and use of clean and certified seeds and treating seeds with seed treating chemicals before sowing for seed borne pests and disease control.
   - Deep summer ploughing is helpful in population reduction of insect pests by exposing them to their natural enemies like birds and to the action of sun and wind. Raking and hoeing of the soil around mango, citrus, guava and other fruit trees help to destroy pupae of fruit trees.
   - Inter-cropping or multiple cropping also helps in reducing the pest population in different crops. All the crops are not preferred by each pest species and certain crops act as repellents, thus keeping the pest species away from preferred crops resulting in reduction of pest incidence.
   - Growing trap crops on the borders or peripheries of fields. Trap crops

are grown to attract some insect pests so that the target crop escapes pest attack. By growing such crops on the border of the fields, pest population develop there which can be either killed by using pesticides or its natural enemies are allowed to develop there for natural control.

Growing of two rows of bold-seeded Indian mustard after every 25 rows of cabbage helps in pest suppression of cabbage. The first row of mustard is sown 15 days prior to the cabbage planting and second row is sown 25 days after planting. Mustard attracts more than 80 percent of the cabbage pests. However, the mustard foliage is to be sprayed with insecticides (Srinivasan and Moorthy 1991; 1992).

African marigold (45 days old) is grown as trap crop for every 16 rows of tomato to synchronize flowering in both the crops for the management of tomato fruit borer, *H. Armigera*. Most of the eggs of borer are laid in marigold flowers or flower buds, and only negligible eggs are laid in tomato (Srinivasan *et al.*, 1994). The limitations of this technology are that the flowering in marigold and tomato has to be synchronized and some tomato rows are to be sacrificed for marigold (Subramaniam, 1997).

- Removal of crowded/dead/broken/diseased branches and destruction of them. It should not be pile them in the orchards which may act as source of pest infestation.
- Crop rotation with the non-preferred hosts is also an effective method of pest management. This strategy is effective for the pests which have narrow host range and low dispersal capacity.
- Cultivation of pest resistant/tolerant varieties also plays a significant role in pest suppression.
- Integrated Nutrient Management (INM) need to be emphasised. Testing of soil for nutrients deficiencies on the basis of which fertilizers should be applied.
- Application of optimum dosage of nitrogen in 2-3 splits avoids build up of insects such as gall midge, leaf folder, BPH and WBPH.
- Proper water management should be done. The amount of moisture in soil has a profound effect on the survival of many insect pests. Flooding of fields has been recommended for reducing the attack of whitegrub, termite etc. Alternate wetting and drying to avoid water stagnation is important suppression of brown plant hopper in rice.
- Proper weed management. It is well-known fact that most of weeds compete with crop for micro nutrients also harbor many pests.

3. **Mechanical control:** Different mechanical controls in pest management are as follows:
   - Hand-picking and destruction of egg masses, larvae, pupae and adults of insect pests and infested plant parts wherever possible also effective for pest management under certain conditions.
   - Installation of bird perches in the field for allowing birds to sit and feed on insects and their immature stages *viz.,* eggs, larvae and pupae.
   - Use of different kind of traps (Light trap, yellow sticky trap, yellow pan trap etc.) for attracting and destruction of trapped insects. Pheromone traps are used for mating disruption, monitoring of pest levels and mass trapping of insect pests.
   - Use of rope for dislodging larvae feeding on leaves *e.g.,* caseworm and leaf folders in rice.
   - Clipping and pruning of infested plant parts also helps in reducing the pest population.
   - Use of pheromone traps for mass trapping and killing.

4. **Biological control:** Natural enemies of insect pests play a key role in reducing the levels of pest populations below those causing economic injury. Both natural and applied biological control tactics are important in successful management of pest populations. DeBach (1964) defined biological control as "the action of parasites, predators, or pathogens in maintaining another organism's population density at a longer average than would occur in their absence." In IPM programmes, native natural enemy populations are conserved, and non-native agents may be released with utmost caution. Trichogramma spp. are the most popular parasitoids being applied on a number of host crops. A number of microorganisms such as *Trichoderma* spp., *Verticillium* spp., *Aspergillus* spp., *Bacillus* spp. and *Pseudomonas* spp. that attack and suppress the plant pathogens have been exploited as biological control agents. Inundative releases of the egg parasitoid, *Trichogramma brasilensis* @2,40,000/ha are recommended for the control of fruit borer. Some of the common bio-control agents are

   Predators: Predators may be insects or other insectivorous animals each of which consumes many insect preys during its lifetime. Predators are often large, active, and conspicuous in their behavior, and they are therefore more readily recognized than are parasites and pathogens. Predators catch and eat their prey. Some common predatory arthropods include ladybird beetles, carabid (ground) beetles, staphylinid (rove) beetles, syrphid (hover) flies, lacewings, minute pirate bugs and spiders.

Parasitoids: These are the organisms which lay eggs in or on the bodies of their hosts and complete their life cycles on host bodies. When the egg of the parasite hatches, the young parasitic larva feeds on the host (the pest) and kills it. A parasitoid may be of different type depending on the host developmental stage in or on which it completes its life cycle. For example, egg, larval, pupal, adult, egg-larval and larval pupal parasitoids. Examples are different species of *Trichogramma, Apanteles, Bracon, Chelonus,* etc. In paddy, release of *Trichogramma* spp. (paddy ecosystem adapted strain) may be useful against the stem borer, *S. incertulas* and rice leaf folder complex, *C. medinalis.* The inundative release of exotic parasite *T. japonicum* @ 20,000 per acre was effective in reducing stem borer infestation in rice (Gupta *et al.*, 1987).

Pathogens: These are micro-organisms which infest and cause diseases in their hosts as a result of which hosts are killed. Major groups of pathogens are fungi, viruses and bacteria. Some nematodes also cause diseases in some insect pests.

Fungi: *Hirsutella thompsonii, Beauveria bassiana, Metarhizium anisopliae etc*

Virus: Nuclear polyhedrosis virus (NPV) and granulosis viruses.

Bacteria: *Different strains of the bacterium, Bacillus thuringiensis commonly known as Bt.*

5. **Microbial Control:** Microbial control of insects is achieved through the inundative application of allowable formulations of insect-pathogenic bacteria (*e.g., Bacillus thuringiensis*), insect-pathogenic fungi (*e.g., Beauveria bassiana),* or insect viruses (NPV). Fungal pathogens such as *Beauveria bassiana* against rice hispa (Hazarika and Puzari, 1997), *Pandora delphacis* against BPH (Narayanasamy, 1995) have been found promising. The sprays of Ha NPV at 250 larval equivalents/ha, has been found to be effective in controlling fruit borer.

There are three broad approaches to bio-logical control

(i) **Classical biological control (also called importation of natural enemies):** It involves the importation, screening, and release of effective natural enemies to permanently establish in new areas. Classical biological control usually targets introduced (non-native) pests, most of which arrive in the new area without the natural enemies that control their populations in their native lands. Native pests that are not adequately controlled by existing natural enemies may also be the target of classical biological control.

(ii) **Augmentative biological control (or augmentation of natural enemies):** In this process, the population of natural enemies already present in the area is increased. This is done by releasing either laboratory reared or field collected bioagents. The goal of augmentative biological control is to temporarily in-crease the number of natural enemies and, therefore, the level of biological control of the target pest.

(iii) **Conservation biological control (or conservation of natural enemies):** It improves the effectiveness of natural enemies through natural enemies' friendly farming practices that provide necessary re-sources for their survival and protect them from toxins and other adverse conditions. These conservation practices will benefit all natural enemies, whether they are native, successfully established through classical biological control, or released for augmentative biological control.

Some example of exotic natural enemies introduced in India

S. No.	Natural Enemies(Order:Family)	Pests/crops
1.	*Aphelinus mali* (Hymenoptera: Aphelinidae)	Introduced for the control of woolly aphid, *Eriosoma lanigerum* on apple.
2.	*Encarsia perniciosi* (Hymenoptera: Aphelinidae)	introduced for the biological suppression of San Jose scale, *Quadraspidiotus perniciosus*
3.	*Rodolia cardinalis* (Coleoptera: Coccinellidae)	Released against cottony cushion scale, *Icerya purchasi* in the Nilgiris (Tamil Nadu) in 1930, upper Palni hills (Tamil Nadu) in 1941
4.	*Cryptolaemus montrouzieri* (Coleoptera: Coccinellidae)	Introduced in June 1898 and presently established on a variety of mealybugs infesting fruit crops, coffee, ornamentals, etc. in south India.
5.	*Leptomastix dactylopii* (Hymenoptera: Encyrtidae)	Introduced for the control of *Planococcus citri*
6.	*Zygogramma bicolorata* (Coleoptera; Chrysomelidae)	Released for the control of *Parthenium hysterophorus*

General advantages of biological control include:

- Biological control can be cost effective in the long run.
- Biological control is less costly and cheaper than any other methods.
- Bio-control agents give protection to the crop throughout the crop period.
- They do not cause toxicity to the plants.
- Self-perpetuation at little or no cost following the initial effort.
- Application of bio-control agents is safer to the environment and to the person who applies them.

- They do not cause residual problem.
- General inability of pests to develop resistance to BC agents.

General Limitations of Biological Control include-

- It's a slow process. It takes a lot of time and patience for the biological agents to work their magic on a pest population, whereas other methods like pesticides work provide immediate results.
- The pest population will continue to exist at a level determined by the properties of the pest, its natural enemies and of the habitat they occupy.
- The effectiveness of natural enemies must be considered relative to economic thresholds.
- The research necessary in seeking a biological control solution to a problem is often demanding in terms of scientific and technical staff, funds, and time and a solution cannot be guaranteed in advance.

6. **Regulatory approaches:** Regulatory or legislative control involves enactment of laws to regulate the entry, establishment and spread of pests.In this process, regulatory rules framed by Govt. are brought into force under which seeds and infested plant materials are not allowed to enter the country or from one part to other parts of the country. These are known as quarantine methods and are of two types i.e. domestic and foreign quarantine.

   Plant quarantine methods: There are number of plant quarantine methods which are used separately or collectively to prevent or retard the introduction and establishment of exotic pests and pathogens. The components of plant quarantine activities are:

   - **Complete embargoes:** It involves absolute prohibition or exclusion of specified plants and plant products from a country infected or infested with highly destructive pests or diseases that could be transmitted by the plant or plant products under consideration and against which no effective plant quarantine treatment can be applied or is not available for application.
   - **Partial embargoes:** When a pest or disease of quarantine importance to an importing country is known to occur only in well-defined area of the exporting country and an effectively operating internal plant quarantine service exists that is able to contain the pest or disease within this area.
   - **Inspection and treatment at point of origin:** It involves the inspection and treatment of a given commodity when it originates from a country where pest/disease of quarantine importance to importing country is known to occur.

- **Inspection and certification at point of origin:** It involves pre-shipment inspection by the importing country in cooperation with exporting country and certification in accordance with quarantine requirements of importing country.
- **Inspection at the point of entry:** It involves inspection of plant material immediately upon arrival at the prescribed port of entry and if necessary subject to treatment before the same related.
- **Utilization of post entry plant quarantine facilities:** It involves growing of introduced plant propagating material under isolated or confined conditions.

7. **Insect sex pheromones:** Sex pheromones have been found effective in the management of some pests. They control the insect through capture and annihilation by either mass trapping or disrupting mating communication. The pheromones are likely to play an important role in rice IPM strategies in future (Krishnaiah *et al.,* 1998). However, sex pheromones are species specific and are not useful in situations where two or three insect pests occur simultaneously.
8. **Chemical:** Chemical control is one of the effective and quickest methods of reducing insect pest population. Pesticides are used to keep the pest populations below economically damaging levels when the pests cannot be controlled by other means. Pesticides include both the synthetic pesticides and plant-derived 19 pesticides. Synthetic pesticides include a wide range of man-made chemicals. These are easy to use, fast-acting and relatively inexpensive. Ideally, pesticides should be used as a last resort in IPM programmes because of their potential negative effect on the environment. In IPM programme indiscriminate spraying is eliminated and only judicious use of pesticides is employed, minimizing damage to biological control agents and the environment. Appropriate chemical control strategy involves the right choice of active ingredient, suitable formulation and application techniques on the basis of pest biology and crop phenology. The knowledge regarding the most susceptible stage of the pest, quantitative data on pest incidence and significance of particular pest populations on yield loss is also crucial for economic and successful pest control. Further, understanding of the potential hazards of pesticides to the users, consumers and environment is essential. Botanical pesticides can be applied as raw crushed plant leaves, extracts of plant parts, and chemicals purified from the plants. Pyrethrum, neem, tobbaco, garlic, and pongamia formulations are some examples of botanicals. Botanicals are generally less harmful to the environment, because of their quick degrading property. They are less hazardous to transport. The major

advantage is that these can be formulated on-farm by the farmers themselves.

- ETL and pest defender ratio must be observed
- Relatively safer pesticides should be selected *e.g.,* neem based and bio-pesticides.
- If pest is present in strips or isolated patches, whole field should not be sprayed.
- If use of a pesticide is necessary, select the one that is most effective but least toxic to non-target organisms or least persistent in the environment, whichever is more important in that location.

Constraints in IPM implementation

Institutional constraints: IPM requires an interdisciplinary, multi-functional approach to solving pest problems. Inadequate interaction between research and extension agencies is a constraint.

Informational constraints: The lack of IPM information which could be used by farmers and by extension workers is still a major constraint in many countries.

Sociological constraints: Some farmers feel it is risky to adopt IPM compared to use of pesticides alone. Our farmers are habituated to using more pesticides. Low awareness and innovativeness of extension personnel and target groups.

Economic constraints

- Lack of funds for training of farmers and extension workers on the use of IPM is also a constraint in implementation of IPM.
- Problem of timely and adequate supply of quality inputs, including biocontrol agents and biopesticides is a major problem.
- Non-availability of location-specific IPM modules for many crops is also a constraint.

Advantages of integrated pest management

IPM provides multiple benefits for society and the environment. Some of the benefits of IPM are as follows:

- Improved crop profitability due to better pest control measures and appropriate use of crop protection products.
- Stable, reliable and quality crop yields.
- Reduced potential for problems of pest resistance or resurgence.

- IPM strategies reduce/slower the risk of development of resistance to insecticides.
- Potential reduced use of pesticides, which reduces chances of environmental contamination and worker health problems.
- Increased consumer confidence in the safety and quality of food and fiber products.

Essentials for implementation of IPM

- Availability of location-specific IPM modules, which are ecologically sound, economically viable and socially acceptable.
- High level of target group participation.
- Area-wide dissemination strategy.
- Removal of obstacles in dissemination of IPM.
- Measuring, evaluating and publicizing the impacts of IPM.

Conclusion

The most obvious and apparently environment-friendly alternative to pesticides is to follow Integrated Pest Management. Many plant species have been reported to possess pesticidal and pest growth inhibiting properties, but their potential remains untapped by the industry. A successful IPM programme requires time, money, patience, short-and long-term planning, flexibility and commitment.

Reference

DeBach, P. (1964). The scope of biological control. p. 3-20. InBiological Control of Insect Pests and Weeds (P. DeBach, editor). Chapman and Hall Ltd., London. 844 pp.

Gupta, M., P.P. Rao and A.D. Pawar. (1987). Role of parasites and predators in rice ecosystem. Proceedings of the National Symposium on Integrated Pest Control: Progress and Perspectives, (Eds. N. Mohan Das and George Koshy). pp. 545.

Hazarika, L.K. and K.C. Puzari. (1997). Field efficacy of white muscardine fungus (Beauveria bassiana) on rice hispa (Dicladispa armigera). *Indian Journal of Agriculture Science* 67(10): 463-465.

Krishnaiah, K., Syed Zainulabedin, A. Ganeswara Rao, D.V.S.S. Kumar and N. Rama Gopala Verma. (1998). Pheromone monitoring system of rice yellow stem borer, Scirpophaga incertulas Walker. *Indian Journal of Plant Protection* 26(2): 99-106.

Narayanasamy, P. (1995). Mycoinsecticide : A novel biopesticide in Indian scenario. Biotechnology and Development Review, Jan.-June, 1995 pp.15-20.

Pedigo, L.P. (1991): Entomology and pest management. Macmillan Publ. Comp., New York. 356 pp.

Srinvasan, K. and P.N. Krishna Moorthy. (1991). Indian mustard as a trap crop for management of major lepidopterous pests on cabbage. *Tropical Pest Management* 37: 26-32.

Srinvasan, K. and P.N. Krishna Moorthy. (1992). Development and adoption of integrated pest management for major pests of cabbage using Indian mustard as trap crop: diamond back moth and other crucifer pests. Proceedings of the Second International Workshop, Tainan, Taiwan, 10-14 December 1990. Asian Vegetable Research and Development Centre, AVRDC Publication Centre, AVRDC Publication No. 92-368. pp. 511-521.

Srinivasan, K., P.N. Krishna Moorthy and T.N. Raviprasad. (1994). African marigold as a trap crop for the management of the fruit borer; Helicoverpa armigera on tomato. *International Journal of Pest Management* 40 (1): 56-63.

Stejskal, V. (2001): A new concept of economic injury level that includes penalization of damage to quality and safety of agricultural products. *Plant Prot. Sci.* 37, 151–156.

Stern, V. M., R. F. Smith, R. van den Bosch, and K. S. Hagen. (1959). The integrated control concept. *Hilgardia* 29:81101.

Subramaniam, K.V. (1997). Critical factors in adoption of integrated pest management technologies: An economic analysis. In: First National Symposium in Pest Management in Horticulture Crops. IIHR, Bangalore. Oct 15-17.

12

Sustainable Plant Disease Management

Sibdas Baskey **and** ***Satyajit Hembram***

Introduction to sustainable disease management

Plants and pathogens are involved in ongoing interactions over millennia that have been modified by coevolutionary processes to limit the spatial extent and temporal duration of disease epidemics (Zhan *et al.* 2015). These interactions are disrupted by modern agricultural practices, such as intensified monoculture using high yielding varieties. Agricultural intensification and monoculture coupled with globalization disrupt the coevolutionary dynamics of plant-pathogen interactions which escalating the evolutionary rate of pathogens and increasing the risk of plant disease epidemics. These activities, when supplemented with high resource inputs, create unique conditions which is conducive to widespread plant disease epidemics and rapid evolution of pathogen. Disease management requires a significant shift to include eco-evolutionary principles in the design of adaptive disease management strategy for minimizing the evolutionary potential of plant pathogens by reducing their genetic variation, stabilizing their evolutionary dynamics, and preventing dissemination of pathogen variants carrying new infectivity or resistance to pesticides (Zhan *et al.* 2014). As environmental and ecological issues continue to impact on agriculture, all the crop-specific and regional-specific, Good Agricultural Practices (GAP) to be developed for crop production, must be economically feasible, ecologically sound, environmentally safe and socially acceptable. Numerous non-chemical methods for management of crop diseases such as pathogen-free seeds, disease resistance, crop rotation, plant extracts, organic amendments and biological control agents are considered less harmful than chemical pesticides and therefore, offer great potential for application in modern agriculture (Marshal *et al.* 2008). No single method can provide satisfactory management of crop diseases. Integration of all the effective and eco-friendly measures in accordance with the dynamics of the agro-ecosystem management would be the best strategy for efficient plant disease management.

Importance of sustainable disease management

Plant diseases result when a susceptible host and a disease causing pathogen meet in a favorable environment. If any one of these three conditions were not met, there would be no disease. The disease triangle model of plant disease is an interaction between the host, the pathogen and the environment (Francl *et al.* 2001). Plant pathogens are influenced by both the host plant (leaf age, shoot age, flower and fruit set etc) and the environment (Temperature, Relative humidity, Leaf wetness and Soil moisture etc). Any host can be susceptible, if the environmental conditions are conducive for infection and if a virulent pathogen is present. One should take these factors into account prior to fixing the timing of chemical pesticide application as well as other disease management strategies. It is important to remember that a set schedule cannot be the only permanent stable useful tool in the dynamic nature, how the host, pathogen and environment interact, it depends on the disease triangle reaction. Although a set schedule might work on particular occasion, the dynamic nature of plant disease would make such a schedule only occasionally useful, at best. To understand and ultimately develop an effective schedule to manage plant diseases, it's useful to remember the plant disease triangle. Keeping plants healthy is the easiest way to prevent disease. Disease management practices must be tailored to the plant species being grown and the diseases that are present. If not, these diseases could develop into a serious threat if environmental conditions change and become conducive to disease (Agrios,1997).

Why sustainable plant disease management?

Modern agriculture is changing fast with globalization and with it the landscape through which disease spreads. This imposes new demands on our understanding of influence the invasion and persistence of new pathogenic strains, how and why they out-compete resident pathogens and how to promote durable management methods for coping up the demand of food for global population in a sustainable way. This requires an understanding of what influences the variability of pathogen between one location and another and from one season to another, and how this impinges upon local, national and international crop loss. But this management method should be environmentally safe and economically feasible and it may be the product of unification of all the available knowledge of plant disease management.

Approach to sustainable plant disease management

Sustainable disease management is basically the unifying concept that connects the wide range of individual disease management technique; these can broadly be categorized as physical method, Cultural method, Biological method and Chemical method that focus on the interactions of microorganisms, environment and plants (Huang *et al.*2009).

Cultural management can be included for reducing the amount of initial or primary inoculums, reducing the rate of spread of an established disease, or plating a crop at a site that is not favourable to pathogens because of its altitude, temperature, or water availability. It is basically preventive rather than curative, it needs long range planning and it depends on detailed knowledge of bio-ecology of the crop-pathogen-natural management-environment relationship, most of which were poorly understood. The results were very variable and it was often difficult to evaluate their effectiveness. But today, due to advancement of science, we have better understanding of crop as well as weather conditions which is useful enough to use cultural methods more effectively to manage the disease. Some important principles of cultural management are (a) reducing the levels of inoculums, (b) reducing the rate of spread of disease, (c) intercropping and mixed cropping, and (d) mulching and soil amendments. Practices that reduce the initial levels of inoculums include selecting appropriate planting materials, destruction of crop residues, elimination of living plants that carry pathogens, and crop rotation. Selection of appropriate planting material can involve planting resistant cultivars, planting a number of mixed cultivars, using certified seed and ensuring that disease is not spread on vegetative propagating materials or on equipment.

In Monoculture agriculture system, due dense population with genetic homogeneity, once a disease becomes established in such a crop community, it can rapidly be spread to epidemic proportion. Hence, Monoculture should be avoided. Destruction of crop residues, which can harbour many pathogens, by burning, burying or removing is an important practice performed between cropping seasons. Elimination of living plants that carry pathogens can include both remnant or diseased crop plants and also wild plants or weeds that can act as alternative hosts between seasons. Crop rotation refers to the successive planting of different crops in the same area, sometimes with a fallow, or resting period in between crops. The survival period of the pathogen, thus reducing or eliminating inoculums in the interval before a susceptible crop is planted there again. It tends not to be effective against pathogens that survive for a long time in the soil. The rate of disease spread can be reduced by controlling the spacing of plants, humidity, moisture levels and amount of sunlight. These factors are more easily controlled in a greenhouse environment, and some can be manipulated in the field also. In general, wider spacing of the plants reduces the speed with which disease moves between them, and reducing moisture levels can inhibit infection by pathogens. The period of time that the plants and soil are wet can be reduced by watering in the morning, spacing plants further apart, pruning or rouging to reduce plant canopy cover, and orienting rows of plants with relation to the prevailing wind. Tillage practices have indirect effects on the spread of plant pathogens, although, some form of inoculums can be spread extensively during tillage. However, Tillage can bury pathogens prevailing at the top soil into the deeper, where they are less likely to cause disease.

Sowing practices, such as changing of time, depth and direction of sowing and changing the density of crop can protect plants from pathogens, to which, they are susceptible only at certain stages of their growth. Changing the time of sowing can exploit weather conditions that are unfavourable to the pathogen, thus reducing crop losses. Suppose a disease favours a particular weather, if we manipulate the sowing time in such a way, that plants becomes much older at that time, when favourable weather condition will be there. Hence, the plant can escape the disease, as because; older plants are much more resistant to diseases than that of at young age. In similar way, viral diseases can be managed by manipulating the vector population. Viruses are carried by some insect vectors and thus spread the disease from one place to another. If we sow our crop in such a way that, when the vector population will be higher, our crop will be much older and thus spread of viral disease will be less.

Crop density and disease incidence are usually correlated, mainly due to the ease with which inoculums are transfer between plants when they are close together and their leaves and roots are able to touch. Microclimate created by densely planted crop fields induces pathogen infection. Temperatures are more uniform, humidity is increased and leaves stay wet for longer time, all these factors favours the development of disease. Crop density can be manipulated by sowing at proper spacing, thinning, and pruning. The practice of planting more than one crop in alternating rows, or intercropping, can reduce disease by increasing the distance between plants of same species, and creating a physical barrier between plants of the same species. It may also reduce the movement of pathogens in the field, if the crops in the same field have different heights. For implementation of successful intercropping, depends on partly on the combination of crop chosen, since, some combinations can actually make disease worse by providing an alternative host or a stimulus that encourages germination of inoculums on the neighbouring species. So, to reduce this problem one should have a proper knowledge of the host range of the disease causing pathogen. Two crops in cropping system should not be taken from same family as same family crops may serve as an alternative host or collateral host of the disease causing pathogen of the main crop. Since, the two crops are not from the same family, there is less chance, that both the crop will be infected by the same major pathogen of the main crop. Root system of both the crops is also an important factor. It is better to choose two crops, one of which has a shallow root system and another has a deep root system.

Mulches are used to conserve moisture and organic matter and reduce erosion in the soil. Straw and sawdust are generally used as organic mulches. If crop residues are used as mulching materials, it can be a food source and create inducive environment for disease development and increase the incidence of a disease. Crop residues can also influence disease incidence by altering immediate environment. Nutrient enrichment of soil, decrease in soil temperature, water retention, weed inhibition are fluctuated by mulching material, ultimately, affects the disease development. Spread of the soil borne diseases that rely on splash dispersal can be reduced, but it may encourage other diseases, for which the altered environment is favourable, can be enhanced. The addition of organic matter to the soil can, sometimes, reduce disease influence by increasing the activity of predatory micro-organism, although it can also increase the incidence of the diseases caused by pathogens that thrive on the amendment.

Polythene mulching is one of the good practices to reduce the soil borne primary inoculums of the pathogen. But it is very costly in nature. The soil should be covered with transparent polythene sheet in the summer months before planting of the crop. The sunlight enters through the transparent polythene sheet and increases the soil temperature. As temperature inside the soil becomes higher and most pathogens are killed in such a high temperature. The polythene sheets should be kept in the field for a minimum period of 15 days.

Hot Water Treatment of planting material is also an effective practice to reduce the primary inoculums load. It is mainly used for transplanted crops. Dipping planting material in hot water for 10 minutes at 50 to 55° C and plated like, banana corms, onion bulbs, citrus root stock, rhizomes, potato tuber seeds.

Bio-fumigation is another agronomic practice in which certain plants having volatile chemicals (allelochemicals) is being incorporated into the soil for a minimum period of 30 days, prior to undertake a cropping programme. Mode of action of biofumigation crops is that the fumigant action of volatile compounds that are released during the biodegradation of organic matter suppresses plant pathogens. Compounds in biofumigation are Glucosinolates (GSLs-glucose-and sulphur-containing organic anions), ITCs (Methyl isothiocyantaes). In this process, it creates a powerful soil sterilizing volatile agents which can easily kill the harmful soil living micro-organism; those are very difficult to manage in deed. Certain *Brassica* cultivars, due to the property of having volatile natural gas-Isothiocyanate (ITC), are being used as organic fumigant in zinger cultivation in west Bengal. Biofumigation crops are *Brassica oleracea* (Broccoli, cabbage, cauliflower, kale), *Brassica rapa* (Turnip), *Brassica napus* (Canola, Rapeseed), *Brassica juncea* (Indian mustard), *Raphanus sativus* (Radish), *Sinapis alba* (White mustard). Green manuring crop like Sunhemp also releases organic acids which reduce survival of pathogenic propagules. It enhances the nutrient supply and encourages the beneficial microbes.

Pathogen survives during off season or unfavourable weather condition in the soil by their spore or resting structure. When the weather conditions become favourable, pathogen becomes active and cause disease in the crop. To reduce the primary inoculums load in the soil, soil solarisation is done. In Soil Solarisation, top 6 inches soil will heat up to 52-60 °C. It is done for 4 weeks during hot summer months. It enhances the soil temperature by 5-10 °C. Clear plastic film of 100 gauge (25 um) is generally used, followed by application of organic amendments. It is effective enough to manage soil borne pathogens, weeds, insects etc. Besides, the crop field is ploughed deeply in summer season. Deeper soils are brought on the top of the soil, where the soil can get direct sunlight.It is kept as such for scorching sunlight for a period of 15 days to 30 days. Thus, the deeper soil inhabitant micro-organism come out to the top soil surface and get exposure to the direct sunlight, subsequently, pathogenic micro-organism get killed. Then, we can plant our main crop safely.

Flooding has both side effects, it may destroy the fungal propagule and it also has the ability to bring foreign pathogenic propagule to the field. It has variable success in disease management, depending on the pathogen present. So, it needs careful handling, keeping the all the points in mind. Most form of irrigation has detrimental effects, rather than beneficial. It prevents desiccation of propagule during dry season and increase the load of primary inoculums. It also facilitates splash-dispersal of pathogenic organism. However, in some case, it also actively reduces the level of inoculums and retard disease development through alternate drying and wetting. The idea behind, is that, we are giving proper condition for the germination of the spore or resting structure of the pathogen by wetting, as most of the plant pathogenic micro-organism are water loving. When pathogen gets it favourable condition, it starts germination, then, starts its active growth and development. At that point of time, during its active vegetative growth, it is very much vulnerable to death, if we able to create dry condition in field by applying alternate dry and wetting irrigation. So, Pathogens can be killed after germination from primary inoculums and can be broken their multiplication process by this method which ultimately reduce the primary inoculums load in the main crop.

The removal of diseased plants can prevent further spread of the disease. Effective roguing requires the removal of infected plants part as soon as possible, after symptoms are seen. The infected plants as well as those are immediately surrounding plants, are to be uprooted and destroyed either by burning or burying under the soil, actually, it is an attempt to halt the disease from spread. Regular field monitoring is the only way to make this methods effective one.

A soil having well balanced supply of nutrients can make a healthy plant which may able withstand the attack of plant pathogenic micro-organism than that of unhealthy one. The application of nitronegenous fertilizer delays the maturity of the crop by prolonging the vegetative growth phase, make the plant fleshy and greener and it can increase the risk of infection by the pathogen, since the plants are more susceptible during their vegetative phase. High nitrogen levels in plants influences the production of secondary metabolites, with various outcomes of disease development. Nitrogenous fertilizer should be applied in 3 split doses, half dose at the time of planting and rest two doses, when the crop becomes older. When nitrogenous fertilizer is applied in the soil, it changes the environment and activity of soil-micro-organism and changes the micro-climate of crop canopy due to increased in canopy cover ad crowding foliage. The addition phosphorous fertilizer affects differently in different crops and diseases. Sometimes encourage the diseases and sometimes inhibit the diseases. Potassium generally inhibits the disease developments, counteracts some of the disadvantages of nitrogenous fertilizers and promotes the healing of crop plants, all of which reduces the disease in the plants.

Trap crops of susceptible plants can be grown on the sideline or on the field to use as trap to the plant pathogenic organism. Trap crop, as it is infected easily by the pathogen, after infection; destroy it before the completion of life cycle of the pathogen. Thus, it reduces the inoculums load in the field and for the main crop. Decoy crops are mainly used to stimulate the germination of the pathogenic propagules, but pathogens will be unable to establish but destroy the decoy-host by infection. Organic amendments of plant origin (Botanicals), contains many triterpenoid compounds- Azadirachtin, having the property of insecticidal, fungicidal, bactericidal and nematicidal also can be used. It is used as Neem cake @ 186 kg/acre, Castor cake @ 10 kg/tree and Plant powders or Neem seed powder @ 1 g/transplant.

Organic amendments (Animal Origin) is also used for improving soil nutrient status and to encourage beneficial micro-organism. Poultry manure : 250g/sq m and Vermi compost @ 1kg/sq m are used in solarised bed one week prior to sowing of seed bed.

Chitinous wastes are used to encourage beneficial fungi. Crushed shells of crustaceans (Crab and shrimp) stimulate growth of nematophagous fungi.

In the total life time of a plant, it interacts with several organisms which ultimately effect the crop production. The different types of interactions are mutualism, photocooperation, neutralism, commensalism, competition, amensalism, parasitism and predation. Biological control can be considered as a net positive result arising from a variety of specific and non specific interactions. Interaction between pathogen and biological control agent helps us to manipulate the soil and the environment to create a condition for conducive for successful bio control measures. Bio-control is eco-friendly in nature. It offers a different mode of action from agrochemicals pesticides and can be applied in rotation with the pesticides to reduce possibility of developing pathogen resistance.

Bio-control agents can be used in the situation where currently no control measurements are available or where the conventional pesticides cannot be used due to concern of residual effects. Bio-control gives a long term management effects. Self perpetuation occurs in case of bio agents. It is ecologically safe and cheap for long term use. Biological control agents, such as *Trichoderma viridi, Trichoderma herzianum, Trichoderma virens, Bacillus subtilis, Pseudomonas fluorescens* can be used to manage the soil borne diseases. *Trichoderma* is most useful for all types of plants and vegetables and it can be used in damping of caused by *Pythium sp. Phytophthora sp.* Root rot caused by *Pellicularia filamentosa*, seedling blight caused by *Pythium sp.* collar root caused by *Pellicularia rolsii,* dry rot caused by *Macrophomina phaseoli*, charcoal rot caused by *Macrophomina phaseoli*, black curf caused by *Rhizoctonia solani*, foot rot of pepper and vetel vine and capsule rot of several crops, dutch elm disease on elm's honey fungus (*Armillaria mellea*) on a range of tree species. *Trichoderma* sp. is also effective against the rots on a wide range of crops caused by *Fusarium, Rhizoctonia, Pythium, Sclerotinia* and *Sclerotium. Trichoderma* is compatible with organic manure. It is compatible with other bio fertilizer like *Rhizobium* etc. For successful disease management, deliver and establishment of *Trichoderma* to the site is a very important factor. Method of application of *Trichoderma* varies in the field like seed treatment, seedling dip, soil application, foliar application, soil drenching and wound dressing. *Pseudomonas fluorescens* is non-pathogenic saprophytes that colonize soil, water and plant surfaces. It suppress plant diseases by producing a number of secondary metabolites including antibiotics, hydrogen cyanide, siderophores.It has unique ability to enter into the vascular system of plants and reach various parts of the plants act as systematic bio-control management agents against various fungal and bacterial diseases. Rapid colonization of rhizosphere by *Pseudomonas fluorescens* is an important factor in plant disease management; it is also known as competitive exclusion of pathogen. *Pseudomonas fluorescens* can be used in the crop like chilli, tomato, potato, cucumber etc.

Sustainable disease management typically involves an integration of various individual component actions, which include planting material or seeds that do not harbour harmful pathogens, using plant varieties with genetic resistance to disease, implementing cultural practices that inhibit pathogen development and/or improve a plant's ability to withstand pathogen attack, deploying agricultural chemicals and/or biological control agents to inhibit pathogen and disease development.

How to develop a sustainable disease management strategy

An effective disease management plan can be formulated by keeping the records of all the parameters of operational activities related to host, pathogen and environment in a systematic manner and that information can be used for formulation of effective strategic management in upcoming season. It should be formulated region and crop specific for a particular period and occasionally, this model needs to be revalidated for better crop production. All the records regarding the interaction among Host, Pathogen and Environment should be documented for development of sustainable disease management models. Principally, sustainable disease management may be achieved through the manipulation of modern agricultural practices to maximize the defense ability of plants, but a successful outcome requires due consideration of, and proactive response to, the ability of pathogens to evolve. No one strategy for disease management necessarily works for all pathogens or for one pathogen in all environments. Disease management programs having high efficacy in short-term disease management are not necessarily the most effective with regard to minimizing pathogen evolutionary potential in the long-term (Zhan *et al.* 2013). It is important to design plant disease management programs that can manage both disease epidemics within growing seasons and pathogen evolution by generating environments conducive for host defense but with unfavorable conditions for pathogen survival, reproduction, and dispersion. Sustainable disease management programs that combine strategies to reduce pathogen population sizes, limit pathogen dispersal, and force pathogens to endure fluctuating, diversifying selection forces have the best prospects for reducing the emergence of new infectivity and new agrochemical resistance as well as for limiting the evolution of increased aggressiveness (Zhan, *et al.*, 2015, Trivedi *et al.*, 2016). Such an integrated approach would have good potential for sustainable disease management.

Key information should be recorded for the development of sustainable disease management model, Such as:

Plants: Include scientific names and cultivars: Record healthiness of the plant, any mechanical injuries and unusual characteristics.

Selection of quality planting materials: Use healthy plants for planting, infected planting material should be discarded. Seeds should always be collected from reliable sources.

Planting method: Planting methods needs to be followed as recommended by agronomist to reduce the plant health problems.

Planting depth: planting depth should be followed as per the recommendation for each species. Planting plants too deeply predisposes them to crown rot and can result prevent flowering. Planting can cause them to topple during watering if planting is done in too shallow depth.

Spacing: Proper spacing of plants should be maintained. Crowded plants prevent good air movement, which favors foliar diseases. Close spacing also causes plants to compete with each other for sunlight and nutrients and becomes susceptible to disease. With vigorous plants, the canopy that's created around the plant favors different plant diseases like downy mildews.

Date of sowing: It should be recorded to assess the on crop growth and yield.

Soil Test: All the soil parameters need to be tested from nearby soil testing laboratory (texture, pH etc). If not done previously, the growing medium should be checked for nutrients and pH, and then monitored regularly. As always, monitor watering levels to prevent damping-off.

Weather parameters: Relative humidity, Temperature, Rainfall, include unusual events.

Irrigation: Irrigation should be given in proper way, keeping in mind that plant should not be water lodged. However, it is important to remember that water is for the roots; avoid overhead irrigation, and when watering, water plants in the morning so foliage can dry before night time.

Watering schedule: It should be prepared consideration critical growth stage of the crops. Quality of water should also be tested and recorded like pH, salts etc.

Application of fertilizer: Judicious application of fertilizer should be done. Too much fertilizer predisposes plants to disease and prevents many flowering plants from flowering.

Field Sanitation: Utmost care should be given to maintain field sanitation to avoid diseases. Sustainable disease management incorporates multiple techniques to reduce disease problems. A corner-stone to sustainable disease management is sanitation. Clean tools, benches, and equipment to remove soil and plant debris prior to sterilization because disinfectants do not penetrate these well. Soak clean tools in a bucket of disinfectant and dry on clean newspaper or in an empty, clean container. Hang them up so that they do not become contaminated with soil or plant debris. Equipment must be kept free of soil-borne plant pathogens. Containers that are reused should be disinfested by submerging in a liquid disinfectant for 10 minutes and then allowed to air dry.

Crop rotation: Some crops (*e.g.,* barley) have been continuously grown on the same land for decades. Control through rotation among different species does not necessarily work for many soil borne pathogens that have long-lived resting structures such as sclerotia. sclerotia of Sclerotinia can survive up to 20 years in the soil before germination. For such situations, creating a series of isogenic lines differing in R genes and using them in an R-gene rotation could help avoid disease problems. Thus, for both foliar diseases of barley (Turkington *et al.* 2005) and blackleg of rape (Marcroft et.al, 2012), sequentially planting different varieties of the same species on the same land substantially reduced disease severity as compared to continuous cultivation of the same varieties.

Roguing: Regular removal of infected leaves or plant parts during the growing season minimizes the buildup and spread of disease causing organisms. Remove infected plants promptly to minimize disease.

Resistant variety: Use disease resistant varieties or cultivar to avoid disease. Increase diversity in field, it reduces disease problems.

Disease Incidence and Severity: Record the date of occurrence, Incidence, severity, symptoms, and signs.

Chemicals used: Note the name of any pesticides or fertilizer used on or near the plants or near the general site. Note the name of Formulation, rate and method of application, and weather conditions at the time of application as well as the time of day the material was applied.

Management calendar: Try to maintain crop management calendar and records all the operational activities.

These above mentioned information regarding the host, pathogen and environment aspects can be used as baseline data to formulate disease management strategies in future years and to see what does and doesn't work and how environmental conditions may have helped or hindered with disease management strategy. Careful end-of-the season, evaluations of plans need to be adjusted and hopefully improved to increase effectiveness, reduce chemical pesticide usage and ultimately, improve both sustainable crop production and reduce the disease infestation. However, effective and efficient sustainable disease development strategies depends on the following criteria, such as:

(a) ***Prevention:*** You have to keep in mind the phrase "Prevention is better than cure". Try to restrict the entry of pathogens into fields through planting materials, irrigation water, workers, and tools.
(b) ***Monitoring the field:*** Prime concerned to be given for monitoring and engage in regular field visit to identify disease symptoms and plant insect vectors.
(c) ***Accurate disease diagnosis:*** Identification of the causal organism for a disease is most important as most biological and chemical management options available are pathogen-specific.
(d) ***Development of acceptable disease thresholds limits:*** Try to measure and understand the effect of a disease incidence and severity and yield loss. For example, 10% disease incidence because of a specific pathogen (Root rot) may not cause a significant yield loss in a crop, in which case chemical management may be an unnecessary expense.
(e) ***Optimal selection of management options:*** Identify the sustainable disease management plan depending upon the disease, crop, and field history and try to revalidate the same for better result.

Few common disease problems in the crop fields and general management

Good agricultural practices (GAP) can play a key role in managing the plant disease effectively but it should be region and crop specific. It should be validated by a regular interval basis. These approaches may include steps to adopt precision farming for optimizing quality of the crop and safety for the environment and human beings.

Fungal disease: Root rot

Symptoms: The symptoms of root rot are easier to spot above ground. Common symptoms are gradual decline, stunted or Poor growth, wilted, yellowed or browned leaves, branch dieback, thinning of the canopy. The fungus grows up from the roots in the inner bark and causes cankers or sunken dead areas.

Management: Proper watering and avoid over watering, well drained soil, choose disease resistant varieties or cultivar, Rouging or pruning out the infected plant parts. Always be sure to disinfect any tools before using them again. If these underlying problems are not addressed, chemical or biological treatments will not be effective. Evaluation of salts and pH (some pathogens favor alkaline growing conditions).Adequate organic matter should be applied for growing plants. Fumigants (chloropicrin or methyl bromide etc) can reduce the level of the infection. Biological control agents, such as *Trichoderma viridi, Trichoderma herzianum,Trichoderma virens, Bacillus subtilis*, can be used to manage the soil borne diseases. *Trichoderma* is most useful for all types of plants and vegetables and it can be used in damping of caused by *Pythium* sp. *Phytophthora sp.* Root rot caused by *Pellicularia filamentosa*, seedling blight caused by *Pythium sp.* Collar root caused by *Pellicularia rolsii*, dry rot caused by *Macrophomina phaseoli,* Charcoal rot caused by *Macrophomina phaseoli,* black curf caused by *Rhizoctonia solani*, foot rot of pepper and vetel vine and capsule rot of several crops, Dutch elm disease on elm's honey fungus (*Armillaria mellea*) on a range of tree species. *Trichoderma sp.* is also effective against the rots on a wide range of crops caused by *Fusarium, Rhizoctonia, Pythium, Sclerotinia* and *Sclerotium*. T*richoderma* formulations are available in the market. These formulations contain 3 – 106 CFU per 1g of carrier material. Tacl powder is generally used as carrier for making the powder formulation of *Trichoderma.* Seed coating of *Trichoderma* is one of the easy and effective methods of delivering the antagonist for the management of seed or soil borne diseases. Seeds are coated with dry powder of *Trichoderma* just before sowing. Dry powder of antagonist is used at 3 to 10 g per kg seed based on the size of the seed. Propagules of bio control agents germinate on the seed surface and colonize roots of germinated seedlings and rhizosphere. Seedling roots can be treated with spore or cell suspension of antagonists either by drenching with the bio-control-agents in nursery bed or by dipping the seedling roots in bio-control-agents suspension before transplanting. This method is used for vegetable crops where transplanting is generally practiced. There are reports about the reduction of sheath blight disease of rice by root dip of seedling before transplanting into the main field. Root dipping in antagonist's suspension not only reduces the disease incidence and severity but also seedling growth in tomato, chilli and brinjal. Rate of *Trichoderma* powder depends on the methods of application, however, during seed treatment 6-10 g of *Trichoderma* powder per kg of seed is being mixed before sowing and during nursery treatment 10-25 g of *Trichoderma* powder per 100 m^2 of nursery bed. At the time of soil treatment 5 kg of *Trichoderma* powder per hector is used and for plant treatment 10 g of *Trichoderma* powder is mixed in a litre of water and then, soil is being drenched near stem region. No single fungicide will kill or inhibit the growth of all the fungi. In fact, some chemical eliminates fungi that compete with the disease-causing fungus. Before any action is taken or Fungicide application, requires accurate diagnosis of the pathogen (*Armillaria, Rhizoctonia, Pythium, Phytophthora, Fusarium*, etc.), so determine what type of root and crown rot is present prior to fungicide application. As for example, *Streptomyces scabies* the cause of potato scab disease is suppressed by the population of *Bacillus subtilis.* Club root disease of cabbage and cauliflower caused by *Plasmodiophora brassicae* can be managed by applying 10gm of *Pseudomonas fluorescens* per kg of seeds as seed treatment, and as soil application at the rate of 2.5 kg *Pseudomonas fluorescens* mixed with 50 kg of FYM and then, apply to the soil before planting. Besides, *Pseudomonas fluorescens* can also be used during seedling root dip practices as it is done by dipping the seedling in the solution containing *Pseudomonas fluorescens* formulation of 5 g/litre of water for 30 minutes. Damping off of vegetables can be prevented through seed treatment by using *Pseudomonas fluorescens* formulation at the rate of 10 kg per kg of seeds 24 hours before sowing. Formulation of *Pseudomonas fluorescens* at the rate of 2.5 kg/ha mixed with well decomposed FYM and apply to the main field which gives good result for the management of damping off disease of vegetables.

Viral disease and management

Symptoms: If plant viral disease is suspected, careful attention to plant appearance can give a good clue. All most all virus disease seems to cause some degree of reduction in yield and the length of life of virus infected plants is usually shortened. The most obvious symptoms of virus infected plants are usually those appearing on the leaves but some viruses may cause striking symptoms on the stem, fruit and roots. The typical symptoms of viral disease are mosaic leaf pattern, crinkled leaves, yellowed leaves, mottling, malformation, plant stunting or dwarfing, Ring spots, streaking, blistering, leaf roll etc.

Management: Inspect plants regularly and remove plants that are suspected of being infected. Monitor pest populations and control the insect vectors (thrips, whiteflies, aphids, etc.) as they are the active transmitters of the viruses from weed and other plant sources. Select virus tolerant variety if possible. Remove weeds from surrounding areas. Use trap crop in the boarder of the crop field, thereby reducing insect vector's access to the plants. The best way is to avoid the disease. If the viral diseases are prevailing in the crop field, adopt crop rotation to avoid the availability of the same host. Planting materials should be collected from reliable source.

Bacterial disease and management

Symptoms: Water soaking is common symptom of many bacterial diseases. The spots look like drops of water and are caused by water accumulation in tissues as a direct result of bacterial infection. It occurs early in the development of bacterial diseases. In later stage of the lesion development, it appears like "greesy looking" spot. It is being occasionally observed that bacterial "ooze" or exudate coming out of water-soaked lesions in the early morning from underside of the leaf. These droplets (Ooze) are a sign of the pathogen, being composed mostly of bacterial cells. Wilting is also another symptom of bacterial infection which leads to death of the plant.

Management: Disease free planting material, Disease resistant variety, crop rotation can be useful to manage the bacterial disease. Remove infected plants immediately. As bacteria require leaf surface water to spread, keep foliage dry through adequate spacing and proper watering. Copper based compound are effective at minimizing inoculums build up but copper resistant bacteria have arisen. Antibiotics are used occasionally in human medicine, have been used for combating plant diseases in few cases. Resistance to streptomycin has been quite common, although no resistance has been detected to tetracycline in plant pathogens (Vidaver, 2002). No new antibiotics are likely to be available for use against plant pathogenic bacteria.

Conclusion

Sustainable disease management should be formulated region-wise and crop-specific and model should be upgraded in regular basis, then only, we can find effective disease management model for that particular area. However, it allows us to blend all the cultural, physical, biological and chemical measure to manipulate the interaction among the plant, pathogen and environment to minimize the disease. Sustainable disease management doesn't mean not using fungicides. It does mean that use the chemical pesticides sustainably. Whereas, using the wrong synthetic fungicide can pollute the environment; it also wastes employee time, money and worsens the condition of the crop health. However, Integration of ecological and evolutionarily attractive sustainable disease management strategies with economic practicalities is a big issue; mechanisms that fail to increase on-farm profitability are not likely to be adopted. The increasing use of remote sensing coupled with planting, harvesting, and spraying equipment will broaden disease management options by making more precise treatments a greater reality. Simulation modeling of the interactive effects of particular disease management strategies on both epidemiological and evolutionary dynamics of pathogens will provide an increasingly important guide to future directions but must include the ecological realism. Understanding interactions at the interface between host and pathogen systems will provide means of reducing disease occurrence in crops and will maximize opportunities to affect the evolutionary trajectories of pathogen population structure. Increasing global information regarding the emergence of novel pathotypes can support efforts in disease resistant breeding aimed at encountering the increasingly rapid dispersal of pathogens around the globe.

Reference

Agrios, G. N. 1997. Plant Pathology. Fourth edition. Academic Press, New York.

Beckerman,J. 2008. Sustainable Disease Management. Grower Talk.1-2

Francl, L. J. 2001. The disease triangle: a plant pathological paradigm revisited. Plant Health Instructor. doi: 10.1094/PHI-T-2001-0517-01

Huang, H. C., and Wu, M. T. 2009.Plant disease management in the era of energy conservation. *Plant Pathol. Bull.* 18: 1-12.

Marcroft, S.J, Van de Wouw, A.P, Salisbury, P.A, Potter, T.D.and Howlett, B.J. 2012. Rotation of canola (Brassica napus) cultivars with different complements of blackleg resistance genes decreases disease severity. *Plant Pathol.* 61:934–44

Marshall B, Newton AC, Zhan J. 2009. Evolution of pathogen aggressiveness under cultivar mixtures. *Plant Pathol.* 58:378–88

Trivedi, P, Trivedi C, Grinyer J, Anderson IC and Singh BK 2016 Harnessing Host-Vector Microbiome for Sustainable Plant Disease Management of Phloem-Limited Bacteria. *Front. Plant Sci.* 7:1423.doi: 10.3389/fpls.2016.01423

Turkington, T.K., Xi, K., Tewari, J.P, Lee, HK, Clayton, G.W. and Harker, KN. 2005. Cultivar rotation as a strategy to reduce leaf diseases under barley monoculture. *Can. J. Plant Pathol.* 27:283–90

Vidaver,Anne K.2002. Uses of antimicrobials in plant agriculture. *Clin. Inf. Dis.*34 (Suppl 3) S107-S110.

Zhan, J. and McDonald BA. 2013. Experimental measures of pathogen competition and relative fitness. Annu. Rev. *Phytopathol.* 51:131–53

Zhan, J, Peter H. Thrall, Julien Papa, Lianhui Xie, and Jeremy J. Burdon 2015. Playing on a Pathogen's Weakness: Using Evolution to Guide Sustainable Plant Disease Control Strategies.*Annual Review of Phytopathology*. 53: 1-19

Zhan, J., Thrall, PH, and Burdon JJ. 2014. Achieving sustainable plant disease management through evolutionary principles. *Trends Plant Sci.* 19:570–75

13

Sustainable Weed Management

Partha Sarathi Patra, Pabitra Adhikary, Shyamal Kheroar Bratati Kanjilal* and *Subhra Shil

Introduction

Weeds are the most costly category of agricultural pests, causing more yield losses and added labor costs than either insect pests or crop disease. Weeds are a constant fact of life in annual row crops, vegetables, and other horticultural crops. In India, the manual method of weed control is quite popular and effective. The usage of herbicides in India and elsewhere in the world is increasing due to possible benefits to farmers. At the same time, the continuous use of herbicides of the same weed shift, herbicide resistance in weeds and environmental pollutions. Application of herbicides also kills species of bacteria, fungi and protozoa thereby upsetting the balance of pathogens and beneficial micro organisms. The complexity of these situations has resulted in a need to develop a holistic sustainable eco-friendly weed management programme throughout the farming period (Gnanavel and Natarajan, 2014). Sustainable development of agriculture is concerned about the conservation of land, water, plant and animal genetic resources, is environmentally non-degrading, technically appropriate, economically viable and socially acceptable. Sustainable weed management is the use of weed control methods that are socially acceptable, environmentally benign and cost-effective.

Objectives of sustainable weed management

There are several basic objectives of the sustainable weed management. The main objectives are;

- To make best use of the resources available for weed control
- To develop cultivation methods that manage weeds and improve soil quality and to determine the impact of weed management systems
- To minimize use of non-renewable resources like herbicides and to use of renewable energy and recycled mineral resources
- To protect the health and safety of farm workers and animals, local communities and society from the application of chemicals

- To protect and enhance the environment and natural resources
- To protect the economic viability of farming operations
- To provide sufficient financial reward to the farmer to enable continued production and contribute to the well-being of the community
- To produce sufficient high-quality and safe food and
- To build on available weed control technology, knowledge and skills in ways that suit local conditions and capacity.

Approaches involved in sustainable weed management

There are three different approaches are involved for sustainable weed management. The different approaches are reviewed in the context of the cultural, mechanical and biological methods.

Cultural Approaches

Proper crop stand: Crop population, spatial arrangement, right method and time of sowing, adequate seed rate and the choice of cultivar (variety) are essential to limit the weed growth. Narrow row widths and a higher seeding density will reduce the biomass of later-emerging weeds by reducing the amount of light available for weeds located below the crop canopy. Similarly, fast growing cultivars can have a competitive edge over the weeds. Planting pattern is a cost effective technique that modifies the crop canopy structure and micro-climate enhances crop competitiveness in weed suppression, improves the resource use efficiency and maximizes crop productivity (Sumathi. *et al.*, 2010). Closer spacing, early planting and increasing the fertilizer rates are observed to increase crop yields and reduce weed populations in barley and wheat under small farming systems of semi-arid regions (Abu-Irmaileh, 2000).

- **Green manure *in situ***: A practice of ploughing or turning into the soil undecomposed green manure crops in the same field where the crop is grown. Green manure crops are commonly associated with organic agriculture, and are considered essential for annual cropping systems that wish to be sustainable. Green manures usually improve the soil health and soil protection. In addition to soil improvement, green manuring is also used for weed suppression in cropping systems. Green manure crop, *Sesbania aculeata* in the preceding off-season and ploughing in situ before puddling reduced the weed counts and increased the weed control index in the succeeding rice crops due to smothering effect of green manure on the emergence and growth of weeds (Gnanavel and Kathiresan, 2002).

- **Intercropping:** Intercropping and cover cropping are practices that increase diversity in the cropping system and enhance the utilization of resources such as light, heat and water. These practices can also help to suppress weeds and increase the likelihood of being able to reduce herbicide use in the cropping system. Intercropping and cover cropping systems tend to suppress weeds better than sole cropping systems (Liebman and Dyck, 1993).
- **Crop rotation** Crop rotation is an important component of sustainable weed management. Crop rotation is a planned sequence of crops growing in the same field year after year. Rotating crops adds diversity to the cropping system, increasing the sustainability of the system. Crop rotation provides the foundation for long-term weed management. Planting a wide variety of crops with varied characteristics reduces the specific weed species. The crops like sorghum, maize, barley, rye, sweet clover, sunflower, rapeseed, soybean, alfalfa, cowpeas and hemp has smothering effect on various weed species through crop interference.
- **Crop nutrition:** Nutrition is a very important aspect in any cropping system. Fertilizer use has definite influence on weed emergence, weed persistence, weed dormancy, weed dynamics, weed growth and weed dispersion attributes. It is necessary to study such interactions in depth to develop safe and pragmatic approaches for weed management in sustainable agriculture (Bajwa, 2014). Proper management of crop nutrients assures sustainable weed management through provision of suitable inputs for crop functioning. It promotes a healthy competition in which participation of plants is strong and thus better growth habits are developed.

  Weeds show a variety of responses to different fertilizers under different tillage systems depending upon rate and method of application (Lindsey *et al.*, 2013). Sensitivity of weeds to fertilization depends on their responding ability in terms of weed growth and nitrogen assimilation. Nutrient availability influences weed crop competition (Evans *et al.*, 2003) however, the response varies with weed species type, crop type and nutrient status of soil (Weaver *et al.*, 1992) reported nutrient availability as one of dynamic approaches that can influence duration and extent of competition. Moreover, environmental factors influencing the efficacy and efficiency of fertilizers affect weed dynamics and distribution. N influenced the germination, emergence and competitiveness of different weeds depending on soil type and climatic conditions (Sweeney *et al.*, 2008).
- **Manage the Soil Weed Seed Bank:** The weed seed bank is the reserve of viable weed seeds present in the soil, and consists of new weed seeds recently shed and older seeds that have persisted from previous years. The weed seed bank includes the tubers, bulbs, rhizomes, and other vegetative

structures through which invasive perennial weeds propagate. Another source for the weed seed bank is the inadvertent arrival of new weed seeds in organic amendments or crop seed mixture, in soil carried on footwear or farm equipment, or via wildlife, irrigation, or floodwaters. The weed seed bank can be reduced by cutting or pulling weeds before they set seed, and by mowing or tilling fields as soon as crop harvest is complete. Removal of large weeds from established crops before the weeds set seed is a good way to reduce seed bank deposits. Competitive cash and cover crops can reduce weed seed deposits. Till the soil several inches deep several weeks before crop planting, watch closely, and cultivate when a flush of weeds emerges. Cultivate again immediately before planting, this time going very shallowly (0.5-1 inch) to avoid stimulating more weed seeds to sprout with the crop.

Mechanical Approaches

- **Tillage:** Tillage and cultivation are the most traditional means of weed management in sustainable agriculture. Both expose bare ground, which is an invitation for weeds to grow. Bare ground also encourages soil erosion, speeds organic matter decomposition, disturbs soil biology, increases water runoff, decreases water infiltration, damages soil structure, and costs money to maintain (for fuel, machinery and labor). Tillage systems tend to discourage most biennial and perennial weed species, leaving annual weeds as the primary problem. Exceptions to this are several weeds with especially resilient underground rhizome structures such as johnson grass, field bindweed, and quack grass. Plowing of fields to bring up the rhizomes and roots has been used to control bindweed and quack grass. Some specific tillage techniques for weed management include the following:
- **Preplant tillage:** Where weeds such as quack grass or johnson grass exist, spring-tooth harrows and similar tools can be effective in catching and pulling the rhizomes to the soil surface, where they desiccate and die. Discing, by contrast, tends to cut and distribute rhizomes and may make the stand even denser.
- **Blind tillage:** Blind cultivation is a pre-emergent and early post-emergent tillage operation for weed control. It usually employs finger weeders, tine harrows, or rotary hoes. These implements are run across the entire field, including directly over the rows. The large-seeded corn, soybeans, or sunflowers survive with minimal damage, while small-seeded weeds are easily uprooted and killed. For corn, the first pass should be made between three and five days after planting and a second at the spike stage. Blind cultivation may continue until the crop is about five inches tall.
- **Off-season ploughing:** Ploughing operations carried out in the off-season with the help of tractors or bullock drawn implements known as off-season

ploughing, before the crops are sown or transplanted. Off-season ploughing was very effective in reducing the weed population in succeeding rice crop as tubers and weed seeds are exposed to scorching sun and a highly unfavorable environment, with eventual destruction of their perennation (IRRI, 1981).

- **Soil solarization:** Solarization during the hot summer months can increase soil temperature to levels that kill many disease-causing organisms, nematodes, and weed seed and seedlings. It is a method of hydrothermal disinfection accomplished by covering moist soil with transparent polyethylene (TPE) film during the hot summer months. Soil solarization increases soil temperatures by 8 to 12 °C over the corresponding non-mulched soil (Hosmani and Meti, 1993).The possible mechanisms of weed control by soil solarization are breaking dormancy of weed seeds and solar scorching of emerged weeds and direct killing of weed seeds by heat.
- **Stale seed bed:** The stale seedbed is a useful weed control technique which involves creating a seedbed some weeks before seed is due to be sown. Preparation of such seedbed makes sure that any weed seeds that have been disturbed and brought to the soil surface during cultivation will thus have a chance to germinate, and can then be hoed off or eliminated with the use of a flame weeder before sowing of the actual crop is carried out. The technique can be utilized in early spring, when the weather is still too cold for proper seed germination. Several passes are made with a rototiller or plow, and then weed seeds are allowed to germinate as weather permits. By tilling, the farmer increases the chance of weed seed germination by the same method as one would for favorable vegetable/crops: The fine soil allows weed seed to grow rapidly by allowing the seed to open and the roots to spread easier than in compacted soil. Deep tilling will also bring dormant seed to the surface for germination. Some species of plant are known for seeds that can lay deeply buried in the soil for years before favorable conditions allow germination. Timing is important, however, weed seeds must be tilled/howed or otherwise destroyed before they themselves can create new seeds. By destroying them early, the farmer eliminates most of that season's annual weeds. Turning the dead weeds back into the soil also increases soil nutrient content, although this difference is slight.
- **Mulching:** Mulches are coverings placed on the surface of the soil. Mulching smothers the weeds by excluding light and providing a physical barrier to impede their emergence. Any material such as straw, plant residues, leaves, loose soil or plastic film can be used as a mulching material. These mulches are most effective against annual weeds. Black plastic mulch or landscape fabric (weed barrier) can block out many perennial weeds, although some weeds emerge through planting holes, nut sedges and a few other weeds

can penetrate the mulch itself (Schonbeck, 2011). With the exception of the synthetic mulches, all of these strategies to minimize weed niches also add organic matter, build soil quality, and/or enhance farm biodiversity.

- **Pre-emergent flaming:** Pre-emergent flaming may be done after seeding, and in some crops post-emergent flaming may be done as well. Flaming is often used as a band treatment for the crop row, and usually combined with interrow cultivation. The burners are offset to reduce turbulence and to avoid concentrating too much heat on the corn. Water shields are available on some flame weeder models. Uniform seedbed preparation and uniform tractor speed are important elements in flaming. Hot and dry weather appears to increase the efficacy of flaming (Cramer., 1990). Flaming has generally proved most successful on young broadleaf weeds. It is reportedly less successful on grasses, as the seedlings develop a protective sheath around the growing tip when they are about 1 inch tall (Tanya., 1994).

Biological Approaches

- **Allelopathy:** The concept of allelopathy is receiving increased attention in the search for weed control strategies. It is an alternative and organic approach to weed control that uses chemicals that are excreted from a plant to cause either direct or indirect harm to weeds by negatively affecting their germination, growth, or development. Nearby weeds can be affected by allelopathic chemicals entering the rhizophere from the roots or the aerial parts of the crop plant. Allelochemicals or natural compounds have more benefits over synthetic compounds as they have novel structure and short half-life, therefore considered safe of environmental toxic (Duke *et al.*, 2002). Crop residues from cover crops, such as fall rye, or other organic mulches can also be used to suppress weeds through such allelopathic interactions (Michael, 2001). The list of allelopathic crops and weeds to interfere with different weeds are given in the Tables 1 and 2.

Table 1: Allelopathic crops to interfere with different weeds

Crop	Weeds
Sorghum bicolor	*Abutilon theoprasti, Amaranthus retroflexusPortulaca oleracea*
Oryza sativa	*Monochoria vaginalis, Echinochloa crussgalli*
Triticum aestivum	*A. retroflexus*
Triticum speltoides	*Avena fatua*
Capsicum annum	*Solanum nigram, A. retroflexus, Chenopodium album*
Secale cereale	*E. crussgalli, Epilobium ciliatum*
Helianthus annuus	*E.colonum*

Table 2: Allelopathic weeds to interfere with other weeds

Weeds	Weeds
Acalypha indica, Trianthema portulacastrum, Parthenuim hysterophorus, Sesbania grandifolia, Ottelia alismoides	*Eichhornia crassipes*
Argemone maxicana	*Cyperus rotundus*
Eupatorium adenophorus	*Lantana camara*
Mikania micrantha	*Cyperus sp., Imperata cylindrica*
Imperata cylindrica	*Echhinochloa colonum, Phalaris minor, Portulaca oleracea*

- **Bio-fertilizers:** Bio-fertilizers accelerate certain microbial processes in the soil which augment the extent of availability of nutrients in a form easily assimilated by plants. *Azolla* is a free floating water fern that floats in water and fixes atmospheric nitrogen in association with nitrogen fixing blue green algae *Anabaena azollae*. Dual culturing of azolla in rice fields had the added benefit of suppressing weed growth besides fixing atmospheric nitrogen. Since it formed a mat over the surface, it reduced the entry of sunlight and aeration into soil thereby suppressing weed growth. The addition of azolla in rice fields suppressed the weeds of *Eichinochloa crusgalli* and *Cyperus difformis* and the degree of suppression increased with increase in per cent of azolla cover and water depth (Sivakumar *et al.*, 1999).
- **Insect bio-control agents:** Weed control through living organisms is an effective weed management practice. A large number of predators, pathogens and other plant competitors of weeds are exploited to kill or suppress the weeds. Bio-control of weeds is the deliberate use of natural enemies to reduce the densities of the weeds economically or aesthetically tolerable limits.

Table 3: List of weed species controlled by insect agents

Bioagent	Weeds
Beetles (*Octotoma scabripennis and Uroplata giraldi*)	*Lantana camara*
Scale insect (*Dactylopius tomentosus*)	Prickly-pear weed -*Opuntia*
Flea beetle (*Agasicles hygrophyla*)	*Alternanthera philoxeroides*
Fish (Common carp, Chinese carp)	Aquatic weeds
Mammals (Manetee or sea-cow)	Water hyacinth
Snails (Marisa sp and other fresh water snails)	Submerged weeds like coontail and algae
Fungi (*Rhizoctinia* blight)	Water hyacinth
Mites (*Tetranychus* sp)	Prickly pear
Plants (Cowpea as intercrop in sorghum)	Reduces the growth of weeds in sorghum

- **Bio-herbicides:** Weeds can be controlled by pathogens like fungi, bacteria, viruses and virus like agents. Among the classes of plant pathogens, fungi have been used to a larger extent than bacteria and virus or nematode pathogens.

Table 4: List of microorganisms used in bio-herbicides and their target weeds and ecosystems

Microorganism	Target weed	Ecosystem
Colletotrichum gleosporioides	Aeschynomene viriginica, *Malva pusilla, Cuscuta* spp., *Hakea sericea*	Rice, soybean, Wheat, horticultural crops
Puccinia canaliculata	*Cyperus esculentus*	Rice, horticultural crops
Cercospora rodmani	*Eichhornia crassipes*	Water ways, impoundments
*Fusarium solani f.*sp. *cucurbitae*	*Cucurbita texana*	Cotton, soybean
Trichoderma virens	Several	Row and horticultural crops
Streptomyces hygroscopicus	General vegetation	Row and horticultural crops
Pseudomonas syringae pv. tagetis	Composite weeds	Maize, soybean
Pseudomonas syringae pv. phaseolicola	*Pueraria lobata*	Non-crop lands, pastures
Ralstonia solanacearum	*Solanum nigrum*	Non-crop lands, pastures
Pseudomonas fluorescens D7	*Bromus tectorum*	Cereal grain crops

- **Herbicide resistant crops:** Herbicide resistance is the inherited ability of the plant to survive and reproduce following exposure to a dose of herbicide that would normally be lethal to the wild type. In a plant, resistance may occur naturally due to selection or it may be induced through such techniques as genetic engineering. Herbicide resistant maize, canola, cotton and soybean accounted for 77% of the GM crop hectares in 2001. However, sugarbeet, wheat, and as many as 14 other crops have transgenic herbicide resistant cultivars that may be commercially available in the near future (Gnanavel and Natarajan, 2014).

Reference

Bajwa, A. A. 2014. Sustainable weed management in conservation agriculture. *Crop Protection* 65:105-113.

Duke, S.O., Dayan F.F. and Rimando, R.M. Schrader, K.K. Aliotta, G. Oliva A. and Romagni G. 2002. Chemicals from nature for weed management. *Weed Sci.,* 50: 138-151.

Cramer, C. 1990. Turbocharge your cultivator. New Farm. March–April. p. 27–30, 35.

Tanya, D. 1994. Non-toxic weed control. The IPM Practitioner. October. p. 20.

Gnanavel, I. and S. K. Natarajan. 2014. Eco-Friendly Weed Control Options For Sustainable Agriculture-A Review. *Agri. Review,* 35 (3): 172-183.

Lindsey, L.E., Darryl, D., Warncke, K.S., Wesley, J.E., 2013. Fertilizer and population affect nitrogen assimilation of common lambsquarters (Chenopodium album) and redroot pigweed (Amaranthus retroflexus). *Weed Science.* 61: 131-135.

Michael, B. 2001. Does allelopathy offer real promise for practical weed management and for explaining rhizoshere interactions involving higher plants? *Plant and Soil.* 232: 31-39.

Sweeney, A.E., Renner, K.A., Laboski, C., Davis, A., 2008. Effect of nitrogen fertilizer on weed emergence and growth. *Weed Sci.* 56: 714-721.

Sivakumar, C., Kathiresan RM. and Kalyanasundaram D. 1999. Effect of azolla on yield and weed suppression in rice. In: Abstr. Of Papers, 8th Bienn. Conf. ISWS, Banaras Hindu Univ., Varanasi.p.6.

Weaver, S.E., Kropff, M.J., Groeneveld, R.M.W., 1992. Use of ecophysiological models for crop-weed interference: the critical period of weed interference. *Weed Science.* 40: 302-307.

Schonbeck, M. 2011. Principles of Sustainable Weed Management in Organic Cropping Systems. *The Virginia Biological Farmer* 3: 1-20.

IIRI. 1981. International Rice Research Institute. Long term effect of reduced tillage. In: Ann Rep.1980.Laguna, Philippines, p. 229-230.

Hosmani, M.M. and Meti, S.S. 1993. Non-chemical means of weed management in crop production. Integrated weed management for sustainable agriculture, Volume 1, pp. 299-305. Proceedings of a conference held at the Department of Agronomy, CCS Haryana Agricultural Universtiy, Hisar, Haryana, India, 18-20 November 1993.

14

Sustainable Livestock Farming

Dilip Kumar Hajra and *Nonigopal Shit*

Introduction to sustainable livestock

Sustainable development has been defined by FAO as "the management and conservation of the natural resource base, and the orientation of technological and institutional change in such a manner as to ensure the attainment and continued satisfaction of human needs for present and future generations. In other words, the development that meets up the needs of the present without compromising the ability of future generations to meet their own needs. Such sustainable development (in the agriculture, forestry and fisheries sectors) conserves land, water, plant and animal genetic resources, is environmentally non-degrading, technically appropriate, economically viable and socially acceptable". (FAO Council, 1989). Hence, sustainable livestock farming is "the efficient production of safe, high quality products, in a way that protects and improves the natural environment, the social and economic conditions of farmers, their employees and local communities, and safeguards the health and welfare of all farmed species".

Importance of livestock

Livestock sector employs over 11 million rural poor and women in principal status and eight million in subsidiary status, which is about 5 per cent of total working force in India. According to estimates of CSO (2009), the value of output from livestock and fisheries sector together was about Rs. 2, 82,779 crore during 2007-08, which is 31.6 per cent of the value of output from agricultural and allied sectors. The contribution of these factors in the total GDP during 2007-08 was 5.21 per cent.

During 2009-10, the country produced 112.5 million tonnes of milk 59.8 billion eggs, 43.2 million kg of wool and 4.0 million tonnes of meat from the organized sector (Economic Survey of India, 2010-11).

The livestock and fisheries sector contributed over 4.07 per cent to the total GDP during 2008-09, which is 29.7 per cent of the value of output from agricultural and allied sectors.

The livestock sector is one of the fastest growing parts of the agricultural economy, the FAO report underlines. Globally, livestock contributes 15 percent of total food energy and 25 percent of dietary protein. Products from livestock provide essential micronutrients that are not easily obtained from other plant food products.

Why sustainable production?

Addressing malnutrition and overconsumption as well as tackling waste are crucial for sustainability of food system. For the ever increasing population of the globe it requires to increase livestock production to meet demand for animal protein.

In order to answer to the question inherent to address the opportunities and challenges in this sector sustainability is perhaps the only way out. In terms of opportunities, chickens have the best rate for converting feed into meat. Cattle are efficient in producing human edible proteins per unit of land where only grass and forage can grow. Furthermore, animals provide manure and thus nutrients as well as organic matter that are recycled in crop production. Both, opportunities and challenges underline the importance of balancing resource-efficient production and consumption for achieving food security in a sustainable manner.

Approach to sustainable livestock production

A systems approach requires attention to detail and adopting innovative practices to move farm businesses towards delivering better on sustainability goals. Sustainable livestock farming refers to a system based on three pillars: Environmental security, Social accountability and Economic practicality

Environmental security

Reducing greenhouse gas emissions and resource use is paramount. We need to reduce the amount of land, energy, and water needed to produce the same or higher amount of food. Hence, we will need to minimize the environmental impact per unit of product and leave more opportunities for enhanced ecosystems and biodiversity. An environmentally sound livestock system should make every effort for the following:

- Efficient use of natural resources per unit of animal, whilst maintaining or increasing production.
- Reduce emission intensity of greenhouse gas (especially methane and nitrous oxide) via improved production efficiency, and decrease air, water, and soil pollution (ammonia, nitrate, phosphorus, etc.).

- Use of Implement and practices to improve air quality.
- Manage the manure in a way that ensure recycling of nutrients and energy, and minimize release of gases with high global warming potential.
- Fodder and grass land management for efficient grazing systems.
- Use energy-saving devices and practices and foster the use of renewable energies.
- Ensure the maintenance or improvement of soil health and fertility.
- Maintain or enhance biodiversity, respect ecosystems and the cultural landscape.
- Reduce waste and maximizing opportunities to reuse and recycle it.

Livestock sector and climate change

Improved technologies can play a key role in mitigating climate change. These approaches may include steps to adopt precision livestock farming for optimizing feed quality and digestibility, breeding for reduced methane emission, vaccines and potential methane inhibitors to subside methane production etc. Future a systems approach may be employed in order to curb GHG emission from the farm complex.

Social accountability

The social responsibility component includes components such as:

- Improving and maintaining the societies and communities where livestock food products are produced.
- Safeguarding the health, well-being, and social rights of farm families and associated workers.
- Improving and safeguarding animal health and welfare.

Measures for small and marginal farmers are an important aspect in the country like India. Farmers and workers need access to training and education as well as social assistance. Furthermore, they need the space in order to grow their business. Farm animals depend on human care and it is society's responsibility to respect their welfare, including health. Internationally recognized Five Freedoms for farm animals are: (1) free from hunger and thirst, (2) discomfort, (3) pain, injury and disease, (4) fear and stress and (5) they should be free to express normal behaviour. Key to getting as close as possible to this ideal state are the knowledge, skills, and personal qualities of the caretaker. These are the so called 'Three Essentials of Stockmanship'. Animal welfare is vital for a sustainable livestock production, which takes into account the three pillars of sustainability. At the livestock production level, this can be achieved through, for example, good husbandry practices, by preventing and controlling animal diseases, and by promoting the responsible use of veterinary medicines, including antimicrobials. Greater investment in safe and scientifically sound technology and innovation can do much to improve food sustainability and efficiency of farm production.

Economic practicality

Food production must be economically viable: the farmer and other food chain stakeholders must be able to prosper and to sustain investment, while consumers need access to quality food that is affordable. From an economic standpoint, sustainable livestock farming systems are therefore characterized by elements such as:

- Enabling economically viable food production along the food chain, while accomplishing social and ecological goals.
- Ensuring that farm operations obtain a fair share of the profits achieved in the food chain.
- Supporting the ability of livestock producers to invest in sustainable improvements.
- Adopting innovation and approaches that help farmers deal with market volatility and hence the prices of products and inputs.

Policy choices should focus on achieving an economically viable livestock sector for farm operations of all sizes, through continuous improvement, innovation, capacity building, knowledge transfer, and translation of research into practice.

Sustainable poultry production its issues and aspects

Poultry species, especially the chickens are the most efficient converters of feed into meat of all land-based livestock species. Research conducted on the environmental impacts of Australian chicken meat have confirmed that chicken meat production generates low levels of greenhouse gas emissions and results in modest levels of energy and water use. However, many sustainability aspects need to be addressed in the current poultry sector. Leinonen *et al.* (2012) reported that feed seems to be the factor in poultry production that caused the largest environmental impact. Poultry adjust well into a diverse farming system, with locally sourced feed. Most of the backyard varieties of chickens with slower growth rate get access to outdoor areas, lead to higher animal welfare, and better use of marginal areas. There is potential for producing poultry in diverse rural and urban settings, where smallholders, as well as large scale animal systems can be integrated into the household and local food chains, and poultry can utilize food and plant byproducts and produce manure and protein rich food for the family and local community in return. The theoretical issues and aspects of sustainable poultry production are discussed here under:

Environmental aspects

Livestock activities have been estimated to contribute about 18% of total anthropogenic greenhouse gas emissions (Steinfeld *et al.,* 2006). This is connected to the production of feed, N fertiliser use, processing, transport and includes deforestation. However, poultry is the livestock species with generally the least impact in terms of land size, water use, environmental stress and footprints both for poultry meat and eggs (De Vries and De Boer, 2010). Excessive nutrients in huge amounts of waste from intensive poultry farms are the main source of serious environmental pollution of air, water and soil. Harmful nitrogen (0.8–1.8 g per hen/day) and phosphorus compounds (78–156 g/layer/year) in poultry manure cause eutrophication of rivers, lakes and ground waters. The contribution of farm animals to ammonia production is high and accounts for 80–85% of global ammonia emissions (Bouwman *et al.,* 1997). Ammonia emissions are recognized as one of the main causes of increased atmospheric acidification (Sapek, 1995). Ideally, the animals are sustained from products on the farm. The manure of the animals contributes to soil fertility. This cycle does not exist in industrial animal production. The manure is handled like 'waste' or byproducts. Beside all, use of antibiotics in poultry operation has become great concern in the face of growing resistance strain of bacteria.

Resource management

Poultry species generally reared on concentrated feed. Concentrated feed has a larger water footprint than roughage, but a higher feed conversion ratio (FCR). The entire process, from the site of production of feed ingredients to the preparation of ration needs to be managed strategically to reduce environmental impacts.

There has been an underline fear of loss of breed diversity in the wake of modern intensive poultry farming practices. Human kind has domesticated 30 species of animals, which accounts for about 8,000 registered breeds, including hundreds of poultry breeds. Most of these breeds are kept by smallholder farmers throughout the world. The current breeds in industrial systems are almost exclusively single purpose breeds, either broilers or layers. Losing a breed means loss in the part of genetic diversity and the possibilities for adapting to different environments in the face of climate changes (FAO, 2015).

Social aspects

Several studies have pointed out two different categories of social aspects for humans: 1) working conditions and health of the humans (owners or workers), and 2) the fairness of employment contracts, *e.g.,* health insurance, decent salaries, and collective bargaining rights. Intensive poultry production often leads to serious health problems in birds, which presents a serious threat to food safety. Crowding of hens and often unhygienic production conditions are leading to a rapid spread of infectious diseases, including zoonoses. Although generally do not cause any public-health concern, odours can represent a strong local problem that is frequently reported by farms' neighbours as the most disturbing environmental impact.

Animal welfare

Animal husbandry systems which cause suffering to the animals are unacceptable from an ethical point of view. Some systems and methods of rearing cause suffering, *e.g.,* foot pad dermatitis, Cage layer fatigue etc., which may be associated with high production stress. No doubt, the system compromise with animal welfare, and cause abnormal behavior. In line with the Code of Recommendations for the Welfare of Livestock, published by the Farm Animal Welfare Council, farmed birds should be guaranteed:

- freedom from hunger and thirst,
- freedom from discomfort,
- freedom from pain, injury or disease,
- freedom to express normal behaviour,
- freedom from fear and distress.

Increasing cage floor area per bird and feed access, and determining minimum (acceptable) cage dimensions will definitely improve bird welfare. One of the most popular housing alternatives to cages is the litter floor system, which is considered more avian friendly. In Europe, free-range and organic egg production systems have become more popular in recent years (Herbut and Walczak, 2003). Hence, to improve the welfare constrains, there is a needs to regulate the system in the industrial poultry production system.

Farm management

Taking environmental issues into account in all management strategies at the farm level can reduce the environmental impacts at the level of production.

Control of odour as well as flies and mosquitoes

- Minimizing the surface of manure in contact with air: frequent collection of litter (once a week in dry seasons and twice a week in rainy seasons), closed storage (bags or closed sheds).
- Cooling animal manure, achieved as a positive side effect of cooling the animal houses – cooling systems can be equipped with biofilters and air scrubbers that trap odours from the ventilation airflow.
- Lowering litter's water content – achieved by the incorporation of hydrophilic substances such as hashes, rice husk, peanut husk, dust or sawdust.
- Applying deodorant products to feed or directly to animal houses
- Building wind protection structures.

Animal waste management

- Storing manure in closed buildings or bags in order to reduce the emissions of gases and the risk of environmental contamination
- Composting manure potentially reduces or even eliminates certain pathogens, fly-larvae and improves the handling characteristics.
- Drying (with machine or by spreading out) – minimizes the moisture content, inhibits chemical reactions and thus reduces emissions. The best way to prevent ammonia emissions from poultry litter and manure is to reduce microbial decomposition, which can be accomplished by drying the freshly produced manure as soon as possible.
- Manure application – manure must be applied at the correct time of year to prevent losses to surface water, groundwater and the atmosphere, and to optimize the utilization of manure nutrients by growing plants.
- Disposal of the dead-bird must comply with legally accepted practices including, composing, incineration and burial. A contingency plan should be in place for disposal of large numbers of dead birds in the event of disease outbreaks. The burial pits should be at least 3 metres above the maximum groundwater table.

Nutrition management

Nutritional management aims to reduce pollution load by limiting excess nutrient intake and/or improving the nutrient utilization efficacy of the animal. It reduces the pollution load of soil, water and air, allows improvement to feed conversion ratios.

Relevant measures include

- Formulating feeds according to the stage of production to reduce the amount of excretion; options in this category include phase feeding, split-sex feeding or feed formulation on an available-nutrient basis.
- Use of low-protein diets supplemented with amino acids, and low-phosphorus diets with highly digestible inorganic phosphates.
- Improving feed digestibility and nutrient bioavailability through the use of dietary enzymes such as phytase, and highly digestible synthetic amino acids and trace minerals.
- Using good quality, uncontaminated feed containing no extra level of copper, zinc, and other additives than is necessary for animal health.

Feed production

The key to reducing the negative environmental impacts associated with intensive agriculture for feed production lies in increasing efficiency, *i.e.*, increasing production while reducing the use of inputs that adversely affect the environment. The negative effects of feed production can be greatly reduced with appropriate use of conservation agriculture (*e.g.*, minimum tillage), integrated pest management (IPM) and targeted fertilizer inputs.

Conclusion

Compared to other livestock species, poultry performs well from an environmental perspective. Poultry's feed conversion ratio represents a major contribution to the profitability of the industry in terms of reduced feed inputs as well as in waste output. Technologies exist that have the potential to produce substantial reductions in environmental impacts. The problem lies in awareness apart from cost involvement in using those mitigation facilities. Economic incentives and regulatory mechanism within a framework of market forces should be sufficient to create a sustainable environment for the poultry operations at large.

Reference

Bouwman, A.F., Lee D.S., Asman, W.A.H., Dentener, F.J., Van der Hoek, K.W., Olivier, J. G. J. (1997). A global high-resolution emission inventory for ammonia, *Global Biogeochemical Cycles*, (11) 561-587

De Vries, M. and De Boer, I.J.M. (2010). Comparing environmental impacts for livestock products: A review of life cycle assessments. *Livestock Science*, 128: 1-11.

FAO (2015) Coping with Climate Change – the roles of genetic resources for food and agriculture. Commision on genetic resources for food and agriculture. Pp. 110. Downloaded from http://www.fao.org/3/a-i3866e.pdf (downloaded 9th Feb 2015 and I just tested it 5th Oct. 2015).

Herbut E. and Walczak J. (2003). Warunki ekologicznej produkcji zwierzęcej z uwzględnieniem istniejącego stanu prawnego w Polsce. Ann. Warsaw. Agricult. Univ. SGGW *Anim. Sci.*, 40: 3–12.

Leinonen, I., Williams, A.G., Wiseman, J., Guy, J. and Kyriazakis, I. (2012). Predicting the environmental impact of chicken systems in the UK through a life cycle assessment: broiler production systems. *Poultry Science* 91: 8-25.

Sapek A. (1995). Emisja amoniaku z produkcji rolnej. *Post. Nauk Roln.*, 2: 3–23.

Steinfeld, H., Gerber, P., Wassenaar, T., Castel, V., Rosales, M. and De Haan, C. (2006). Livestock's long shadow-environmental issues and options. FAO, Rome, Italy, pp. 390.

Sheep and goat production

Almost three billion people, or almost half of the world's population, live on less than US$2 per day and poverty is largely but by no means only a phenomenon of the rural areas. Effective poverty alleviation measures can only be successful if the livelihoods of the rural poor can be improved. Considering the possibility of equal distribution of livestock than agriculture lands to the marginal farmers, precisely small ruminants such as sheep and goat could be prioritized for both subsistence and economic reasons. They also play socio-cultural roles for the wellbeing of rural households, such as food supply, source of income, asset saving, source of employment, soil fertility, livelihoods, transport, agricultural traction, agricultural diversification and sustainable agricultural production (Moyo *et al.,* 2010). Their contributions towards the use of crop by-products and locally grown vegetation, integration with other farming enterprises including soil fertility, use of household wastes make it precious for all farming situations. Besides, small ruminants

1. Are less likely to damage and compact soils
2. Their grazing pReference enable them to feed on vegetations that other species of domestic animals tend to refuse.
3. Easy to handle and less costly to manage and maintain the species
4. Ready market with high demand of products
5. Disease resistance ultimately reduce the cost of production.

Present scenario and impacts of small ruminants

Because of small size, easy handling and early return with less risk of disease, small ruminants are focused for the sustainable rural livelihoods. Further, women and children in addition to their assignment can look after these small ruminants and can fetch lucrative price from them. Presently about 65.07 million sheep (*Ovis aries*) and 135.2 million goat (*Capra hircus*) population contributing 485 and 943 metric million tons of meat respectively in this country (DAHD&F, 2017). Based on the agro-climatic condition, there are about 40 sheep and 20 goat breeds have been identified in this country and breed concern is important for the sustainability.

Income and food security

There are about 1.57 million sheep and 15.01 million goats (ARDWB, 2009) in West Bengal and among the major economically important livestock playing an important role in the livelihood of resource-poor marginal and land less farmers. Though, door step marketing channel fetch low economic benefit, farmers earn lucrative remuneration on demand by selling their animals. Apart from the food security, small ruminants also increase the animal diversity of food and cash sources. The national mission for protein supplement (NMPS) is implemented against malnutrition on goat for the societal development. Application of Information Technology to strengthen small farmers as well as investment for value addition has the power to enhance livelihood opportunity from this sector.

Empowering women and children

Women and children are usually the managers, if not actually the owners, of small ruminants. They derive some socio-economic benefits from this activity, not only in terms of access to food and cash but also through more subtle changes in their status. This may arise from the empowerment conferred on managers of livestock.

Use feedstuffs not suitable for human

Many small ruminants exist on by-products from crop production on tree fodder (leaves, twigs and fruits) and on household wastes. The use of these by products for livestock feed confers a value on them that they would not otherwise have. They also contribute to the production of food and other animal products from what would otherwise be waste. This increases the efficiency of the household and farm system and effectively recycles waste materials

Recycling nutrients in the soil

Small ruminants play an important role in recycling nutrients in the soil. The use of crop wastes as feed facilitates the return of nutrients to the soil via the production of manure and urine. The form and amount of dung and urine does not generally pose burden in the management operation. The manure is composted over a longer period of time. This enables the breakdown of roughage and the decomposition of weed seeds that lose their germinative power. Harmful substances in the first stage of decomposition are also eliminated.

Inputs to small ruminant production systems

Using the right species and breed

The basic unit of production in all livestock systems is the animal. Selecting or using the best genetic resource for the function, the animal is expected to fulfil and for the production environment is of critical importance. Suitable breed are available for different agro-climatic condition throughout the length and breadth of the country which maintains right balance between productivity and health, fitness. The local livestock are adaptable to the local climate, system stresses and productivity requirements. Unhealthy crossbreeding programme seems apparently advantageous but adaptability to the local climate and functional to the production is important for sustainability. The tendency to choose supposedly highly productive animals, while largely ignoring disease and climatic adaptation, is widespread, but the costs of coping with new disease or environmental conditions add considerably to the costs of production.

Sources for replenishing stock

All livestock systems require stock to be replaced at some time in the production cycle and that is to be made appropriately with cost effective way. However, it is necessary to go outside the home environment to avoid inbreeding depression and would be the best practice if arranged locally with same breed type. Care must be taken not to use highly inbred stock that may result from the lack of introduction of new genes into the local population. Without considering local environment and adaptation, introduction of exotic and imported stock for their high production potential would be the worst practice in replenishing stock. The following care should be taken during selection of the animals during replenishing the stock

1. The animal exhibiting favourable traits could be identified with reasonable accuracy.
2. Incorporation of too many traits for selection will lower the efficiency of selection in individual traits. Therefore, a minimum number of traits should be selected at a time.
3. Selection pressure should be applied continuously, or there will be rapid regression towards average.
4. Only select for economically important traits. Selecting for a trait which has a short time popular appeal will result in losing the opportunity of selecting for other economically important traits.

Feeds and fodders to the goat and sheep

A prime objective of sheep and goat production through small holder systems is to make them as self sustaining as possible. Assure stall feeding concept for these small ruminants will never bring profit as well as sustainability for rural households. It is also true optimization of growth cum production will never be possible without concentrate. Free grazing at least 6-8 hours can fulfil 70% of their daily requirements. Therefore, natural grazing land or induced practice for green roughage is to be ensured for reproduction as well as to reduce the cost of production. Small stock keepers may be encouraged to have locally grown or processed feed resources at a realistic cost if available that will satisfy the need of concentrate.

Table 1: Average growth and feeding chart for kids and lambs

Age (days)	Weight (kg)	Requirements of feed		
		Milk (gm)	Kid Starter (gm)	Green fodders
0-5	1.5-2.0	Colostrum	-	
6-30	2.0-3.0	300-500	Small quantity	-
30-60	3.0-5.0	400-500	50-100	Small quantity
60-90	5.0-7.0	350-500	100-150	250
90-120	7.5-10.0	-	200-250	250
120-180	10.0-12.0	-	250-300	750

Plantation of multipurpose trees or shrubs *i.e.*, jack fruit, subabool, banyan etc. and or encourage to practice techniques such as intercropping, alley cropping, foraging in their surroundings to provide the right balance of animal nutrients during odd seasons with scarce of green fodders. Hybrid Napier, Para grass, Fodder variety maize, Barseem, Lucern, Cow pea, Rice bean etc are common fodder used for small ruminants. Some un-conventional greens such as bamboo leaves, jack fruit leaves, Tapioca leaves, Mulberry leaves etc. and cakes *i.e.,* saal seed cake, mahua cake, karanja cake etc. may be recommended in the formulation of concentrate for sheep and goats.

Locally arranged maize or maize crust, rice bran, wheat husk, arable crop products and by-products of oil extraction, the source of protein, will make keepers' confidence for low cost feed formulation in addition to small quantity of pre-mixture. An adult non-pregnant ewe requires 93 kcal ME/kg$W^{0.75}$ for maintenance with an additional allowance of 50% over maintenance is generally provided to pregnant ewes. The lactating ewes' needs still higher amount of energy and that is of 102 kcal ME/kg$W^{0.75}$ as estimated for maintenance requirement (DIPA, 2002). The work done in the country on grazing with supplementation indicated that in addition to free grazing supplementation of limited quantity of concentrate (1.5-2.0% of the body weight) will provide marketable finishing weight of 25 and 30kg at 6 and 9 months of age, respectively.

Table 2: Ingredients and proportion for formulation of low cost feed

Kids/Lambs		Growers		Adults	
Constituents	**Percent**	**Constituents**	**Percent**	**Constituents**	**Percent**
Broken maize	22.00	Broken maize	37.00	Broken maize	32.00
Broken grams	20.00	Broken grams	15.00	Broken grams	18.00
Cakes	35.00	Cakes	25.00	Cakes	22.00
Wheat husk	20.00	Wheat husk	20.00	Wheat husk	25.00
Min. Mixture	02.50	Min. Mixture	02.50	Min. Mixture	02.50
Salt	00.50	Salt	00.50	Salt	00.50

Small ruminants are provided *ad-libitum* clean water essential for proper functioning of the body including thermoregulatory mechanism. Though they usually satisfy their water intake from feed consumed and drinking water, it depends on the level of lactation, environmental temperature and humidity, water content of the forage consume, amount of exercise and mineral content of the diet including salt. Sub-optimal water intake will result initially in the reduced feed intake followed by reduced performance and gradual starvation. Specially goat amongst the most efficient domestic animals in the use of water and possess a low rate of water turnover per unit body weight. Small ruminants have the ability to conserve the body water by reducing losses in the urine and faeces (goats) and can withstand water deprivation up to 3 days (sheep). An adult animal requires average 2 litres of water per day during winter and 3.5-4.0 litres during summer.

Using local materials to make livestock housing and equipment

As both sheep and goat are reared almost for the same purpose, so requirements of building units are more or less the same for. They do not require any elaborate buildings rather it should be clean, dry with a provision of outside temporary shed to provide shelter during monsoon and winter. The floor should preferably be elevated in low lying and heavy rainfall areas and may be made of wood in temperate Himalayan regions. Animals of different ages should better to maintain separately. Sheds with long axis running east-west provide a cooler environment underneath than the one with a north-south direction. An attempt to cover sheds to protect the animals from inclement weather may result in an increase of humidity up to 90% and ammonia concentration to 20 ppm. So good ventilation is important in hot-dry and warm-humid climate to promote the heat loss from animals with increased convection and radiation.

Local materials preferably should be used whenever possible. The rural habitual practice for the animal shed refers low cost and readily available local materials, such as wood, bamboos, sun dried bricks, thatch from local grasses and bush poles etc. In view of the sustainability, constructional details such as stud height (3m), pitch angle (30-45°), floor, wall, gutter and its slope (1 in 12), space requirement etc. should be scientific to reduce the physiological stress and augment production. Apart from the shelter, some farmstead construction such as quarantine shelter, shearing hall, ram/buck shed, kidding/lambing shed and other accessory building for the routine farm operation are required for systemic management with efficient data record.

Table 3: Space requirement for the different category of the animals

Category of the animals	Age	Space requirement (m^2)
Male and Female	Adult (above 1 year)	1.25-1.50
Lactating and pregnant		2.00
Bucks/Ram		2.00
Kids to young	7-90 days	0.50-0.60
	90-180 days	0.70-0.90
	180-365 days	1.00

Keeping livestock healthy

The small ruminants, Sheep and Goat, except few are less susceptible to deadly diseases. The flexibility of their immunity may change according to the seasonal environmental variation and make them resistant. However, maintenance of animals in a healthy state is important for sustainable development. Animal health is crucial not only for the production may also have implications for reproduction and human health as bears some zoonotic hazards like anthrax, brucellosis, tuberculosis, helminthiasis etc.

Table 4: The common bacterial and viral disease occurs in sheep and goat

Diseases	Sheep	Goat
Viral diseases	Sheep pox, Cotagious ecthyma, Fooft and Mouth Disesase, Blue tongue	Peste-de-petits ruminant (PPR), Goat pox, Foot and Mouth Disease (FMD),
Bacterial disease	Black leg, Enterotoxaemia, Tetanus, Pasteurellosis, Brucellosis	Enterotoxaemia, Anthrax, Mastitis, Brucellosis, Black quarter, Colibacillosis, Haemorrhagic septicaemia (HS), Tetanus
Mycoplasma	Contagious caprine pluoro pneumonia (CCPP)	Contagious caprine pluoro pneumonia (CCPP)
Metabolic	Diarrhoea, Tympany	Diarrhoea, Tympany, Ketosis
Worm infestation	External: Ticks, Mites Louse Internal: Round worm, Flukes and Tape worm	External: Ticks, Mites Louse Internal: Round worm, Flukes and Tape worm
Protozoa	Coccidiosis, Babesiosis, Theileriosis,	Coccidiosis, Babesiosis, Theileriosis,
Deficiency disease	Condition meant for the deficiency of minerals and vitamins in the species *i.e.*, Rickets, Hypocalcaemia, Night blindness etc.	

Disease occurrence depends on several factors including nutritional status and age of the animals. On seasonal variation, animals become prone to various diseases which indirectly affect the production potential. The availability of affordable medicines and the routine (prophylactic) treatment of animals for diseases known to be present in an area are important aspects of livestock management. Immunization against viral diseases are only way to escape the animal from illness and need to be done as per recommendation for healthy population. Hence, some bacterial diseases like HS, anthrax, BQ etc. also needs to be secured by vaccination. A routine prophylactic treatment with proper doses of anthemintics is advisable to keep the animals free from worm infection.

Table 5: Vaccination schedule of Sheep

Disease	Age and booster doses	Route	Remarks
Foot and mouth disease	6-8 weeks; repeat every 6-9 months	s/c or i/m depending on the vaccine	Nov./Dec. and May/ June
Hemorrhagic Septicaemia	3-4 months; repeat annually	1 ml s/c	May/June
Sheep pox	3 months	s/c	Sept./Oct.
Tetanus	Tetanus toxoid	0.5-1 ml s/c or i/m	Any time of the year

Anthrax	4-6 months; repeat annually	0.5 ml s/c at tail fold	In endemic areas only
Enterotoxaemia	3-4 months, repeat after 15 days and then annually.	2.5 ml s/c	First two doses before august

Table 6: Vaccination schedule of Goat

Disease	Age and booster doses	Route	Remarks
Foot and mouth disease	6-8 weeks, repeat every 6-9 months	s/c or i/m depending on the vaccine	At winter and rainy
Enterotoxaemia	3-4 months, repeat after 15 days and then annually.	2.5 ml s/c	First two doses before August
Hemorrhagic Septicaemia	3-4 months, repeat annually	1ml s/c	May/June
Anthrax	4-6 months, repeat annually	0.5 ml s/c at tail fold	In endemic areas only
Tetanus	3-4 months, repeat at 6 months and then annually	0.5-1 ml s/c or i/m	Any time of the year

Support services for sustainable small ruminant production

(a) Sustainable services for all

Small stock farmers need access to a range of inputs and marketing services. When such services are planned, their long-term delivery and sustainability must be considered. A major aspect of sustainability is the involvement of smallholders, particularly women, in planning and managing the services. Livestock producers have usually had less support than crop producers and women are often neglected even though it is frequently recognised that they are extremely important to the agricultural sector. Now almost all schemes from government or NGOs under poverty alleviation programmes intend to empower the women strength. The rural force (women) is encouraged through SHGs to be clustered to initiate income generation for better livelihood. The capacity building programme (CBP) at various stages is tried to build their confidence for scientific animal husbandry practices. However, one of the greatest deterrents to the growth of small stock production is spreading of animal diseases that result huge loss of production. Irrespective of the seasonal variation and other factors, supplying appropriate breeds, proper management such as improving levels of feeding, ensuring timely and focussed veterinary prophylaxis etc. and in-time immunization is thought to overcome these constrains. Once these aspects of livestock keeping have been addressed, other weaknesses such as poor marketing and pricing arrangements, or the lack of indigenous knowledge in the sector may become apparent.

(b) Access to technical and commercial information, loans and credit

Small ruminant producers differ in many ways from keepers of intensive livestock or large stock. Because their enterprises are small, they are not very visible and their needs are often overlooked or underestimated. Consequently, sheep and goat keepers may find it difficult to gain access to technical information concerning livestock, livestock services, loans, credit facilities, savings mechanisms and marketing possibilities. As a result, the likelihood of making a success of their enterprise and achieving the potential benefits of their stock may be limited. In view of the large number of small ruminants and their potential contribution to the local and national economy, it is essential that policy-makers address the issues that prevent access to such services by small stock keepers. Formation of Farmers' cluster to maintain all possible records including channel of marketing is necessary to emphasize this small enterprise. Authorized organizations dealing animal husbandry practices needs to re-orient their service design so that small keepers get enriched with knowledge and skills through proper training and become eligible to have financial assistance from the loan sanction institutions. The provision of services, credit and physical inputs must be sensitive to labour division between women and men in livestock production. Particular attention should be paid to the ability of women to access such services and steps taken to overcome any barriers they may face.

(c) Veterinary services

Like other livestock species, the ailing condition of the small ruminants must be handled in a cost effective way in order to create a framework for successful livestock keeping. This may involve a combination of national veterinary services, particularly for diseases that need a national or even a transnational eradication policy coupled with locally delivered interventions. The traditional approach of government veterinary service comprising a top down structure of regional and district offices, has generally been unable to deliver services to local farmers because of inadequate funding, poor motivation amongst staff and a failure of trust between farmer and veterinarian. The main need is often simple prophylaxis programmes that can be taught and delivered at the local level. In each locality it should be possible to determine what are the most likely disease threats and train farmers and para-vets to deal with them. In view of the close involvement of women and of children in the management of small ruminants, policy-makers and extension managers should ensure that they are provided with training in primary animal health care.

(d) Markets for produce

Much of the output from small-scale livestock production is consumed in the home or sold or bartered with friends and neighbours. Encouraging and supporting small holder access to more formal markets is an additional means of generating cash and providing a wider outlet for surplus produce. Products need to be marketed in a reliable and cost effective way. Assistance may take the form of market information about market opportunities, prices and quality requirements, transaction mechanisms between farmers and buyers, transport to more distant markets, transporting goods to market in smaller or larger lots, product advertising, creation of local markets and the provision of information to potential buyers about the existence of markets. The creation of specialized markets, the use of agricultural shows, the setting up of food fairs as well as the more traditional food markets are possible outlets for surplus produce. More innovative methods to supply processing factories with bulk produce may be an option worth pursuing for farmer groups.

(e) Potential for establishing cooperatives and women's groups

Livestock producer cooperatives and women's groups can provide a range of benefits to members including delivering technological messages, sharing best practices, joint purchasing of inputs, collective value-adding activities, product differentiation through labelling, risk reduction, securing higher prices and economies of scale through collective marketing and providing access to credit and savings mechanisms.

Challenges and opportunities

There are technical constraints which work against successful improvements and sustainability in small ruminant production and productivity:

1. The genetic potential of indigenous livestock in need of improvement
2. The widespread distribution of livestock diseases
3. Inadequate feed supplies and poor animal nutrition
4. Poor marketing infrastructure and arrangements

Non-technical constraints against sustainable sheep and goat production

1. An almost universal absence of clear livestock development policies and strategies
2. The pace of privatization and the possibilities of cost recovery for goods and services

Small ruminants provide a very suitable base for increasing the supply of animal products. High value essential dietary proteins for people are an important output of small ruminants. There is rapidly rising demand for livestock products especially from the urban populations that are escalating throughout the world. Increasing population pressure, uncertain rainfall, declining soil fertility, and traditional techniques continue to inhibit increases in animal output. Comprehensive and coherent programmes for development of small ruminant production would seem to be the most appropriate way to confront the complex biological, technical and climatic factors, and associated policy issues with which these extremely useful animals have to contend. Strategic options to improve small ruminant production may include:

1. prioritizing the development of the small-scale and traditional sector
2. re-orient the role of governments in creating an appropriate enabling environment
3. making more appropriate production technology available through research and extension
4. improving resource management by smallholder farmers
5. increasing beneficiary participation in efforts and rewards

Reference

Animal Production, Diseases and Treatment. By Nilotpal Ghosh. 2nd Edition, 2008. Kalyani Publishers, New Delhi

Animal Husbandry for Income Generation. West Bengal Livestock Processing Development Corporation Ltd. ARD Department, Govt. of West Bengal.

Goat rearing practices, Directorate of Research, Extension and Farms, WBUAFS, Kolkata. 4th Edition, 2011.

Handbook of Animal Husbandry, Directorate of Information and Publications of Agriculture, ICAR. 3rd Edition, 2005. New Delhi

Livestock Production Management. Edited by N.S.R. Sastry and C.K. Thomas. 2005, Kalyani Publishers, New Delhi

Prani Palan Sahayika, Directorate of Animal Husbandry and Veterinary Services, Govt. of West Bengal. 2nd Edition, 2008.

Dairy Farming

Post white revolution, the growth curve of Indian dairy industry has taken upward growth trajectory in terms of total milk production as well as in per capita milk availability. It was 51.4 million tons in 1990 and about 127 million tons in 2011-12 and 291 gm/day respectively. Few reports suggest that with current growth rate of approx. 3%-4% it is thought to grow to 185 million tons and become a $24 billion (₹1,44,000 Crore) organised industry by 2020 and $140 billion (₹8,40,000 Crore) including unorganised sector. However, various issues in connection to sustainability should be examined critically in the face of newer challenges like global warming. As per the joint publication of the International Dairy Federation (IDF) and the Food and Agriculture

Organization of the United Nations (2004) the Principles and Practices for Sustainable Dairy Farming should be planned in such a way which would be able to answer the following question.

(a) For Sustainable Dairy Farming Systems

- Are breeds suited to the local climate?
- Are animals kept under good conditions, which will in turn deliver good products?
- Is milk protected from any contamination from milking to delivery?

(b) Economic sustainability

- Is it possible to increase production?
- Is the farm system diverse enough?
- Is there access to market information?

(c) Social Sustainability

- Are workers treated fairly?
- Is training a priority?

(d) Local community/economy

Is there a positive impact in the local community from the farm system?

(e) Environmental sustainability

- How soil fertility is is maintained?
- Is soil erosion an issue?
- Is the water source for irrigation sustainable?
- Are the impacts of fertilisers and pesticides considered?
- Are there natural habitats on farm?
- Are rare species of plant/animal threatened by growing the crop?
- What are the major energy inputs?
- How can their impact on climate change be reduced?
- How wastes are safely handled?
- How are the technologies used to reduce, reuse, recycle, dispose them?

Farm Size

Unlike other major milk producing countries, the growth story in India is driven largely by small scale dairy farmers. Around 80% of Indian cattle belong to farmers having herd size of 1 to 2 animals, producing on an average less than 4 Kg milk per day. In the recent past, there have been initiatives in the direction of establishing large scale dairy farms (with more than 1,000 animals) by private organizations. However, most of the players have been able to achieve limited success due to high operating costs, low productivity, large investment and lack of adequate knowledge to manage large scale farms.

Major Challenges Faced by Small Holder Dairy Farmers in India

Following characteristics define the dynamics of small holder dairy farming systems in India.

- Scattered land and small number of cattle holdings make provisions of relevant services and technology transfer a herculean task and are big impediment towards investments in infrastructure development.
- Access to institutional credit for small holder is difficult and makes farmers vulnerable to moneylenders' trap.
- Rapid shrinkage and degradation of grazing land has resulted in green fodder shortage, increasing the risk of nutritional deficiency, occurrence of diseases coupled with loss of productivity.
- Improper genetic evaluation and lack of record keeping have resulted into generation of inferior quality progenies.
- Lack of access to organized marketing channel.
- Gradual migration of manpower to other non-farm occupations.

Feed and Fodder Management

Availability of quality feed and fodder is increasingly becoming a challenge. Hence, it is critical to develop farmer's expertise in effective utilization of locally available feed resources. Low-cost technologies may be applied to maximize the nutritional value of crop residues and to provide some green.

(a) Urea molasses treatment of straw and other low quality forage.

(b) Preparation of ration or feed block for animal using locally available feed resources.

(c) Practicing intercropping and multi-cropping with cereal crops, relay cropping to ensure year round fodder supply.

(d) Cultivation of azola (*Azolla caroliniana*), a good supplement of protein in animal feed.

(e) Production of fodder using low cost hydroponics technology in the farmers' backyard will reduce the use of grain.

(f) Balanced ration formulation by using locally available resources.

(g) Encouraging fodder production systems which will also ensure the restoration of ecological balance in fragile and degraded ecosystem.

Keep animals healthy

With education, farmers could improve husbandry which would result in more stability in the production system. For better biosecurity measures the following measures may be applied in the dairy operation.

- Have secure boundaries/fencing.
- Direct contact of visitors with animals should be avoided.
- Fallen stock should be disposed properly.
- Disinfectant should be used for undertaking cleaning and disinfecting of vehicles and facilities.

Animal welfare & lodging

Sufficient feed and water should be provided every day. Enough space should be given to avoid competition among animals and also to.

- Ensure that all animals are free from discomfort.
- Provide adequate space allowances and dry bedding.
- Protect the animals from adverse weather conditions.
- Have an effective herd health management programme in place.
- Protect the animals against lameness and sufferings.

Waste

The farm shall continuously reduce, reuse and recycle the quantity of waste and by-products that it generates. In particular, organic waste may be composted on the farm and reused for soil conditioning where there is no risk of disease carry-over. This is to ensure that animal and human wastes are stored and managed to minimize the risk of environmental pollution.

Role of Information Technology (IT)

Information technology can be harnessed for efficient management of small scale dairy farms in India. It is important to invest in the information systems to make strategic decisions on optimization of the dairy supply chain and the costs involved. IT finds many applications for information management of demand-supply aggregation of the milk and milk products across production and consumption clusters. Small holder farmers do not have direct access to basic financial services including insurance. IT can play a pivotal role in bringing large chunk of small holder farmers under the ambit of financial services and government schemes.

Access to Finance

To strengthen the dairy sector for rural employment and revenue generation, easy access to institutional credit and instant payments for sale is critical. Financial institutions need to design innovative systems. This would pave the way for financial literacy of the farmers and reduce their dependence on moneylenders for credit requirements. Rural financial institutions need to focus on widening their reach to small farmers.

Extension Services and Skill Development

Dairying needs skill. Skill itself can be helpful in enhancing family income and provide a sustainable source of livelihood. Augmenting knowledge and skill levels of the workforce is essential to enhance productivity, boost innovation, manage finance, mitigate risks and improve decision making ability which will enable sustainable dairy farming. Extension services from different corners need to be converged for a desired goal. Market led practices are to be encouraged to increase resilience in the smallholder dairy farming situation.

Climate Change and Sustainability

Climate change scenario demands adopting green technologies. Harnessing solar or other alternative sources of energy for various heat based applications in farm. India is endowed with a vast solar energy potential. Small holder dairy farmers may be supported with cold chain, bulk milk coolers and harnessing the alternate sources of energy. Efficient collection and utilization of waste for vermin-compost/bio-manure production and establishment of community biogas plants will further benefit the small and marginal farmers.

Opportunities in Indian dairy sector

- The Indian dairy market is amongst the largest and fastest growing markets in the world.
- Per capita availability of milk in India has reached 322 grams per day, higher than the world average of 293.7 grams per day.
- Changing lifestyle pattern, increasing purchasing power and increasing health consciousness are the key growth drivers for milk and high value milk products in India.

Key Fiscal Incentives

- 100% FDI permitted through automatic route in food processing sector.
- Concessional rate of Customs Duty applicable on imported equipment under the project import benefits.
- High rate of Income Tax deductions on capital expenditure are allowed for setting up and operating cold chain or warehouse for storage of agriculture produce.
- Complete exemption of income tax was made available to new food processing, preservation and packaging units for the first 5 years of operation.

Conclusion

A good management of a farming system constitutes the grassroots of the system's economic, environmental and social sustainability. Strategic application of financial and social schemes related to dairying in a convergence mode for making the sector better livelihood option is a key issue for the small farmers in India. Skill development has to be strengthened for Applying the principles & practices of scientific approach to the whole farm system with the philosophy of continuous improvement. Switching to cheaper, renewable and cleaner sources of energy in dairy will open new avenues for entrepreneurs to invest in the sector along with connecting all the stakeholders involved in the chain.

Reference

Bickell, S. L., Durmic, Z., Blache, D., Vercoe, P. E. & Martin, G. B. in Updates on Ruminant Production and Medicine Proc. 26th World Buiatrics Congress, Santiago, Chile (eds Wittwer, F. *et al.*) 317–325 (Andros Impresores, 2010).

Department of Animal Husbandry, Dairying and Fisheries (DAHD), Basic Animal Husbandry, Dairying and Fisheries Statistics, Agricultural and Processed Food Products Export Development Authority (APEDA), National Dairy Development Board and other relevant websites

Eisler, M.C. and Lee, M.R.F. (2014). Steps to sustainable livestock, Nature, (507), 32-34.

Food and Agribusiness Research Management (FARM), YES BANK (2015). Making Indian Dairy Farming Competitive – The Small Farmer Perspective

Scientific approach for sustainable pig production

As per 19th Livestock census, India's pig population is 10.294 million which amounts to about 2% of the country's entire livestock. The statistics reveals 2.456 million exotic and 7.837 million indigenous pigs in India which decreased by 7.54% over the previous census. The highest share of country's pig population is recorded in Assam (15.89%) followed by Uttar Pradesh (12.96%), Jharkhand (9.35%), Bihar (6.31%) and West Bengal (6.30%). However, the number of Pigs per 1000 households in rural areas was highest in Arunachal Pradesh (2221) followed by Nagaland (1424), Mizoram (1220), Meghalaya (1124) and Manipur (657). In addition to protein suppelemt for human, pigs are often one of the main sources of cash income in rural areas and provide manure for cropping. Swine rearing promotes better self-reliance along with greater food security to urban households (Mougeot, 2000) and increases incomes.

North East India: the epicentre for pork production

In states of North-East India, almost every household has a small piggery unit. There is a huge demand for pigs because people of these states prefer pork than other meats. Majority of the pigs in this hilly region are reared in intensive production system and fed with homemade cooked feed including kitchen waste and locally available plants (Kumaresan *et al.*, 2009). Moanaro *et al.* (2011) reported pig rearing under a low-input production system in the Nagaland State of India where nonconventional feed resources contribute majority of the feedstuffs offered to the animals. Small land holders in Nagaland are adopting more profitable and less labour intensive backyard pig production to increase cash returns, and accumulate capital in banks. Families usually keep an average of 1-2 indigenous or crossbred pigs for fattening with zero to minimum inputs in terms of family labour and feeding (Patra *et al.*, 2014). In Mizoram and Sikkim, it has been reported that crossbred pigs were preferred for better growth performance, higher weight gain, larger litter size and greater back fat thickness (Nath *et al.*, 2013).

Why to encourage for pig farming

In comparison to other livestock species, Pig rearing compared to other domestic species has higher potential to contribute more economic gain for small, marginal farmers or rural poors. In addition, it has following advantages:

1. Better feed conversion efficiency of pigs as compared to other meat producing animals except broiler birds.
2. Higher fecundity in pigs: Sows produces 6-12.
3. Pigs reach sexual maturity by an age of 8-9 months and can give birth of 6-12 piglets in each farrowing.
4. Pigs have shorter generation interval and offers quick returns since the market weight of 60-90 kg can be achieved in a period of 7-10 months.
5. Can survive and grow on wide variety of feed stuff viz. grains, vegetables, fruits, fodder, sugarcane, kitchen waste.
6. Offers employment and income generation opportunities to seasonally employed rural farmers for better livelihood and starts with little initial investment

7. Pig products ranges from primary commodities such as pork to processed food like sausages, smoked hams to cooked salted ears etc. readily supplements animal proteins

Selection of breeding stock

Selection of individual animals from a herd is more important than the selection of a particular breed. Each producer at the time of setting up his herd should purchase his animals from a reliable disease free herd and should obtain as much information of the animals as possible. Once the herd is established the selection of the gilts and boars for replacement in the breeding herd should be based on the types and performance. Besides, the following status should also be taken into consideration

1. Size of litters
2. Strength and vigour of litters
3. Milking ability
4. Temperament

S. No.	Selection of Gilt	Selection of Boar
1.	Selection of gilts for the breeding should be weigh about 90 kg	The boar should be purchased from a breeder or a farm maintaining adequate information on its performance
2.	Select gilts from sows which have consistently farrowed and weaned large litters	The boar should be selected from a dam which has consistently farrowed and weaned high litters
3.	They have reached market in minimum time and have desirable market type	A good boar will weigh 90 kg in about 5-6 months will be of good type and will be strong on feet and legs
4.	It would be desirable to choose gilts whose litter mates and other fullsibs have given good performance in daily weight gain and feed conversion efficiency	The feed conversion from weaning to 90 kg weight would be the most desirable

Feeding Management

1. Most economical ingredients should be selected
2. Grains-maize, sorghum, oat, other millets, wheat and rice should form the basic ingredients as energy riched source
3. Protein supplements-oil cakes and fishmeal and meat meal
4. No vitamin supplements is necessary if the pigs are allowed to pasture or are fed fresh green legumes. Vitamin B 12 supplement would be necessary if little or no animal protein is fed. Mineral supplements should be provided

Table 7: The composition of the concentrate feed for variousage groups pigs

Ingredients	Creeper 14-56 day	Grower till 40kg	Finisher 40-90 kg	Pregnanat & Nursing
Maize or sorghum or broken wheat, broken rice and barley in convenient combinations	65	50	50	50
Oil cakes (groundnut oil cake, soya bean oil-cake, sesame oil cake, linseed oilcake	14	18	20	20
Molasses	5	5	5	5
Wheat bran or rice bran	10	15	25	18
Fishmeal or meat meal or cooked offal, skim milk powder dairy wastes	5	5	3	5
Mineral mixture	1	1.5	1.5	1.5
Salt	-	0.5	0.5	0.5

Housing and management

Adequate housing and equipment for raising pigs are necessary to provide shelter against inclement weather, prevent diseases, control parasites and save labour. The flooring should have a rough finish and should be of a regular masonary type. Proper drains should be provided so that the effluents are disposed off easily. Walls should be 1.2-1.5m high from the floor. For the purposes of farrowing some of the pens could be converted into farrowing pens by providing guard rails made up of G.I pipes of 5cm diameter, along the walls, 20-25 cm from the ground and the wall. In addition to guard rails, creep space can be provided for the piglets along the wall by making a partition or in one of the corners with separate entrances for the piglets. This space usually of 0.75 m x 2.4m area. Shade helps in preventing deaths and increasing production efficiency during hot weather. While it is desirable to plant trees in the neighbourhood of pens for reducing the intensity of heat.

Table 8: Floor area and water requirement for pigs

Class of animals	Covered floor area/ animal (m^2)	Open yard area/ animal (m^2)	Water required (L)
Boar	6.25-7.5	8.8-12.0	45.5
Farrowing	7.5-9.0	8.8-12.0	18-22
Winner	0.96-1.8	8.8-12.0	3.5-4.0
Dry sow	1.8-2.7	1.4-1.8	4.5-5.0

Breeding management of sow

The average length of oestrous cycle in pigs is 21 days. The oestrous symptoms last for five to seven days beginning with vulvar swelling and vaginal discharge. In true oestrous there is frequent urination, reduced appetite, mounting and standing for service detected by the erection of ears and immobility when normal pressure is applied to the back. The application of pressure on the back is used to determine the correct breeding time. Animals with a predisposition for weak oestrous should be brought near the boar to exhibit heat symptoms a little more clearly. Best time for breeding is during the latter half of the first day or early on the second day of oestrous. In many cases the gilts and sows continue to exhibit the standing heat on the next day. In these cases the animals should be rebred and the interval in the case of rebreeding should be 12-14 hours. This procedure will ensure a high conception rate in the herd. Sows come into oestrous one to four days after farrowing but they should not be bred at this time. Sows may also come into heat two to ten days after weaning and may be bred at this time. But better results can be obtained by breeding them in the second post lactational oestrous. The animals which have been bred should be observed for the appearance of subsequent oestrous. If sows not conceived even after successful mating with a boar in two continousoestrous cycles it is desirable to cull them from the herd.

Care and management of pregnant animals

The gestation period of sow is of 114 days. Pregnant animals should be housed in groups in separate enclosures and should not be mixed with new animals to avoid fighting which at times may result in abortion. It would also be advisable to house pregnant gilts and sows in separate groups during gestation. About 3 m^2 of dry housing should be available for each sow. The pregnant animals should be allowed to move about every day in the morning on a free range or a pasture if available. A pasture area is presumed to be clean if a cultivated crop was raised.

Management at farrowing

Farrowing time is the critical time in pig production. Death rate is high during farrowing and the first week after farrowing. Sows may be farrowed in pens equipped with guard rails and a creep space in farrowing crates or in farrowing stalls. A pen equipped with guard rails and a creep space is adequate. The pen should be maintained at 24°C to 28°C until the piglets are three or four days old and at 18°C to 22°C until the piglets are approximately six weeks old. The heat lamps should be hung 45 cm from the floor and suitably protected. The farrowing pens should be thoroughly cleaned before the sow is brought in. This will prevent a large number of diseases of piglets.The sow should brought to the farrowing pen atleast one week prior to farrowing so that it becomes familiar with the surroundings. She should be washed thoroughly before being brought to the farrowing pen. The feed ration should be made bulky by substituting one-third of the regular ration with wheat bran. The amount of ration fed should also be reduced by one third till the sow farrows. The sow should be watched closely for determining the approximate time of farrowing and feed should not be given 12 hours before farrowing.

Care during farrowing

An attendant should be on hand when the sow farrows. Otherwise many piglets will die. It takes generally 2 to 4 hours for complete farrowing to take place. The piglets should be removed as they are farrowed and kept warm in the creep space until farrowing is complete. Each piglet should be cleaned of all mucus to ensure that the breathing passages are clear. The navel cord should be tied 2-5 cm away from the navel, cut with a disinfected pairof scissors and the stumps painted with iodine. Piglets should be allowed to be nursed after birth. In about 2 days they settle down to their individual teats. They nurse 8-10 times in 24 hours in the initial period. Trampling by the sow should be prevented during the first two weeks.

Weaning and raising of piglets

Normal weaning age of piglets is at 8 weeks of age. The sow should be separated from the piglets for a few hours each day to prevent stress of weaning and feed is reduced gradually. The piglets should be dewormed after 2 weeks of weaning. The piglets should be gradually shifted from 18 percent protein creep feed to 16 percent grower ration over a period of two weeks. Group of 20 piglets of more or less the same age should be housed in each pen.

Piglets anaemia

Anaemia is a common nutritional disease in piglets.This condition can be prevented and cured by supplying iron either orally or by injection. Oral administraion consists of spraying or swabbing the sow's udder with a saturated solution of ferrous sulphate (0.5 kg of ferrous sulphate in 10 litres of hot water). This solution must be applied daily from birth until the piglets start eating creep feed. Intra muscular infection of iron –dextran compounds is the more effective method of preventing anaemia.

Prevention and control of swine diseases

1. All the newly born piglets should be vaccinated atleast against Swine fever at the age of 2 months.
2. The piglets should be dewormed regularly once in every there months. Worms from pigs can cause disease in human. The eggs of these parasites infect new hosts and can stay in the ground or the pigsty for several years.

3. The next commonly occurring ailment is skin infection, which may be caused by several organisms like bacteria, lice, ticks, mites and results in thickening and crusting of the skin. Mange occurs around the head, ears, legs and tail but will spread over the body if not treated. These parasite may carry other infections to the animals. Mange is controlled by spraying, dipping or painting the infected areas with a suitable preparation. The pen and shelter should also be thoroughly cleaned out and washed down.
4. Once, the unwanted symptoms are noticed, veterinarian should be contacted for treatment.

From production house to consumer's table, pig production system in India is mostly a neglected rather a hate story in most part except in the North Eastern states. No doubt, the pig in the larger part of the country is reared in most unhygienic condition, the glimpse of which is visible besides railway tracks. However, there are innumerable untold story on pig farming and a successful livelihood enterprise. In the changing global atmosphere, sustainable food production in general needs to be addressed and the issue is more crucial for the densely inhabited country like India. Policy and initiatives from all corners are to be framed to change the pig production scenario which itself may change the mindset of entire country.

Reference

Kumaresan, A., Bujarbaruah, K.M., Pathak, K.A., Das, A. and Bardoloi, R.K. (2009). Integrated resource-driven pig production systems in a mountainous area of Northeast India: production practices and pig performance. Tropical Animal Health and Production, 41 (7): 1187–1196

Luc J.A. Mougeot., (2000) Urban Agriculture: Definition, Presence, Potentials and Risks, and Policy Challenges. Cities Feeding People Series Report 31. International Development Research Centre (IDRC).

Moanaro, Ngullie E., Walling I., Krose M. and Bhatt B.P. 2011. Traditional Animal Husbandry Practices in Tribal States of Eastern Himalaya, India: A Case Study. Indian Journal of Animal Nutrition, 28: 23-28.

Patra, M.K., Sonuwara Begum, and Bidyut, C. Deka.(2014) Problems and Prospects of Traditional Pig Farming for Tribal Livelihood in Nagaland. Indian Res. J. Ext. Edu. 14 (4): 6-11.

Nath, B.G., Pathak, P.K., Ngachan, S.V., Tripathi, A.K. and Mohanty, A.K. (2013) Characterization of smallholder pig production system: productive and reproductive performances of local and crossbred pigs in Sikkim Himalayan region. Tropical Animal Health and Prodn. 45: 1513-8.

15

Extension Management for Sustainable Agriculture

K. Pradhan

Introduction

With India's population to catapult to a gigantic figure of 1.63 billion by 2050, it rests upon the shoulders of the country's dwindling land resources to cater to the food demand to the surging human race with an eco-friendly manner. This requires a mammoth effort from the food producing machinery of the country-an effort which has to be inclusive of the technological innovation and extension in this field and to the fullest possible extent. Coping with the short-run challenges to food security posed by food price volatility is indeed a daunting task. But the serious concern is the long-term challenges of avoiding a perpetual food crisis under conditions of global warming. The decline of contributions of Indian Agriculture to the National GDP and an apparent stagnation of the growth rate in agriculture, non remunerative agricultural system, ever increasing population pressure, shrinking land resources, global warming and climate change, indiscriminate use of agricultural inputs, over exploitation of natural resource base and environmental quality have made the situation more vulnerable towards food security. In this backdrop, agricultural knowledge and technology accumulation, transfer, application, and diffusion are key to sustainable development and food security in an eco-friendly manner. The knowledge and technology revolution is critically different from the past industrial revolution in that it is based upon a shift of wealth creating assets from physical things to intangible resources based on knowledge and technologies. Thus, effective management and transfer of knowledge and technologies are believed to be the most critical capability of individuals, organizations, and nations in the globalized 21st knowledge society. Extension could play a pivotal role in fostering sustainability through its educational programs but there has been a growing realization that traditional extension models have not been sufficiently effective in promoting socialization of sustainable agricultural practices. Since sustainable agriculture is a knowledge-intensive system, it requires a new kind of knowledge, which differs from other forms on the basis of conventional agricultural practices. In fact, conventional extension system cannot accomplish sustainability in agriculture; because today's agricultural extension must consider environmental implications, social issues, and overall economic growth within the agriculture sector. In such a situation, the present conceptual discussion shall be envisaging the dictum and destiny of evolving extension approaches by cherishing a clear focus on extension management perspectives towards sustainable development and environment protection and management so as to the well established extension approaches can be restructured and applied in the niche of extension management towards sustainability.

Methodology

The exploratory or formulative research study was carried out to prepare the entire content of the Chapter. The major emphasis was given on the discovery of ideas, insights and theoretical construct. The survey of concerning literature was the main base to formulate the theoretical construct. The researchers reviewed and built upon the work already done by the previous scientists. The incorporation of researcher's idea, knowledge, opinion and discussion with the experts, related to the field helped to envisage the conceptual framework as well as to derive a conclusion of the entire Chapter.

Conceptual framework

Our existing industrial-age, organizational structures are derived from such institutions where the framework sees the manager as "the boss" and everyone else as "the subordinate." Extension can no longer be organized and managed this traditional way if the goal is to retain and motivate our best agents and specialists, for two reasons. One is that people's work-related expectations have changed with a much higher value placed on the intrinsic rewards of achievement, personal growth, challenge, and less on traditional rewards like promotion. And another one is the qualitative knowledge work. Extension management is more concerned with doing things right rather than in doing the right things. Managers must begin to shift their focus from command and control to creating a culture of productivity-one which challenges, rather than reinforces, established practices. In knowledge work, the workers are the "bosses" and the administrators and managers are facilitators. So, better conceptualization of extension management the key components like knowledge and technology transfer, technology generation and socialization to be addressed.

Knowledge and Technology transfer

Theoreticians and practitioners define the concepts of knowledge and technology transfer in many different ways. There is usually agreement, however, that (1) knowledge and technology is not just "thing," and (2) that transfer requires a profoundly human endeavour (Gibson and Smilor, 1991). Transfer is the movement of knowledge and technology via some channel from one individual or organization to another. The transfer of knowledge and technology is a particularly difficult type of communication if that it often requires collaborative activity between two or more individuals or functional units who are separated by structural, cultural, and organizational boundaries. Appreciation for the human component in knowledge and technology transfer directs us away from thinking of simply moving knowledge and technology from point "A" to point "B". Instead, think of knowledge and technology transfer as an interactive process with a great deal of back-and-forth exchange among individuals over an extended period of time (Gibson and Smilor, 1991).

Technology generation

Agriculture has been the foundation of Indian economy and has been part of the culture, knowledge system and way of life of Indian society for centuries. Today approximately 80 per cent of the population depends on agriculture for subsistence. Realizing the importance of agriculture as a means of livelihood, for majority of the people, the government began planned interventions in the agriculture sector in the 1950s. Technological inputs particularly the introduction of improved varieties of crops and their trials were the initial outside interventions in agriculture. The focus on the production of crops under the influence of dominant discourse of science and technology has created an inherent conflict between the scientific and indigenous knowledge systems.

Emerging issues in agricultural technology development: Whose knowledge counts?

The knowledge system is organized along disciplinary lines and agricultural commodities. The key knowledge development function of scientists is to improve crop varieties and animals through scientific experiments and methods. The scientists tend to work in isolation rather than in a holistic and integrated way in knowledge production. Although they have developed low input technology such as zero tillage, the basis of the technology generation is guided mainly by their scientific wisdom that originates from the research stations and academic institutions. The present model of technology development does not consider the structure of social relations and system of social practices through which technologies filter and become accessible to people with limited land and other resources. Interpersonal socio-economic differences play important roles in determining who benefits with the technological innovations.

Besides, scientists seem to regard themselves as the formal, legal and authentic source of knowledge. Scientists reinforce the idea that they still have strong sense of themselves as the only legitimate knowledge authorities in agricultural sector. Generally, they did not seem to appreciate the local context specific research carried out by small-scale research agencies such as local farmers' networks and professional NGOs. This indicates the continued preoccupation of bureaucratic power and the scientists' emphasis on technical rationality in generating new knowledge.

The development of new technology by research strategies and methods do not appear sensitive to the needs and constraints of the resource poor farmers. The research seems to focus on those who have land, and a great majority of landless and land-poor farmers are excluded from the targeted group. Another bottleneck seems to be in the area of sharing or dissemination of knowledge with programme partners like the government extension agencies.

Coordination and linkages among the actors of agricultural development is important for agricultural knowledge development, its dissemination and effective use. Most of the developing countries are concerned about rural poverty and malnutrition, unsustainable agricultural production and swelling of urban population. Most of the problem is due to lack of effectiveness of agricultural research. Further, the research has been done which is not relevant to the farming situation and this lack of relevance of research is due to poor linkages between research and its clients, namely farmers and extension. The process of Transfer of Technology (TOT) cannot be examined in isolation. It is long and involves various institutions and different stages from research stations to farmers' field. The effectiveness of TOT is greatly influenced by the linkages between and among those responsible for technology generation; technology transfer (extension sub system); those who actually put to use the technology (client sub-system) and those who act as a catalyst the adoption of modern technology by supplying the inputs (input sub-system). Thus TOT of any improved practice calls for concerted and coordinated efforts and mutual understanding between and among the above four sub-systems involved. There are certain lacunae in the present research-extension system. Due to rapid development in the past, several institutions and agencies are created to support and strengthen the extension system for the common cause of helping the last man. However, most of the programmes suffer due to absence of teamwork and pReference are given to individuals rather than institutional interests. Patience, understanding, proper altitudes and sprit of co-ordination through inner urge are lacking, therefore, involvement and functioning of Non Government Organisations (NGOs) become relevant.

There is lack of an effective monitoring system to identify the impact and learning from the intervention of new technologies. Moreover, the extension agents are also unaware of the technologies generated. In many cases, the technologies generated are detached from the life of the people (culture, institutions and individual farmers). When useful, they seemed to benefit richer farmers. Most of the scientists are operating in the linear model of development paradigm as they think that their own knowledge is superior. They are not always ready to accept mistakes as a process of learning and refining the knowledge system. What is even more critical is that farmers in the study areas perceived the new varieties or seeds as a threat to the extinction of local knowledge systems.

Suitable context of extension management towards sustainability

Sustainable agricultural systems are situation-specific systems (Zhen and Routray, 2003), thus, in order to support them, extension systems should be situation-specific, too (Qamar, 2002). The meaning of situation-specific is that sustainable agriculture systems should be acted on the base of available assets (Van Loon *et al.*, 2005). In this model, we put these assets as the context of extension activities and functions to achieve sustain-ability; because the success of extension systems is related to identify assets, assess their vulnerabilities and then take steps either to enhance or strengthen those assets or to reduce vulnerabilities placing stress on those assets (Worth, 2006; 2002). These assets are human capital, social capital, natural capital, financial capital and built (manufactured) capital. Human capital refers to skills, abilities, education, indigenous knowledge and health of community. Social capital refers to the networks both within and outside the community; sense of place belonging and the extent of opportunities existing for participation in local activities and decision-making. Natural capital refers to the existence and health of environmental assets. Financial capital refers to the financial resources in or available to the community. Built capital refers to community infrastructure (Gasteyer and Flora, 2002). Each form of capital can enhance the productivity of other forms of capital (Flora,1997a; 1997b). In addition, these assets are sub-system for "forces of change" and are interacting to them. The following is the sustainable extension management model, proposed by Allahyari (2008).

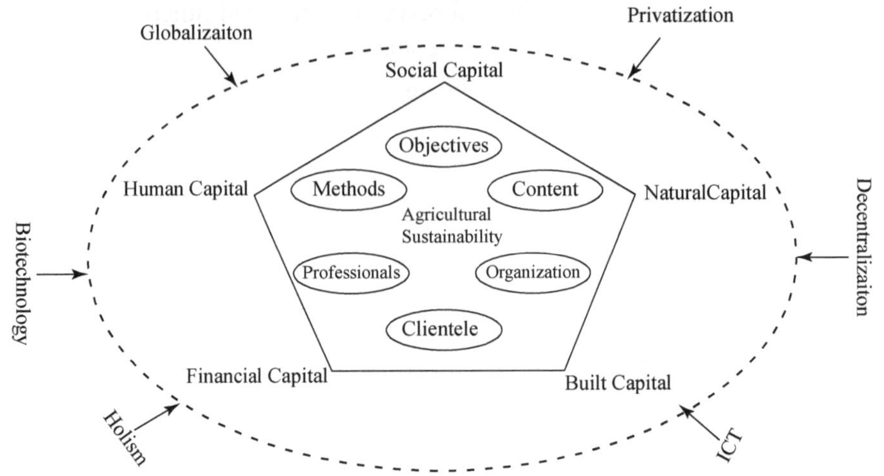

Technology transfer and socialization

In India the "Transfer of Technology"(ToT) model has been the prevalent practice for developing and spreading innovations. It is based on the assumption that a transfer of technology and knowledge from scientists to farmers will trigger development. Applied to agriculture, this model assumes that farmers' problems can be solved by people and institutions that have this 'modern' knowledge. This top-down model creates a rigid hierarchy, which discourages the feed back of information. The major identified constraints of TOT model are constraints in agricultural research system as it is highly reductionist, parochial, and discipline-oriented, the local classification system is neglected, farmers' informal experimentations are not considered as a source of innovation, farmers are mainly seen as the recipients of technical messages but not the originators of either technical knowledge or improved practice, feedback from farmer is a missing link etc.

Development through agricultural innovation in any sphere within the social system is not a linear phenomenon, rather it is multitudes of interactions among the innumerable elements constituting the very social system and leading to an intricate fabrication of multidimensional dynamism. The agricultural scientists of experimental stations, planners and implementing agencies from a techno-cultural complex situation where from flow the ideas, techniques, innovations, components and device, concepts down to the farmers who, again constitute a separate techno-cultural situation. As the concepts and devices are diffused towards the farmers, and then on their part provide the laboratories and planning and implementing agencies with the required feed back for a better way of transferring the needed technology. It is undoubtedly necessary that the diffusion process in the transformation of technology has to be studied. It is often found that socio-cultural and socio-economic situation is also taken into account for finding out a better course for the transfer of technology. The flow of supply of agricultural inputs were increased after the prior assumption that farmers would promptly take to these supplied resources and use them to their fullest advantage to raise their individual productivity and contribute to the increase in agricultural production of the country. But the actual experience in the field, however, showed that farmers were not prompt in adopting these practices as was expected. Many farmers to the surprise of extension worker rejected many of these practices, which had obvious qualities to be of economic benefit to them. In spite of economic advantage, the social and cultural considerations frequently take precedence of adopting a new agricultural practice.

To study impact of Green Revolution the score of extension scientists with a given socio-political context started collecting events of adoption of exotic technologies and ideas as a glory of modernisation and cases of non-adoption had been attributed the failure of individuals in getting themselves tunned with the process of modernisation. And so they are branded as **'unwilling scores of masses'** having low level of education, high skepticism and poor scienticism.

The theory of expansionism had more or less been an explicit or implicit dent in the transfer of technology process as adopter in different front line extension activities like national demonstration, lab to land, operational research project etc. and the process earned a colossal shape through the introduction of Training and Visit system. Extension professional started pilling up reports as to how many had adopted the practices. They were not that interested as to how many rejected the recommendations and what the reasons were therein. Besides documentation in the case of technologies discontinued, dropped outs, reinvented, modified or rectified had been very poor or absent. Thus it was astounding to observe that the extension researchers started in delineating the innovation decision process or diffusion process as to have only one indomitable consequence that is adoption of technology. This could be branded as a **pro-adoption bias** in diffusion researches.

It is discernible to any sensitive mind that the relentless study of transfer of technology had so far refrained our extension scientists as well as scholars from starting the **'technology socialisation process'** as a whole due to its' better implication in the field of ultimate users' by incorporating the two extremes of knowledge systems (generation end and users end) after giving due importance to the prerogative of the ultimate technology users (Acharya *et al*, 2008)

Effective Research Extension Farmer (REF) Linkage

An integrated approach to REF linkage attempts to link all system participants-researchers, extension workers, educators, input suppliers, farmers and others-so that they are jointly involved in the agricultural technology innovation process. The system draws on both modern science and farmers, indigenous knowledge.

In an integrated REF linkage system, the process of technology innovation is viewed as a cycle that offers opportunities for research education-extension-farmer interfaces at the various stages of the process.

An integrated REF linkage approach emphasizes the importance of interactive, mutual learning between formal and informal knowledge/ technology systems and stresses linkages with farmers so that they actively participate in agricultural technology innovation efforts.

REF Linkage and their mechanisms

Linkages are defined as the channels for the two-way flow of knowledge, information and resource between research and its partners. The link can be discussed from two points of view.

- Functional links, which relates to research and technology transfer activities
- Institutional links, which relates to the institutions and personnel that carry out these activities, which aims to form a bridge between research and technology transfer.

There are at least four types of linkages every extension system must develop and maintain. They must have "Enabling linkages" with government cooperatives, or the private sector to give them the right to exist and to provide their financial support. They must have "Functional linkages" with universities and other research systems, "Normative linkages" with colleagues in other related professions, as well as "Diffuse linkages" with farmers and other clientele groups.

Needs for links

- Strong links ensure that:
- Research tackles users priority needs and problems
- Farmers and TOT workers keep up with research development
- Research results from experiment stations are applied to solving farmer's problems
- Available technologies are adopted to with local agro-ecological and socio-economic conditions.

- Successful technologies are promoted and distributed widely to farmers
- Users have access to the information, inputs and services required to support a technology
- Researchers can capitalize on users knowledge and obtain feed back on the relevance and performance of technology

The function linkages are meant to perform

Looking at the problem from a functional perspective to bridge technology generation with technology transfer seven functions have been identified:

- **Input Feedback**: People involved in technology transfer, who are (or should be) in close contact with farmers, can play a role in expressing the farmer's technological needs to researchers. They can also provide information of how new technologies perform under wide spread in on-farm conditions and producers reactions to them.
- **Testing or Adoptive Research**: Technologies must be adopted to local conditions and the needs of particular types of farmers. In addition, those who are expected to transfer these technologies must become familiar with them and convinced of their utility.
- **Transformation of Research Results**: Before experimental results can be effectively transferred, the domain for which they are relevant must be specified, they must be put in a form technology transfer workers can interpret and producers can use, their economic implications must be analysed and certain extrapolations must be made.
- **Regulation**: After a technology is generated and before it is transferred, someone must decide whether it should be transferred. This is the role of regulation.
- **Communication of Research Results**: Researchers must transmit the knowledge generated by their research either to technology transfer workers or, in some cases, directly to producers.
- **Training**: Training has the broader task of providing researchers and technology transfer workers with the different skills they need to interact effectively and maintain proper linkage
- **Services**: Certain technologies are passed from researchers to technology transfer workers in the form of services such as library and laboratory facilities and the production of improved inputs such as seed, breeding cattle, fertilizer, pesticides and agricultural machinery. Although these services are not generally included in discussions of research technology transfer relations, these are significant points of interaction between the two types of institutions.

The need to perform these linkage functions is more or less common to all agricultural technology systems

Improving Linkages

- In order to improve REF linkages, apart from increased investment, better management and an improved policy environment important elements are-
- REF linkage needs to be viewed as an integrated agricultural knowledge and technology system
- A focus on REF linkage functions might include other partners and improved integration.
- Linkages must be formed and strengthened. Policy and mandate changes, as well as changes in "organizational attitudes", may be needed.
- Promotion of informal networking at many levels with an incentive system that rewards collaboration.
- Strengthening and empowering "client groups"

Integrating formal and informal research and development systems

Farmers' efforts to improve their farming and living conditions have resulted in the development of a number of biotechnologies. However, now it is recognized that local people's knowledge needs to be complemented with that of farmers elsewhere and with formal science-based knowledge for improvements. For developing appropriate biotechnologies various groups of people involved in the process must interact with one another effectively. These include farmers and scientists, but also extensionists, NGO staff, representative of farmers and women's organisations, policy-makers, donors, traders, processors, input sales staff and others.

Considering the different backgrounds of these groups and their different perceptions of the problems of small-scale farmers and appropriate solutions, a challenge is to integrate the knowledge of these people.

Effective Linkages

Links are likely to be effective where members of groups

- Share a common overarching goal and sense of mission
- Feel that they depend on each other to successfully accomplish their tasks
- Agree upon the tasks each should perform to get the job done and on their relative importance
- View each other as professionally credible, with the capacity to follow through on agreements

- View each other as partners, not as competitors
- Have adequate opportunities for formal and informal interaction
- Consider that the personal benefits of collaboration outweigh the costs

Finally, the linkages actions required to improve agricultural technology systems may involve shifts in REF linkage priorities, policies, functions and funding, changes in the organization, staffing and management of REF linkage institution; and the development of strong-multilevel links among REF linkage institutions and with farmers who are in a position to demand improved integrated services. Research, extension and farmer need to be seen not as separate entities, but as parts of an integrated agricultural technology system.

Cutting edge extension approaches for proper extension management

The identified extension approaches for proper extension management toward sustainability can be represented as:

Market led extension

With globalization of market, farmers have to transform themselves from mere producer, seller in domestic market to producers cum sellers in wider market sense to best realize the returns for his investments, risks and efforts. This to be achieved, farmers need to know answers to questions like what to produce, when to produce, how much to produce, when and where to sell, at what price and whom to sell his produce. Farmers received most of the production technologies from extension system. Extension system now needs to be oriented with knowledge and skills related to market. An efficient marketing system is essential for the development of the agricultural sector. The marketing system contributes greatly to commercialization of subsistence farmers. Effective linkages of productions system with marketing are required. Agro-processing and other value added activities would play an increasingly important role in the diversification of agriculture. The government provides much of the infrastructure required for efficient marketing. One of the most important is the information and extension services to farmers besides transport & communication facilities, public utility supply, like water, electricity, fiscal and trade administration and public storage, market and abattoir facilities. Extension system now needs to be oriented with knowledge and skills related to market. The supply chain management along with the marketing intelligence are the key components for the market led extension approach. The extension would occur on the basis of the market demand. The major challenges of market led extension are:

- Gigantic size of Public Extension System of the public extension in the country.
- Generation of data on market intelligence would be a huge task by itself.
- The present extension system suffers from several limitations of stationery, mobility, allowances, personal development of cadres etc.
- Infrastructure challenges.
- Storage and preservation of produce
- Minimization post harvest losses.
- Capacity building of extension personnel and farmers.

However the present technology transfer system should be modified and modernised with the market led extension approach for appropriate extension management toward sustainability.

Information Communication technology (ICT)

Information and Communication Technology (ICT) is now the key enabler and a vital component of the new knowledge based economy because it has revolutionized the process of dissemination of knowledge and information. It is a major factor in economic growth and increasing productivity. India is increasingly integrating ICT into its national development plans and adopting strategies for its widespread promotion in all the spheres of economic activities. There is a need to ensure that the benefit of the ICT percolates to all the different socio-economic strata and to the grass roots of the rural India. The requirement of transforming a nation, into a knowledge vibrant e-learning environment is vital. In this context the mission can be taken on the establishment of Village knowledge centre, Telecentre, Community Radio Station, development of software for exercising the right of the villagers to manage their own water resources, weather forecasting, forecasting of cropping pattern to cope the ill effects of climate change, e-governance system, GIS based planning and decision support system or expert system, installation of low cost, self-sustainable and community owned rural Intranet system for promoting the rural entrepreneurship of the youth, providing telemedicine service through net, development of different information portals to provide access to products and services needed by rural households, farmers, and industries, adult literacy. However, the present technology transfer system should be modified and modernised with the help of information communication technology for appropriate extension management toward sustainability.

The information communication technology based expert system is the hand held solution to the technology users and emphasises the strong REF linkage through a single window toward sustainability. An Expert System can be defined as a Knowledge Based System which is basically a computer program designed to simulate the problem-solving behavior of an expert in a narrow domain or discipline. In agriculture, expert systems unite the accumulated expertise of individual disciplines, *e.g.,* plant pathology, entomology, horticulture and agricultural meteorology, into a framework that best addresses the specific, on-site needs of farmers. Expert systems combine the experimental and experiential knowledge with the intuitive reasoning skills of a multitude of specialists to aid farmers in making the best decisions for their crops. The rule-based expert system has five components: (i) the knowledge base, (ii) the database, (iii) the inference engine, (iv) the explanation facilities, and (v) the user interface. Knowledge-based expert system technology has been applied to a variety of agricultural problems since the early eighties. The expert system applied to the problems of diagnosing Soybean diseases (Michalski *et al.*, 1983) is one of the earliest expert systems developed in agriculture. A unique feature of the system is that it uses two types of decision rules namely the rules representing experts' diagnostic knowledge, and the rules obtained through inductive learning from several hundred cases of disease.

Computer based general support and expert system can be helpful in creating awareness and giving timely preventive suggestions. It can also help in insect identification, monitoring, insecticide selection, and disease management. It can be cost effective as users need not wait for experts. By repeated consultations, farmers can develop self skill and awareness for care-taking and expert system can become advisory-cum-training tools. The expert system technology provides a good platform to facilitate the transfer of knowledge from expert to farmers. The expert system technology in the area of horticulture has been developed especially related with decision making, pests, diseases, or disorders diagnosis.

Some examples of available expert system in agriculture are:

(a) **CALEX:** CALEX, the integrated expert system for Cotton production and management, developed in Egypt is user friendly computer program that simulates human problem solving behavior. Growers can use this system to help manage crop production or predicts the effects of any one decision on subsequent events. CALEX contains plant and pest simulation models to diagnose the pest.

(b) **PEKA-SEWIT:** PEKA-SEWIT is an expert system for Oil-Palm disease control diagnosis. It will perform diagnosis process for assigning an infected disease treatment solution and also the preventive control. Agricultural officers and planters who involve directly with oil-palm plantation may use this system as assistance for helping them in managing crop activities especially in disease control [Ramil and Suhaimi (2008)].

Prasad and Babu (2006) interpreted various agricultural expert systems. They mentioned that in order to remain competitive, the modern farmers often relies on agricultural specialist and advisor to provide information for decision making.

(c) **UNU-AES:** UNU-AES is an expert system in agro-forestry management. This system was designed to support land–use officials, research scientists, farmers and other individuals, for maximizing benefits from agri-forestry management techniques.

(d) **Rice-crop-Doctor:** MANAGE (National Institute of Agriculture Extension Management) developed an expert system to diagnose pest and disease for rice crop. This system diagnoses disease and pest and suggest preventive and curative measures. This system considers disease like rice blast, brown spots, sheath blight, rice tungro virus, false smut fungi, bacterial leaf blight and it include pest like stem borer, rice gall midge, brown plant hopper, rice leaf folder, green leaf hopper, gandhi bug etc.

(e) **NAPER-WHEAT:** It is another expert system developed by CLAES for irrigated Wheat management.

(f) **NEPER:** NEPER is one of those systems, which was implemented using a tool developed at Michigan State University on top of the object-oriented language Small Talk. The expert system (NEPER) included two subsystems: the strategic subsystem, and the tactical subsystem. The strategic subsystem consists of six modules namely: variety selection, pre-cultivation, pest control, tillage, planting, irrigation and fertilization, and harvest. The tactical subsystem consists of two modules namely: weed identification and control, and diagnosis and treatment (Baig *et al*, 2005).

(g) **MAIZE:** The MAIZE expert system was developed by Penn State University to help extension agents and agribusiness personnel identify pest management strategies in field corn for their clientele. (Baig *et al*, 2005)

(h) **NAPRA:** NAPRA (National Pesticide Risk Assessment) was developed by the Soil Conservation Service (now the Natural Resource Conservation Service) to assess the relative risk of pesticides contaminating the ground or surface water. Then a joint MAIZE/NAPRA program was designed by linking the output of the NAPRA program with the MAIZE expert system.

(i) **DIAGNOS-4:** Kerala Agricultural University developed an agricultural Expert System for diagnosing pests and diseases of nine major crops of Kerala called 'DIAGNOS-4' which has drawn tremendous attraction from extension personnel.. The user when selects a complaint, a list of symptoms will appear on the screen.

(j) **AGREX:** Centre for Informatics Research and Advancement, Kerala has prepared an Expert System called AGREX to help the agricultural field personnel give timely and correct advice to the farmers. This expert system find extensive use in the areas of fertilizer application, crop protection, irrigation scheduling, and diagnosis of diseases in post-harvest technology of fruits and vegetables.

(k) **Farm advisory system:** Punjab Agricultural University, Ludhiana, has developed the Farm Advisory System to support agri-business management. The conversation between the system and the user is arranged in such a way that the system asks all the questions from user one by one which it needs to give recommendations on the topic of farm management.

(l) **Maize Agri Daksh:** It is the first online expert system developed by IASRI (Indian Agricultural Statistics Research Institute, New Delhi) in collaboration with Maize Research, New Delhi using AgriDaksh. It provides ICT based advisories on Maize crops and allows interaction with experts using internet.

(m) **SEMAGI:** An interactive microcomputer program named SEMAGI has been developed for sunflower to evaluate the potential yield reduction from multispecies weed infestations and from the parasitic weed broomrape and to determine appropriate selection of herbicide. It combines relational database on herbicides, weed and their interactions. SEMAGI provides an economic study of any herbicide treatment selected or introduced by the user, based on herbicide treatment cost, expected yield increase from the weed control treatment and sunflower selling price.

(n) **PLANT:** An excremental expert system developed by University of Illinois at Urbana-Champaign to provide consultation on the diagnosis of soybean disease and also for decision making regarding both crop diseases and insect damages.

(o) **HERB:** It is a computer-based expert system for soybean weed management developed at North Carolina State University, was evaluated for managing weeds under Georgia conditions. The project was initiated in two phases: a) training Cooperative Extension county agents followed by evaluation in six Georgia counties and b) revision, licensing, and distribution across the state. Field evaluations indicated that HERB was not highly accurate for predicting final yield loss because of weed species senescence and environmental extremes later in the growing season. HERB generally provided a reasonable prediction for a positive economic return due to treatment approximately 60% of the time (Monks *et al*, 1995).

Farming System Based Extension (FSBE)

A major challenge in Indian agricultural development in the present decade and beyond lies in the effective involvement of farmers in the extension and research programs. The continued stress is more on developing procedures or methods that encourages farmer's participation in planning and management of above programs. Involving farmers in the process of technology development can develop appropriate recommendations specific to crop situation. This requires major changes in the attitudes, approach and role of researchers and extensionists. As such an approach which provides active participation of farmer-researcher and extensionists in developing a modified or fine tuned technological recommendations specific to crop situation has been attempted through "Farming Situation Based Extension. The activities in this approach would be governed by participatory approach. Participatory extension provides a framework for extension staff to participate with village communities in facilitating development planning and activity implementation. This approach ensures the extension response becomes community driven and assist village communities implement their planned activities with routine monitoring and evaluation of activities and development progress. Importantly, as the name implies, the extension process is seeking maximum participation from women and men from all groups within the target village communities. People's participation implies the active involvement in development of the rural people, particularly disadvantaged groups that form the mass of the rural population and have previously been excluded from the development process. In this context participatory evaluation is also plays a pivotal role. Participatory evaluation is a partnership approach to evaluation in which stakeholders actively engage in developing the evaluation and all phases of its' implementation. Those who have the most at stake in the programme — partners, pro-gram beneficiaries, funders and key decision makers — play active roles. Participation occurs throughout the evaluation process including:

- Identifying relevant questions;
- Planning the evaluation design;
- Selecting appropriate measures and data collection methods;
- Gathering and analyzing data;
- Reaching consensus about findings, conclusions and recommendations;
- Disseminating results and preparing an action plan to improve program performance

Fundamentally, participatory evaluation is about sharing knowledge and building the evaluation skills of program beneficiaries and implementers, funders and others. The process seeks to honour the perspectives, voices, pReference and decisions of the least powerful and most affected stakeholders and program beneficiaries. Ideally, through this process, participants determine the evaluation's focus, design and outcomes within their own socioeconomic, cultural and political environments.

Characteristics of participatory evaluation

- It elicits involvement of local program stakeholders, allowing them to reflect on their own experiences and to learn from them.
- Participatory evaluation allows program managers, field staff and community members to better understand the perspectives of program stakeholders and the dynamics of community programs, which can contribute to improved program implementation.
- Participatory evaluation can increase the capacity and confidence of local program staff and community members to analyze their own needs and programs and to undertake action-planning based on the conclusions of such analysis.
- Through involvement of community and program stakeholders in evaluation, community members can articulate their priorities and criticisms of development program strategies.
- It can contribute to sustainability of program strategies by increasing the sense of ownership on the part of local development staff and community members of the conclusions and recommendations for future action.

Categories of tools for Participatory evaluation

The variety of tools and techniques available for use in participatory evaluation include more recently-developed PRA tools. Most of the PRA tools were originally developed for use in initial community assessments (participatory rural appraisals etc.), however, most of them can also be used for planning, documenting and reporting on program activities. The purpose of these tools is to elicit group discussion, reflection and sharing and to stimulate groups of program stakeholders to formulate conclusions and plans for action.

1. PRA and PLA-related tools and methods
    - Visualized analysis
    - Venn diagrams
    - Pie diagrams

- Matrix ranking
- Transect walks
- Ranking
- Community mapping
- Flow diagrams
- Seasonal calendars
- Welfare classification/wealth ranking

2. Quantitative tools
   - Community surveys
   - Intercept interviews
   - Structured observations
3. Tools derived from the anthropological tradition
   - Participant observation and Oral testimonies
4. Micro-planning tools
   - Logical framework analysis and stakeholder analysis

Methods for participatory evaluation

- Interviews
- Focus group discussions
- Group and team dynamics methods
- Community meetings
- Team review sessions and Lessons learned exercise

However, the present technology transfer system should be modified and modernised with the help of farming system based extension approach for appropriate extension management toward sustainability.

Indigenous Knowledge System Approach

Indigenous knowledge on the flora of India and adjacent region is as old as ancient scriptures, bio-geographical niche, cultural history, natural resource on which the indigenous communities are dependent upon on this subcontinent. The topography coupled with gigantic watershed river system, diverse mountainous, desert, oceanic ecosystems and varied climatic influences and other factors have contributed immensely towards the rich flora of the region with variety of life forms subsistent on them. With the shrinkage of the bio-resources rich areas under indigenous communities and vanishing of knowledge along with indigenous communities, there is urgent need to make joint venture through the application and integration of scientific knowledge through the ways and means the communities understand and prefer most considering the components of indigenous technical knowledge. From the integration of taxonomic knowledge there is need for characterization of traditional methods of identification and classification, value addition and methods of pre and post harvesting of species. Popular beliefs and folklore have now scientific principles for interpretative approach. However, the present technology transfer system should be modified and modernised with the help of the systematic approach of indigenous knowledge and scientific knowledge integration for appropriate extension management toward sustainability.

Sustainable Livelihood Approach (SLA)

The Sustainable Livelihood Approach (SLA) is a way of looking at development in a way that is concerned principally with people. It may be described as a way of thinking about the objectives, scope and priorities for development in order to enhance progress in poverty elimination. This approach seeks to understand people's strengths, including their skills and possession, and how they use these assets to improve the quality of their lives. The SLA is comprised of a development objective, an analytical framework and a third component, at its heart, principles for poverty focused development. In this sense, the capability approach complements the Sustainable Livelihood Approach (SLA) as adopted by many a contemporary research and development organization.
A particular strength of this approach is that it recognizes human agencies and examines the way in which household livelihood strategies are built around protecting, substituting, increasing, and using assets to produce security and achieve other goals. Recently, the NGOs are playing an emerging role in case of strengthening the linkage between research-extension-farmer through SLA. The Non-Government Organizations (NGOs) are most valuable development resource and active in different states and union territories India. The Training and Visit (T&V) system of agricultural extension, employing over a hundred thousand extension personnel, is the most extensive extension system in India serving India's million of farmers. The joint efforts of T&V system and NGOs are likely to show better results in serving their common client, the farmers, that the sum of their individual efforts. This is the premise on which linkages between the organizations can be based. Many of NGOs co-sharing mission opens up a number of areas in which these two systems can establish linkages to their mutual benefit and thus enhance the quality of extension service to the farming community in case of soil and water management, community organization, common property management, research linkage, monitoring

and evaluation and extension training. However, the present technology transfer system should be modified and modernised with the help of the sustainable livelihood approach for appropriate extension management toward sustainability.

Epilogue

India's agriculture has been experiencing some transformation over the decades through substituting her traditional peasant farming with that of surplus generating one and, accordingly, capital-intensive approach is getting entry into the realm of Indian agriculture. While agriculture held an important place, its' efficiency remained at a low level. Sustainable development through agricultural innovation in any sphere within the social system is not a linear phenomenon, rather it is multitudes of interactions among the innumerable elements constituting the very social system and leading to an intricate fabrication of multidimensional dynamism. The agricultural scientists of experimental stations, planners and implementing agencies from a techno cultural complex situation where from flow the ideas, techniques, innovations, components and device, concepts down to the farmers who, again constitute a separate techno-cultural situation. As the concepts and devices are diffused towards the farmers, and then on their part provide the laboratories and planning and implementing agencies with the required feed back for a better way of transferring the needed technology. It is undoubtedly necessary that the diffusion process in the transformation of technology has to be studied. It is often found that socio-cultural and socio-economic situation is also taken into account for finding out a better course for the transfer of technology. Under the aegis of paradigm shift of extension approaches from mere technology transfer to technology socialisation and so on, the need of rejuvenating the newer extension approaches like market led extension, information communication technology application, farming system based extension with the help of participatory approach, integration of indigenous knowledge with scientific knowledge and sustainable livelihood approach with emerging increased role performance of NGOs can replenish the appropriate extension management system towards sustainability.

Reference

Acharya, S.K., Pradhan, K and Adhikary, M.M. (2008) Socialisation of technology in Agriculture (A paradigm shift from TOT). *Agrotech Publishing House*, New Delhi.

Allahyari, M S (2008) Extensionists' attitude toward sustainable agriculture in Iran. *Journal of Applied Science*. 8: 3761-3763.

Baig Fawad, Nawaz Naima and Rahman Saif Ur (2005). Expert Systems for Decision Making in Agriculture Sector. *Journal of Agriculture and Social Science.* 1813-2235/2005/1-2-208-211.

Flora, B C (1997a) Enhancing community capitals: the optimization equation. *Rural Development News.* 21(1): 12-23

Flora, B C (1997b) Building social capital: the importance of entrepreneurial, social infrastructure. *Rural Development News*, 21(2): 76-89.

Gasteyer, S P and Flora, C B (2002) Community participation for conservation and Development of natural resources. *A summary of literature and report of research findings.*

Gibson, David and Smilor, Raymond (1991) Key Variables in Technology Transfer: A Field-Study Based Empirical Analysis. *Journal of Engineering and Technology Management.* 8: 287-312.

Michalski, R., Davis, J., Visht, V. and Sinclair, J. (1983) A computer-based advisory system for diagnosing soybean diseases in Illinois. *Plant Disease.* 67:459-46.

Monks C. Dale, Bridges David C., Woodruff John W., Murphy Tim R. and Berry Daniel J. (1995). Expert System Evaluation and Implementation for Soybean (Glycine max) Weed Management. *Weed Technology.* 9 (3) : 535-540.

Prasad G.N.R. and Babu A Vinaya (2006). A Study of Various Expert Systems in Agriculture. *Georgian Electronic Scientific Journal: Computer Science and Telecommunication.* 11: 4

Qamar, M K (2002) Global trends in Agricultural Extension challenges facing Asia and the Pacific region. *Sustainable Development Department (SD), FAO,* Rome.

Ramil Azizul Azar bin and Suhaimi Nur Suhallin bt. (2008). *Expert System For Oil–Palm Disease Diagnosis (Peka-Sewit)*, www.eprints.uthm.edu.my/1833/.

Van Loon, G W; Patil, S G and Hugar, L B (2005) Agricultural sustainability; strategies for assessment. *Sage publication*, New Delhi.

Worth, S (2002) Sustainable extension not transforming, but renewal. *Proceedings of the 18th* annual conference of AIAEE, Durban, South Africa.

Worth, S (2006) Agriflection: A Learning Model for Agricultural Extension in South Africa. *Journal of Agricultural Education and Extension.* 12(3):179-193.

Zhen, L and Routray, J K (2003) Operational Indicators for MeasuringAgricultural sustainability in Developing Countries. *Environment Management.* 32(1): 34–46

16

Sustainable Horticulture

Subhra Shil* and *Pabitra Adhikary

Our production systems are at a pivotal stage in terms of meeting consumer demand for affordable food while improving sustainability. Current intensive crop production practices designed to maximize yield have created an unstable, fragile, and non-sustainable production system. Such systems are prone to reduced soil quality, frequent insect and disease resurgence, emergence of resistant weed species, reduced food quality, and detrimental effects on the health and well-being of our communities. With increasing awareness among consumers of produce quality, nutritional value, methods of crop production and effects of production techniques on the environment, there has been a strong move to improve the quality of our production systems. Grower response to this change has been supportive. Growers have shown interest in developing resilient and stable production systems through the adoption of production practices that enhance soil health, crop productivity, and improve long-term farm sustainability and profitability. Such practices emphasize the use of natural processes within farming systems, often called 'ecologically sound' practices which build resilience through synergies and complementarities within the field, the farm, and across our landscape and communities.

Sustainable horticulture is nothing but a contemporary approach to horticulture that emphasizes organic/natural approaches and allows an integrated meshing of ideas with synthesized chemicals used sparingly or relegated to the "last resort." High Tech & Low Tech are encouraged to create a long-term viable result. Short-Term techniques that impact the environment negatively and thwart the biodynamic continuum are rejected.

Sustainable vegetable production does not necessarily mean organic vegetable production. Many growers have made a commitment to create a more sustainable farm for their families and communities. But the immediate elimination of all chemical fertilizers and pesticides would be a disaster for many. In transition from a purely chemical-based production to a more environmentally benign one, economics may dictate the use of a pesticide to save the crop. Time is needed for growers, researchers, and extension agents to become familiar with environmentally sound practices and their efficient implementation. (John McGrady, 1992)

Some of the key practices include conservation (or reduced) tillage systems, cover cropping, crop diversity including crop rotations and intercropping, use of suitable crop cultivars, efficient water use, sound nutrient management plans, and integrated pest management and. Sustainable production systems are complex and dynamic and can involve diverse number of production practices.

Sustainable production: practices and technology

Protected cultivation

Protected cultivation is one of the answers to most of the burning issues facing Indian Agriculture including adversities due to uncertain and varying climatic conditions, climate change, improper uses and low productivity of natural resources, nutritional security in topographic and climatic disadvantaged areas, environmental pollution due to pesticide use, etc. It generally refers to providing congenial conditions for better plant growth and enhancing the production level through artificial means. It makes possible by controlling the climatic conditions by providing full or partial covering surrounding the plant so that plant does not experience higher or low temperatures or humidity, while getting enough light for photosynthesis, optimum fertilisation and watering, and other factors for best growth and production. In the present day context, the interventions include polythene covered greenhouses (polyhouses), shadenet houses, insect net houses, low plastic tunnels, row covers, plastic film mulching, etc. Such interventions have potential to enhance the productivity of any crop to several folds along with enhancing quality to very high level which is not expected normally in open conditions. It increases the land, water and fertiliser productivity along with reduction in pesticide use.

Vegetable and flower production is significantly influenced by the seasonality and weather conditions. The extent of their production cause considerable fluctuations in the prices and quality of vegetables. Striking a balance between all-season availability of vegetables and flowers with minimum environmental impact, and still to remain competitive, is a major challenge for the implementation of modern technology of crop production. The crop productivity is influenced by the genetic characteristics of the cultivar, growing environment and management practices. The plant's environment can be specified by five basic factors, namely, light, temperature, relative humidity, carbon dioxide and nutrients. The main purpose of protected cultivation is to create a favourable environment for the sustained growth of plant so as to realize its maximum potential even in adverse climatic conditions. Greenhouses, rain shelters, plastic tunnels, mulches, insect-proof net houses, shade nets etc. are used as protective structures and means depending on the requirements and cost-effectiveness. Besides modifying the plant's environment, these protective structures provide protection against wind, rain and insects. Protected cultivation is relevant to growers in India who have marginal and small land holdings, which helps them to produce more crops each year from their land, particularly during off-season when prices are higher. However, growing vegetables and flowers under protected conditions requires comparatively high input cost and good management practices, which have direct bearing on the economic viability of the production system. Even if the protective structures are cost effective, proper planning, management and attention to details are needed to achieve maximum benefits (Shweta *et al.*, 2014).

Water conservation and sustainable irrigation

Sustainable crop production practices include methods of water conservation and sustainable irrigation. Over-irrigation causes the salinization of soil, which can lead to declines in yield. Additionally, in many agricultural areas, aquifers used for irrigation are depleting rapidly. In sustainable agriculture water conservation practices include low volume irrigation, rainwater catchment, and the planting of drought-resistant crops or crops that have been bred for a particular environment.

Irrigation

Irrigation is one facet of the inputs which need to be carefully managed so that profits are maximised. If there is no water there is no growth. Water is an essential ingredient for growth. Any commercial agricultural enterprise requires water if grown during long dry season. This includes both annual vegetable and melon crops as well as perennial fruit trees. Despite the importance of water to production, there has been minimal care taken to maximise the efficiency of irrigation, and hence, crop yield and quality. The timing, frequency and quantity of application can all have a marked effect on crop yield quality and time of harvest. Minimal use of water for maximum returns is also a sensible aim, due to the cost of extracting water (bore installation costs and energy costs to pump the water to where it is required) and to the limited resources of readily available bore and river water which exist in the Top End. Irrigation practices can be improved if a range of important factors are taken into account. These are: crop type, crop water requirements, climatic conditions, soil type, water quality, method of irrigation (Y. Diczbalis, 1999).

Crop type

The type of crop grown is influences irrigation practice. Annual vegetable and melon crops are high users of water and irrigation management is extremely critical to their productivity during their relatively short life span (10-20 weeks). Perennial fruit tree crops, tend to require less water and management, although critical, is not generally as critical as for annual crops. Some tree crops (mango and cashew) require little or no water for their survival during non-flowering and nonfruiting growth periods, whereas, fruit trees (*e.g.,* carambola, mangosteen, jackfruit and banana) require continuous irrigation throughout the year. Crop type influences rooting depth which determines how much available soil water the plant is able to tap. Effective root depths vary from 15-30 cm for vegetable crops to 80-100 cm for many tree crops (*e.g.,* mango, citrus).

Crop water requirements

The crop water requirement can be calculated using indirect evaporation based models, or directly measured using soil moisture monitoring devices. In theory both methods should be used, the former method allows the grower and irrigation designer to know how much water is likely to be required during the cropping season and how best to deliver the water so that the crop peak demand for water can be met. The latter method allows the grower to monitor water use during crop growth and to make any small adjustments which are required. The simplest and most commonly used evaporation based model uses open pan evaporation data to calculate water requirements. This can be done on a daily basis, or by using long term average data for your particular area, generally collected and stored by the weather bureau. This method is useful because the driving forces, (solar radiation, wind, temperature and humidity) which cause evaporation of water from an open surface of water are also responsible for driving water loss (evapo-transpiration) from the plant and its immediate soil surrounds. Plant water requirements are generally less than open pan evaporation and are expressed as crop factors (ratio of plant water use to evaporation loss). For example, a crop with a crop factor of 0.6 will require 6.0 mm of water when the daily loss of water from an open pan evaporimeter is 10.0 mm. Crop factors for a large range of crops, mainly sub-tropical and temperate, have been measured by agricultural scientists using direct soil measurement techniques. As a rough rule of thumb, most vegetable crops would have a crop factor somewhere between 0.8 to 1.0, whereas tree crops can have crop factors which range from less than 0.5 and 1.0 depending on the time of year and the particular growth stage (flowering, fruiting or vegetative) the crop is in. Tropical lawn grasses, generally grow well if irrigation is based on a crop factor of 0.6.

Soil type

Soil type influences irrigation management due to the ability of soils to store varying quantities of water, depending on their texture. Sandy soils hold the least and clays hold the most. Most soil profiles are made up of various texture classes, hence the water storage capacity depends on the cumulative storage capacities of the various layers within the profile. The terms describing soil water holding capacity have been loosely used and as such deserve some clarification.

Water quality

The quality of irrigation water in the Top End is generally good. In areas where water is being extracted from limestone aquifers the lime (bicarbonate) levels may present some management difficulties, particularly if using overhead irrigation where the leaves are wetted. Most serious water quality related problems are as a result of high salt (sodium chloride) levels in the water. Fortunately this is not a problem in Top End horticultural areas.

Irrigation method

Method of application of irrigation is an important part of the management process. The three broad categories, over-head (high pressure), micro-sprinkler (medium pressure) and drip (low pressure) systems all have to be managed in their own specific way. Technically, drip irrigation systems are the most efficient, in terms of units of water delivered per units of water pumped. Micro-sprinklers are again more efficient than over-head spray irrigation. Pitcher drip irrigation is similar to drip irrigation, but less expensive to install. The pitchers are the round earthen containers used in rural areas for water storage, ranging from 10 to 20 liters in capacity. This kind of irrigation is ideal for spreading plants such as gourd, pumpkin, and melon because few pitchers are needed per unit area. It is also very good irrigation for saplings, promoting deep root growth. The type of system utilised will depend on: water availability, crop type, soil type and the managers preferred option. It is of vital importance that the system is hydraulically correct: *i.e.,* it is able to deliver the same amount of water at the required emitter operating pressure to all areas of the crop or orchard.

Timing of irrigation

The timing of irrigation can be a crucial part of managing an orchard. Some tree species, *e.g.,* mango, respond to a dry period by flowering earlier and more profusely than would otherwise occur if they were continually irrigated. Other species which may respond positively to a dry period are like cashew but the response and the length of dry period required will depend on the species. In summary, efficient irrigation management requires knowledge of your crop, irrigation system, soil and water quality as well as methods of monitoring your soil water status

Hydroponics

Hydroponics is a subset of hydro culture, which is the growing of plants in a soil less medium, or an aquatic based environment. Hydroponic growing uses mineral nutrient solutions to feed the plants in water, without soil.

Hydroponics is a way for growers to ensure the most efficient use in irrigation. By using hydroponics, growers are able to avoid crop failure or large expenses associated with lack of access to ground water. Also, because of the past issues with soil-borne diseases, it is again to the grower's advantage to implement a surface or soilless growing system whether it is raised beds, containers, soilless media or hydroponics. Because of the challenges faced by growers who attempt to exploit resources like water and soil, it is easier to take a more controlled approach like growing in soilless media or using hydroponics which in turn ensures a better crop and profit for the grower (SAEP, 2010).

Hydroponic plant cultivation is widely considered more sustainable than regular farming as it uses around 90% less water. Secondly, hydroponic farming yields more vegetables and herbs in a shorter time. Thirdly, its carbon footprint is minimal as the greens are grown locally. Soilless cultivation minimizes the use of harsh chemicals needed to fertilize soil; it does not depend on the climate or season and can be installed indoors. Also, hydroponic farms physically take less space (and can be even vertical), are fully controlled remotely and often require less human labour. Another commercial advantage of hydroponic farming is adaptability and scaling: you can have either a small mini-greenhouse at home to grow vegetables for your family only, or run a huge production of industrial scale for sale.

Jalkund - A low cost rainwater harvesting structures.

Unavailability of adequate amount of water during the dry season is a serious problem for successful farming in high rainfall areas. This problem can be minimized by rainwater harvesting and its judicious use for crop production. Direct rainfall collection through water catch ponds/pits (Jalkund) can be highly beneficial to farmers for providing irrigation to crops during moisture scarcity conditions during dry seasons. Rainwater can be stored directly in Jalkunds during the rainy season which can be utilized to provide protective irrigation to the crops for successful cultivation. Otherwise, it may cause soil erosion and nutrient loss through runoff, stored water can also be utilized for animal husbandry activities, Piggery, Poultry and Duckery, fish rearing can also be taken up in the harvested water, harvested water can be used for cultivating high value vegetable crops such as Brinjal, Chilly, Tomato, Radish, Amaranthus, Coriander, Cowpea.

Better soil health management

Soil health is a critical component of sustainable agriculture and comprises a number of different growing practices and principles. There is also some evidence that sustainably grown plants may be higher in vital macro-and micronutrients, resulting from increased soil health as a direct consequence of organic growing methods and sustainable practices. Some of these practices are outlined below:

Organic fertilizer use

Inorganic (commercial) fertilizers are synthetically created (or mined) for the purpose of adding nutrients plants need to grow to the soil. The practice of monocropping and the lack of crop rotation on industrial farms result in the greater need for soil augmentation with synthetic fertilizers. Commercial fertilizer use can impair soil health over time, resulting in the need for additional application of inorganic fertilizers. It may also cause soil acidification and soil contamination with heavy metals. In addition, commercial fertilizers are a primary source of water pollution, causing algal blooms and dead zones in bodies of water throughout the US. The production of inorganic fertilizers also requires large quantities of fossil fuels.

Alternatives to synthetic fertilizer use include compost (decomposed organic matter), animal manure, seaweed, and worm castings. Each of these products can help boost soil health through the introduction and maintenance of healthy soil organisms and micronutrients. Organic fertilizers increase soil biodiversity and have been shown to increase the uptake of nutrients by plants. There is also evidence that use of organic fertilizers improves the nutrient value of the plants themselves.

Green manuring

Green manuring is the process of turning of green plants into the soil either by raising them in same field or plants grown elsewhere at the green stage before flowering and incorporated into the soil. Green manuring, defined by Pieters (1927) as 'the practice of enriching the soil by turning under fresh plant material either in situ or brought from a distance' is a widely used practice in organic farming to maintain soil organic matter. Green manure may be plant of grain legume such as pigeon pea, green gram, soybean or ground nut; perennial woody multipurpose legumes trees viz., *Leucaena leucocephala* (Subabul), *Gliricidia sepium* (Gliricidia or Mata Ratón), *Cassia siamea* (Kassod tree) or non-grain legumes like sunn hemp (*Crotalaria sp*), dhaincha (*Sesbania spp*), *Centosemia, Stylosanthes* and *Desmodium* (Palaniappan,1994). Sun hemp (*Crotalaria juncea* L.), dhaincha (*Sesbania aculeata* L.), berseem (*Trifolium alexandrinum*) and mungbean (*Vigna radiata*) are most commonly grown as green manure crops. The use of green manure crops can play an important role in achieving targets of the sustainable agriculture.

Green manuring practices improve soil fertility, add nutrients and organic matter in soil (Sinha *et al.*, 2009, Selvi and Kalpana, 2009) and improves soils physical and biological properties ((Egodawatta *et al.*, 2011; Eriksen, 2005), reduce soil erosion (Schumann *et al.*, 2000), suppress weeds (Blackshaw, *et al.*, 2001), helps to reduce inoculum of soil-borne phytopathogens and other soil inhabiting pests of agricultural crops with no hazardous effects on agro-eco system (Larkin and Griffin, 2007; Kamil *et al.*, 2009),.

Soil testing

Soil testing plays a very important role in diagnosing the physical, chemical and biological properties of the soils by providing the conditions of available nutrients which indicates the fertility and productivity of the soils. It also helps in assessing the capability and suitability of land for agriculture and allied activities and quantifies the soil constraints. Excessive use of fertilizers especially nitrogenous and phosphatic fertilizers leads to environmental pollution such as eutrophication and nitrate toxicity of ground water. Therefore, soil testing provides sound information for recommendation of fertilizers and correct amount of chemical fertilizers to be integrated with organic amendments for improving the soil health and quality in order to increase the productivity per unit area (Lungmuana and Lalnunsiama Colney, 2011).

Crop rotation, intercropping and mulching

Intercropping, crop rotation, and mulching are other sustainable crop production methods that help replenish the soil. Intercropping and crop rotation can improve soil health by introducing plants that fix nitrogen (a process that pulls nitrogen from the air and releases it into the soil) or plants that can be turned under after their growing season is complete to add additional nutrients to the soil. Crop rotation also generally increases yields, while monocropping has been implicated in declines in crop yield and loss of nutrients from the soil. Rotating crops allows soil to "rest," that is, to replenish its vital micronutrients, microbes, and other important components. Mulching can reduce soil erosion and help retain critical soil moisture.

Mulching is covering the soil around the plant with something which beneficial for production. Mulching generally forms a protective layer around the plant for the benefit of its growth and productivity by modifying in the soil microclimate. It also helps in improving the overall plant health. The naturally available mulch materials include fallen leaves, needles, twigs, pieces of bark and other organic materials, etc. Alternatively, gavel, pabbels and different kinds of plastic films can be used as organic mulch. Mulching is also an important practice for new fruit plantation as it helps to conserve moisture in the root zone of the new plant until the roots have grown out into the surrounding soil. The growth rate of trees and shrubs increase when there is no competition for water and nutrients from weed.

Type of Mulch Formerly, merely agricultural waste was used as mulch material. But with the advancement of science and technology, different sorts of polyethylene sheet were included in mulch material. Principally there are two basic type of mulching material.

Organic Mulch

Organic mulch is made up of natural substance such as bark, wood chips, pine needles, dry grasses, paddy straw, dry leaves, saw dust, grass clipping, etc. But organic mulch attracts insects, slugs and the cutworms that eat them. They get decomposed easily and need frequent replacement.

Inorganic Mulch

Inorganic mulches are extensively used in commercial agriculture. Plastic mulch is the most versatile of all inorganic mulches as it does not decompose easily. Polyethylene and polyvinyl chloride films too are used as mulch materials. Plastic film mulches, low density polyethylene (LDPE) or linear low density polyethylene (LLDP), are available in different types of thickness, widths and colours in the market

Advantages of mulching

Moisture conservation: Mulch material forms an insulated layer on soil surface and hinders the entry of sunlight and solar radiation. The impermeability of plastic film prevents water loss due to evaporation

Soil conservation: Mulch protects the fertile top soil against erosion during rainfall and helps in binding the soil particles by plant root system.

Soil temperature: The fluctuation in soil temperature during day night is reduced significantly by application of mulch in basin of the crops. Black mulches generally result in higher root-zone temperatures while silver or white mulches generally maintain cooler root-zone temperatures. It was observed that stretching of the transparent plastic film tightly across the moist soil for 4 weeks resulted in effective soil heating by conduction than by direct transmission of solar radiation.

Soil solarisation: Soil solarisation is done with transparent plastic mulch to prevent the incidence of soil borne disease. Soil moisture and heat retained during solarisation kills the harmful soil borne pathogens.

Weed control: Almost all types of film mulch materials are effective for control of weeds. In general, black plastic mulch is better than other coloured films for control of weeds.

Root system: Mulching stimulates the lateral root growth particularly in nutrient rich upper soil layer. The greater lateral spread with enhance root growth in the upper layers of soil in mulched trees, perhaps due to the presence of higher moisture and comparatively lower soil temperature at upper surface. Experiment conducted at the CISH on mulching in different cultivars of mango and vegetables crops indicated that the area covered by roots in mulched tree was generally more as the unmulched trees. Similarly the growth of root hairs was also more increase in mulched trees resulting enhanced water and nutrient absorption surface of roots some extent and minimized the disorders. In strawberry it protects the delicate fruits from direct soil contact to avoid infection (Singh *et al,* 2010).

Yield: Plastic mulching has tremendous impact on yield enhancement. Orchardists are benefitted by early production with improved quality which fetches higher price in the market. The maximum fruit retention, minimum fruit drop and high yield in mango by mulching was attributed largely to improve availability of soil moisture during fruit growth. The low abscission in the pedicel is also one of the important reasons for minimum fruit drop and maximum fruit yield in mulch tree.

Intercropping is the cultivation of two or more crops at the same time in the same field. There has been an increase in grower interest in using intercropping since it could reduce management inputs that result in sustainable systems more efficiently using an even potentially replenishing natural resources used during crop production for long term management of farmland (George and Jeruto, 2010). Intercropping has been practiced in the developing countries of Central America, Asia and Africa (Allen and Obura, 1983, Altier and Liebman1994, Bekunda and Woomer, 1996 1999, Grossman and Qualles1993, Henzel and Vallis1977, Jodha, 1979, Prabhakar *et al*, 1983). The advantages of intercropping are risk minimization, effective use of available resources, efficient use of labour, increased crop productivity, erosion control and food security (Bekunda and Woomer, 1996, 1999, Jodha1979, Owuor *et al*,2002). There is reduction of insect/mite pest populations due to the diversity of crops grown and reduction of plant diseases because the distance between plants of the same species is increased due to the planting of other crops between them, alteration of more beneficial insects especially when flowering crops are included in the cropping system, increase of total farm production and profitability and reduction of weed population through allelopathy and efficient crop production (Maguguda *et al*,1979). Further, when the intercrop provides a good soil cover, soil temperature will stay relatively low. This prevents burning of the organic matter in the soil and loss of nutrients. It also provides a microclimate that can be favourable for associated crops.

In India bananas are intercropped with potato which has led to good returns (Okigbo, 1979). Mustard has also been intercropped with bananas in India. Other fruit trees have also been intercropped. For example, in an intercrop between citrus mandarin seedlings and cucumber there was a high yield of cucumber fruit and minimal interference in the growth of citrus seedlings possibly due to the low growing nature of the latter (Nataraja and Nairk,1992). In Kenya fruit trees are intercropped with all types of short term crops such as beans, peas, potatoes, maize, millet, exotic and indigenous vegetables when they are still young as a way of attaining food security and income before the trees mature (George and Jeruto, 2010). Vegetables such as cabbage, cucumber, radish, snap bean, and broccoli are intercropped into double rows of field maize planted on raised soil beds. Intercropping two vegetables with different architecture and nutritional value such as beet and okra, pepper and onion has been practised in tropical Asia (Okigbo, 1979).In some areas of Africa, vegetable cropping systems such as maize and cowpea often are practiced (Mead and Willey, 1980).

Intercropping and soil fertility: One important reason for intercropping is the improvement and maintenance of fertility. An example of this is when a cereal crop or tuber crop is intercropped with legumes (beans, peas, ground nuts. After the intercrop is harvested, decaying roots and fallen leaves provide nitrogen and other nutrients for the next crop, legumes also fix nitrogen. The crop residues of the legumes can also be used as fodder, by cutting and carrying them to the animals, or by letting the animals graze the residues in the field. The nutrients in the crop residues can then be recycled when manure is used to fertilize crops. Fertilizers are more efficiently used in an intercropping system, due to the increased amount of humus and the different rooting systems of the crops as well as differences in the amount of nutrients taken up (Rahman *et al,* 2006., Rukazambuga *et al,* 2001.,Sakala *et al*, 2000., Trenbath,1979., Willey and Osiru,1979).

No tilling of soil and reduction of heavy machinery use

Industrial agricultural operations use tilling (plowing) to create rows, loosen soil, and to remove weeds. Sustainable farms use no-till methods or minimize tilling in order to protect the soil. No-to-minimal till methods can reduce soil erosion and compaction, increase aeration (critical for root growth and function), and reduce loss of water and critical nutrients.

Large industrial operations also use heavy machinery to till the soil, to plant, and to harvest. Sustainable producers limit (or eliminate) use of heavy machinery, which conserves non-renewable resources (*e.g.,* oil) and can decrease soil compaction and erosion.

Integrated pest and disease management

Pests in vegetables (insects, diseases, and weed) are part of every horticultural field in the world. It is part of their nature to eat, inhabit, and reproduce, using the vegetables as hosts to complete their life cycle. Pest management strategies such as IPM, or Integrated Pest Management, are therefore critical. Integrated pest management is the basic framework used in vegetable production to decide when and how pests are controlled. The primary goal of IPM is to provide clear pest management guidelines to growers in order to optimize pest control in an economically and ecologically sound manner. IPM integrates habitat modification and cultural, physical, biological, and chemical practices to minimize crop losses. Monitoring, recordkeeping, and life-cycle information on pests and their natural enemies are used to determine when control options are needed to keep pests below an economically damaging threshold. As they move towards greater sustainability, vegetable IPM programs tend to go through three phases, with each stage using and building on previous knowledge and techniques:

1. The pesticide management phase, characterized by establishing economic thresholds, sampling, and spraying as needed.
2. The cultural management phase, based on a thorough understanding of the pest's biology and its relationship to the cropping system. Tactics employed to control pests include delayed planting dates, crop rotation, altering harvest dates, etc.
3. The biological control phase, or "bio-intensive IPM," requires thorough understanding of the biology of natural enemies (in addition to that of the pest) and an ability to measure how effective these agents are in controlling pests. When natural agents do not meet expected goals, "soft" pesticides (non-toxic to non-target organisms) are used, and applications are timed to minimize pesticide exposure of beneficials.

Insect pheromones can be used to monitor, mass-trap, and/or disrupt the mating process of insect pests. Sex pheromones of major lepidopteran pests such as tomato fruit worm (*Helicoverpa armigera*) and common armyworm (*Spodoptera litura*) are widely used as monitoring lures in tropical vegetable production systems. The use of sex pheromone traps as a mass-trapping tool against polyphagous insects is limited. However, such traps are highly effective in reducing the damage and yield losses caused by monophagous insects such as eggplant fruit and shoot borer (*Leucinodes orbonalis*), as demonstrated in the Indo-Gangetic Plains of South Asia by AVRDC-The World Vegetable Center. By definition, pheromones are species-specific, yet "cross-talks" were observed when we attempted to refine sex pheromones for monitoring and mass-trapping legume pod borer (*Maruca vitrata*) and cucumber moth (*Diaphania indica*). In addition to sex pheromones, the possible use of aggregation pheromone as a pest management tool has been attempted in striped flea beetle (*Phyllotreta striolata*) on vegetable brassicas. Recent research indicated that pheromones act synergistically when combined with host plant volatiles, a process validated against *P. striolata* at AVRDC.

Post-harvest management

Sustainable commercial horticultural success depends on satisfying consumer requirements. Fresh fruits and vegetables are important components of human food. However, fresh horticultural produce is highly perishable with some estimates suggesting a postharvest loss of between 30-50% in fruits and vegetables. The losses are due to poor pre-production and postharvest management as well as lack of appropriate processing and marketing facilities. These losses have several adverse impacts on farmer's income, consumer prices and nutritional quality of the produce. Prices of seasonal horticultural crops fluctuate greatly and during the period of maximum availability the prices are not remunerative to the farmer. At other times these commodities are so highly priced that the ordinary consumers find them beyond their purchasing power. Another problem is that fruits and vegetables are not uniformly available and some areas suffer from inadequate supply even when there is a glut in other parts. At present there is considerable gap between the gross production and net availability of fruits and vegetables due to heavy postharvest losses.

Postharvest management is a set of post-production practices that includes: cleaning, washing, selection, grading, disinfection, drying, packing and storage. These eliminate undesirable elements and improve product appearance, as well as ensuring that the product complies with established quality standards for fresh and processed products. Postharvest practices include the management and control of variables such as temperature and

Laboratory, SydneyPostharvest Laboratory Information Sheet. www.postharvest.com.au/. 22 Aug 2012

Jodha, N. S. (1979) Intercropping in traditional farming systems. Proc. Intl. Workshop on intercropping 10-13 Jan 1979 Hyderabad, India.

John McGrady. 1992. Sustainable Vegetable Production:Introduction to the Colloquium. *Hortscience.* 27(7):735-736

Kader AA (1992) Postharvest biology and technology: an overview. In: Kader AA (ed) Postharvest technology of horticultural crops. UC Publication No. 3311. University of California, Division of Agriculture and Natural Resources, Oakland, pp 15–20

Kamil, D. Kumar, R. and Sinha, A. (2009). Effect of green manuring of Crotalaria juncea L on some soil-borne pathogens. *Indian Phytopath.* 62(3):304-309.

Larkin, R.P. and Griffin, T.S. (2007). Control of soilborne potato diseases using Brassica green manures. *Crop Prot.* 26(7):1067-1077

Lathwell, Douglas J., "Legume Green Manures, Principles for Management Based on Recent Research" in TropSoils Bulletin No. 90-01 (Raleigh, North Carolina: Soil Management Collaborative Research Support Program, North Carolina State University, June 1990).

Lungmuana and Lalnunsiama Colney. 2011. Soil testing-an important tool for assessing soil health and quality. *Science Vision.* 11 (4), 214-217

Magaguda, G. T., I. Haque, W. Godfrey, I. Fendu and G. T. Masina (1979) Intercropping studies in Swaziland: Present status and future projections. Proc. Intl. Workshop on intercropping 10-13 Jan., 1979, Hyderabad, India, Pp. 98-104.

Mead, R. and R. W. Wiley (1980) The concept of "Land Equivalent Ratio" and advantages in yields from intercropping *Expt. Agric.* 16: 217-228.

Natarajaa, M. and D. M. Nairk (1992) Competitive effects of short duration, bush type cowpea when intercropped with cotton in Zimbabwe *Expt. Agric.* 28: 41-48

Okigbo, B. N. (1979) Evaluation for plant interactions and productivity in complex mixtures as a basis for improved cropping systems design: In: Proc. Intl. Workshop on intercropping 10-13 Jan., 1979, Hyderbad, India Pp. 350-356.

Owuor, C. Tenywa, J. S., Muwanga S., Woomer P. L. and Esele, P. (2002) Performance of a sorghum-legume intercrops in response to row levels of nitrogen. Program and extended abstracts. FORUM Working Document. No. 5. The Forum on Agriculture Resource Husbandry, Nairobi, Pp. 542-544.

Pieters, J.A. (1927). Green Manuring, Principles and Practice. John Wiley and Sons, New York, 365 p.

Prabhakar, B. S., V. Shulka and K. Srinwa (1983) Nutritional potential of vegetable intercropping system. *Indian J. Hort.* Vol. 2 No. 3 Pp. 258-268.

Rukazambuga, N., C. S. Gold, S. Gowen and P. Ragama (2001) The influence of crop management on banana weevil (cosmopolites sordidus) populations and yield of highland cooking banana in Uganda. Bulletin of Entomological Research Vol. 92: Pp. 413-421. Cambridge University Press.

Sakala, W. D., Cadisch, G. and Giller, K. E. (2000) Interactions between residues of maize and pigeon pea and mineral N fertilizer during decomposition and N mineralization. *Soil Biology and Biochemistry* 32: 699-706.

Schumann, R.A. Meyer, J.H. and Van Antwerpen, R. (2000). A review of green manuring practices in sugarcane production. *Proc. South Africa Sugar Technol. Assess.* 74: 93–100.

Selvi, R.V. and Kalpana, R. (2009). Potentials of green manure in integrated nutrient management for rice – *A review. Agri. Rev.* 30(1):40-47.

Shweta., S. K. Bhatia and M. Malik. 2014. Protected Farming. *Popular Kheti*. **2**(1):74-79

Singh, V.K., Misra, R. and Pandey, S. (2010). Mulching in HorticulturalCrops, PFDC, CISH, Lucknow-227.107 p.20

Sinha, A. Kumar, R. Kamil, D. and Kapur, P. (2009). Release of nitrogen, phosphorus and potassium from decomposing Crotalaria juncea L. in relation to different climatic factors. *Environ. Ecol.* 27(4B):2077-2081.

South African Education and Environment Project. 2010. Sustainable Development in South Africa : Introduction to Basic Concepts

Trenbath, B. R. (1979) Light use efficiency of crops and potential for improvement through intercropping. *Proc. Intl. Workshop on intercropping* 10-13 1979 Hyderabad, India Pp. 200-215.

Vivekanand, B., Kapil S.S. and Remya, L. 2015. Rooftop Micro Organic Farm: A Promising Initiative to Enhance the Vegetable Production. *Indian Horticulture Journal*; **5**(3/4): 114-116,

Wiley, R. W. and D. S. O. Osiru (1972) Studies on mixture of maize and beans (P. vulgaris) with particular reference to plant population *J. Agric. Sci.* 79: 518-528.

Y. Diczbalis. 1999. An Introduction to Irrigation of Horticultural Crops. *Agnote.* Pp: 0157-8243